Radiation Curing
of Polymers II

Special Publication No. 89

Radiation Curing
of Polymers II

The Proceedings of the Third International Symposium
organised by the North West Region of the Industrial
Division of the Royal Society of Chemistry

U.M.I.S.T., Manchester, 12th–14th September 1990

Edited by
D.R. Randell
Ciba-Geigy Additives, Manchester

ISBN 0-85186-377-9

A catalogue record for this book is available from the British Library

Published by The Royal Society of Chemistry,
Thomas Graham House, Science Park, Cambridge
CB4 4WF

Printed in Great Britain by J. W. Arrowsmith Ltd., Bristol

Preface

During the last decade radiation cured polymers have continued to grow in importance not only by expansion within existing coatings applications but also by extension into new fields of application such as ceramics, ink-jet inks and fibres. Indeed the volume rate of growth of radiation cured coating is currently running at 15-20% per annum with the market in West Europe expected to grow from the present to 56,000 tonnes by 1995.

To provide a further update on the rapidly growing science and technology of radiation curing and as a follow-up to the topics of the Second International Symposium (R.S.C. Special Publication No. 64, 1987) the Third International Symposium, organised by the North West Region of the Industrial Division of The Royal Society of Chemistry, was held at U.M.I.S.T., Manchester on 12th-14th September, 1990. The 2.5 day conference attracted 230 delegates from the U.K. and abroad, from industry and academia and was accompanied by an exhibition by manufacturers, consultants and publishers.

Apart from providing an update on the application, chemistry and control aspects of the radiation curing the aim of the meeting was also to provide the newcomer with a basic insight into radiation curing applications.

Accordingly the proceedings contained in this special publication which follow closely the format of the meeting, has five sections covering the background/trends, applications, initiator chemistry, substrate chemistry and analytical, physical chemical and health and safety aspects.

I wish to acknowledge my gratitude to the other members of the organising committee namely,

Professor N.S. Allen (Manchester Polytechnic)
Mr. M. Combey (Ciba-Geigy Industrial Chemicals)
Mr. W.A. Green (International Bio-synthetics Ltd)
Ms. J.A. Hart (Consultant)
Mr. A. Howell (Anchor Chemicals U.K. Ltd)
Dr. M.S. Salim (Harcros Chemicals U.K. Ltd)
Mr. J. Sutcliffe (Crown Industrial Products)
Mr. A. Williamson (R.S.C. N.W. Region Industrial Division)

My sincere thanks also to Mrs. E.S. Wellingham for her diligent and painstaking services as Conference Secretary and to the Chairman of the five sessions of the meeting, Dr. S.P. Pappas (Loctite Corporation, USA), Dr. H.J. Hageman (Akzo Corporate Research), Dr. M.S. Salim, Mr. J. Sutcliffe and Professor N.S. Allen for exercising tight and understanding control of the meeting.

I should also like to express my thanks to Royal Society of Chemistry for agreeing to publish the monograph and especially Mr. A.G. Cubitt, Miss. C.N. Lyall and Mr. B.P. Jackson for support in organising the publication and to Mrs. C.L. Sharp of Ciba-Geigy Additives and Mrs. A. Randell for assistance in checking the scripts and compiling the index.

It is hoped that the monograph, which includes all the papers from the conference, will prove useful to newcomers, as well as experienced practitioners of the radiation curing of polymers.

Contents

Background/Trends

Overview of UV Curable Coatings   3
    M.S. Salim

Electron Beam Curing (EBC)   22
    J. Sutcliffe

Photoinitiators and Photocatalysts for Various
Polymerisation and Crosslinking Processes   46
    H.J. Hageman

Trends in UV Equipment   61
    R.E. Knight

Applications

UV Curable Media for Decal Production in the
Ceramics Industry   73
    P.R. Jackson

The Use of Dynamic Mechanical Analysis (DMA) in
the Assessment of Radiation Curable Optical Fibre
Coatings   84
    J. Mahone

The Use of UV Technology to Produce Thick
Pigmented Coatings   103
    P.G. Garratt

Photoinitiators for Litho Inks   124
    S.J. Wilson

The Use of UV Curable Prepolymers in Ink Jet Inks   132
    A.C. Marshall, M. Sutty, N. Miller, and
    A.L. Hudd

Radiation Curable Adhesives with Auxiliary Cure
Mechanisms   147
    S.P. Pappas and J.G. Woods

Initiator Chemistry

New Optimised Alpha-Aminoketone Photoinitiators   163
    M. Koehler, L. Misev, V. Desobry,
    K. Dietliker, B.M. Bussian, and H. Karfunkel

Photochemistry and Photopolymerisation Activity of
Novel Perester Derivatives of Fluorenone   182
    N.S. Allen, S.J. Hardy, A. Jacobine,
    D.M. Glaser, F. Catalina, S. Navaratnam,
    and B.J. Parsons

Photoredox Induced Cationic Polymerisation of
Divinyl Ethers   199
    P.-E. Sundell and S. Jönsson

UV Photoinitiators in Pigmented Systems   216
    K. Dorfner

Substrate Chemistry

Performance of UV Curing Systems Through
Monomer Selection   247
    R.C.W. Zwanenburg

Trends in Low Viscosity Acrylate Resins for
Radiation Cured Formulations   269
    A. Howell

Cationic UV Curable Coatings: Contribution of
Acrylate Oligomers to Final Film Properties   284
    P.J.-M. Manus

Hybrid Cure of 'Half Acrylates' by Radiation   302
    J.-P. Ravijst

Effects of Light on Cyclic Vinyl Ethers   326
    C. Lowe and A.E. Wade

Contents

Norbornene Resins as Substrate in Thiol-ene
Polymerizations: Novel Non-Acrylate
Photopolymers for Adhesives, Sealants, and
Coatings   342
    A.F. Jacobine, D.M. Glaser, S.T. Nakos,
    M. Masterson, P.J. Grabek, M.A. Rakas,
    D. Mancini, and J.G. Woods

UV and Υ-Irradiation of Polysilastyrene Films
and Fibres   358
    G.C. East, V. Kalyvas, and J.E. McIntyre

**Analytical, Physical Chemical and Health and Safety
Aspects**

Experimental Techniques to Monitor the Degree of
Cure in Radiation-Cured Coatings   379
    A.K. Davies

Use of Infra-red Spectroscopy to Monitor the
Degree of Cure in Acrylate and Epoxy Curable
Systems   400
    R.S. Davidson, K.S. Tranter, and
    S.A. Wilkinson

Measurement of Dissolution Kinetics of Thin
Polymer Films   416
    G.J. Price and J.M. Buley

Health and Safety Aspects of Radiation Curing   430
    B.C. Ross

Subject Index   437

Background/Trends

Overview of UV Curable Coatings

M.S. Salim

HARCROS CHEMICALS (U.K.) LTD., LANKRO HOUSE, P.O. BOX 1, ECCLES, MANCHESTER M30 0BH, UK

1   INTRODUCTION

UV Cured coatings are widely used by industry. Some of these application areas include:

- cork tiles
- flush doors
- cosmetic packaging, record sleeves
- toothpaste tubes
- kitchen and bedroom furniture
- various food packaging
- PVC floor covering
- Car headlamp reflectors as well as various plastic and wooden components in motor vehicles.
- Printed circuit boards e.g. telephone and computers
- office window may be equipped with acoustic or safety glass cured with U.V.
- Book and magazine covers
- TV cabinets, compact discs
- Optical fibre coatings

All these coatings are cured by using UV light (between wavelengths 200-400 nm).

In this paper I shall begin by discussing radiation curing technology in terms of applications, raw materials and equipment that is currently employed and follow this up by discussing possible new applications of this technology in the future. Finally, the European market for Radiation Curing products will be discussed.

1.1   Background to Radiation Cured Coatings

Throughout the 1970s there was a growing awareness that
UV curing represented a commercially viable and efficient
method of curing inks and coatings for wood, paper and board,
metal, plastics and other substrates.

The oil crisis of the mid-seventies was followed by ever
escalating costs of conventional energy sources and petro-
chemical derived solvents. This led to renewed activity
in UV curing as a clean, efficient and cost effective
alternative to energy intensive air drying of conventional
solvent based coatings.

Some of the major benefits to be gained from UV and EB
curing are listed below:

-       Lower energy consumption than thermal curing.
-       smaller space requirements, i.e. more efficient
        compact operating environment.
-       Fast cure speed.
-       Instant start and shut down.
-       Lower overall costs.
-       100% solids coatings.
-       Reduced fire risks - freedom from volatile solvents.
-       Heat sensitive substrates can be printed or coated
        without degradation

Early UV ink and lacquer formulations tended to suffer from
the disadvantage of possessing a relatively high order of
skin irritancy which led to operator resistance to the new
products in some areas of the coating industry.

However, the development of second and now third generation
of products, which possess a lower order of skin irritancy,
has enabled current formulations to become widely accepted
on a commercial scale and radiation curing is now a firmly
established and rapidly growing sector of the coatings
industry.

Harcros, who pioneered the development of the second generation monomers during the Seventies, is now established as a
leading supplier of speciality monomers and oligomers and is
firmly committed to the production of high purity, low
irritancy products under the trade name of "PHOTOMER".

1.2   Radiation Cured Formulation

The typical U.V. and Electron Beam Cured formulations are shown in Figure 1.

U.V.	E.B.C.
Monomer (s)	Monomer (s)
Oligomer (s)	Oligomer (s)
Photoinitiators	
Amine Synergists	
Pigments/Extenders	Pigments/Extenders
Additives	Additives

For E.B.C. coatings, the formulation does not contain Photoinitiators.

1.2.1 Monomers

Monomers can be mono-, di- or trifunctional giving different properties in the cured films. Some of the properties that are important in selecting a suitable monomer for a particular application are:

a)   Viscosity and solubility

Low viscosity and good solubility are important to formulate the end product with the correct application viscosity.

b)   Odour/taint and volatility

Low odour/taint and volatility are important in coatings - especially in packaging formulations.

c)   Functionality

The functionality will effect reactivity, Tg and percentage shrinkage of monomers, as well as the cure rate, flexibility, hardness, durability and adhesion to substrate. In general the greater the function-ality the more reactive is the coating but this reduces flexibility properties as well as adhesion because of the resulting increase in volume shrinkage.

d)   Low Draize reading

This measures skin irritancy and monomers having a
draize <3.5 are used in the UK industry.

Some of the monomers used in radiation cured coatings
include the following:

Tripropylene glycol diacrylate   (Photomer 4061)

$$CH_2 = Ch - \overset{\overset{\displaystyle O}{\|}}{C} - O \left[(C_3H_6O)_3 \right]_3 \overset{\overset{\displaystyle O}{\|}}{C} - CH = CH_2$$

This difunctional monomer has a low viscosity and
gives high reactivity. This is used in formulations
for wood, packaging and plastics.

Trimethylol Propane Ethoxylate Acrylate (Photomer 4149)

$$C_2H_5 - C \left[CH_2O(CH_2CH_2O)_n \overset{\overset{\displaystyle O}{\|}}{C} - CH = CH_2 \right]_3$$

This has a viscosity of approximately 380 cps. and
offers high reactivity. It is used as one of the
monomers in roller coat and litho varnishes to
enhance the cure rates.

Some examples of Monomers that are widely used in
radiation cured coatings are shown in Table I.

TABLE 1 - PHOTOMER 4000 SERIES - MONOMERS

PRODUCT NAME	FUNCTIONALITY	VISCOSITY AT 25°C (CPS)
Photomer 4039	1	33
Photomer 4028	2	1250
Photomer 4061	2	13
Photomer 4127	2	18
Photomer 4193	2	15
Photomer 4094	3	100
Photomer 4149	3	80

1.3   Oligomers

The oligomers used in free radical polymerisation include the following:

-        epoxy acrylates

-        polyester acrylates

-        urethane acrylates

-        emulsions and water thinnable oligomers

-        specialised oligomers

The type that is employed in a particular formulation will depend on the properties required from the cured film.

1.3.1 Epoxy Acrylates

A typical example of an acrylated epoxy would be the reaction product of a bisphenol A epoxide with acrylic acid, e.g.

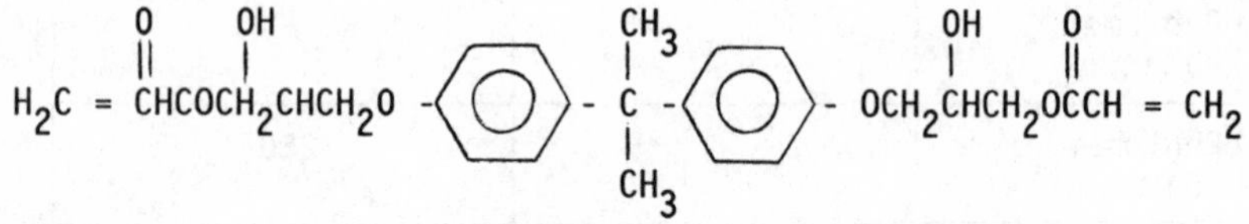

$$H_2C = CHCOCH_2CHCH_2O-\langle O \rangle - \overset{CH_3}{\underset{CH_3}{C}} - \langle O \rangle - OCH_2CHCH_2OCCH = CH_2$$

Some examples of epoxy acrylates and their uses are shown in Table II.

TABLE II - PHOTOMER 3000 SERIES - EPOXY ACRYLATES

PRODUCT	DILUTION	FUNCTIONALITY	APPLICATION AREAS
Photomer 3005	None	3	Flexibiliser - used in varnishes and inks. Improves pigment wetting.
Photomer 3016	None	2	i) Varnishes and inks for paper/board. ii) Metal dec. iii) Wood coatings
Photomer 3038	20% Photomer 4061	2	As for Ph.3016 except UV wet offset litho inks and varnishes.

1.3.2 Polyester Acrylates

A product of the reaction product of acrylic acid with an adipic acid/hexane diol-based polyester is shown below:

$$H_2C = CHCO(CH_2)_6 \left[OC-(CH_2)_4CO(CH_2)_6 \right]_n OCCH = CH_2$$

Examples of Polyester acrylates and their application use is shown in Table III.

TABLE III   PHOTOMER 5000 SERIES - POLYESTER ACRYLATES

PRODUCT NAME	FUNCTIONALITY	VISCOSITY AT 15° (cps)	APPLICATIONS
Photomer 5018	4	600	i) wood sealers ii) varnishes
Photomer 5029	4	5000	As above.

1.3.3 Urethane Acrylates

These are prepared by the reaction of an hydroxyl group of a monomer such as hydroxy propyl acrylate with an isocyanate group e.g. MDI (Diphenyl methane di-isocyanate). When hydroxyl compounds such as polyethers, polyesters and polyols which contain more than one hydroxyl group per molecule are employed then chain lengthening occurs. Typical of this class of oligomer is the following product:

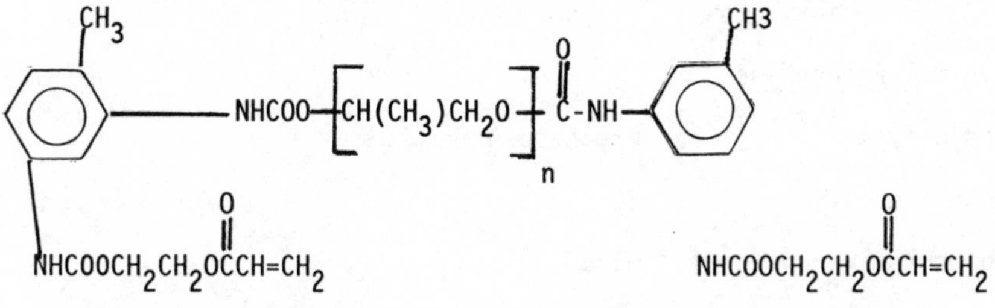

Some of the variables that are present in the design of Urethane acrylates include the following :

i)   Degree of unsaturation - this is controlled by the selection of hydroxyl monomer, molecular weight of the pre-polymer, isocyanate content and the nature and level of the reactive monomer.

ii)   Nature of the hydroxyl monomer.

iii) Nature of the di-isocyanate.

iv)   Viscosity and molecular weight i.e. the higher the
      viscosity the greater the molecular weight. This
      effects the amount of monomer to be added in the
      coatings and this effects application viscosity,
      adhesion, flexibility etc.

Examples of aromatic urethane acrylates are shown in Table IV
and aliphatic urethane acrylate in Table V. Urethane
acrylate used in the electronics application area are shown
in Table VI.

1.3.4   Urethane Acrylates in UV Water Based Coatings

There are two types of UV water based products.

-      UV water thinnable oligomers

-      UV emulsions

UV water thinnable oligomers, such as RCP 4071, are used in
UV silk screen ink formulations for paper/board.   The
urethane acrylate emulsions, such as Photomer EL1303, are
used in UV wood coatings.

RCP 4071 - U.V. water thinnable aliphatic polyurethane
           acrylate.

Typical Properties

Resin Type   Urethane Acrylate

Functionality   2

Molecular weight   c1550

Appearance   Clear Liquid

Colour   Gardner 3

Viscosity at 40°C   1000 P

Flash Point   > 125°C (closed cup)

This product can take up to 70% water.

Photomer EL1303 - is an aqueous dispersion of a UV Curable
Polyether Urethane

TABLE IV

ALIPHATIC URETHANE ACRYLATES

PRODUCT NAME	FUNCTIONALITY	VISCOSITY AT 25°C (POISE)	APPLICATION AREA
Photomer 6129	2		Plastic substrate.
Photomer 6140	2	3250	Flexible substrate.
Photomer 6250	3	1000	PVC tiles etc.

TABLE V

AROMATIC URETHANE ACRYLATES

PRODUCT NAME	FUNCTIONALITY	VISCOSITY AT 25°C (POISE)	APPLICATION AREA
Photomer 6052	2	3000	Varnishes and inks for flexible substrates.
Photomer 6162	2	700	Coatings for flexible substrates.
Photomer 6202	4	7000	Reactive plasticise.
Photomer 6261	6	300	UV litho inks.

TABLE VI

PHOTOMER 6000 SERIES - URETHANE ACRYLATES

ELECTRONICS APPLICATION

PRODUCT NAME	FUNCTIONALITY	VISCOSITY AT 25°C (POISE)	APPLICATION AREA
Photomer 6118	3	200	Solder resist
Photomer 6305	1	50	Etch resist.

Typical Properties

Appearance at 20°C	Translucent mobile liquid with a blue tinge
Type	Anionic
Viscosity at 20°C (Ford 4 cup)	15 secs
% solids content	38.3
pH	7.45

Some of the advantages of UV water based coatings include:

-    Low irritant and odour water as monomer.

-    Absence of monomers reduces level of shrinkage when curing film. The adhesion is improved.

-    Addition of matting agents and waxes to water based products is easy and a reproducible gloss can be obtained.

-    Equipment and spillage can easily be cleaned.

Some of the disadvantages of UV water based coatings include:

-    The water must be removed before UV curing in UV thick coatings. Therefore, additional energy is required.

-    Low solids content products will give low gloss.

1.3.5 Speciality Acrylates

These include products such as Photomer 7020 and Photomer 7031. The former is used as a slip additive in UV varnishes, while the latter is employed as wetting agent on plastic substrates.

1.3.6 Amine Synergists

These act as source for hydrogen abstraction by Photo-initiators such as benzophenone. Examples of commercially available amine synergists are shown in Table VII.

TABLE VII   AMINE SYNERGISTS - PHOTOMER 4000 SERIES

PRODUCT NAME NAME	AMINE VALUE mgKOH/g	ACRYLATE FUNCTIONALITY
Photomer 4116	90-95	2
Photomer 4182	190	1
Photomer 4215	210	0.5

1.4   Photoinitiators

These act as photocatalysts and are receptive to UV light by absorbing the radiant energy, and undergoing a chemical process, resulting in reactive intermediates which are capable of initiating the curing process.

The following process occurs when a UV coating containing photoinitiators is exposed to UV light:

a)   Interaction between the photoinitiator and the UV light occurs.

b)   Free radicals result from the chemical re-arrangement of the photoinitiator molecule.

c)   A chain polymerisation occurs when radicals react with the unsaturated monomers and oligomers.

d)   Free radical chain polymerisation continues via free radical chain mechanism until a solid polymer matrix forms.

Some of the photoinitiators used are listed below:

a)   <u>Benzoin butyl ether</u>

This follows the Norrish I type cleavage reaction.

b)   Benzil Ketals

Irgacure 651 (ex Ciba-Geigy) is widely used in
radiation cured coatings.

Benzil dimethyl ketal.

This photoinitiator also undergoes a Norrish I
type[15] fragmentation when irradiated.

The mechanism is as follows:

As the temperature increases, a side reaction occurs
resulting in the formation of methyl benzoate.

The resulting odour can cause problems in radiation
cured coatings for food wrapping and packaging.

c) __Hydrogen abstraction photoinitiator__

The most commonly used in radiation cured coatings is benzophenone.

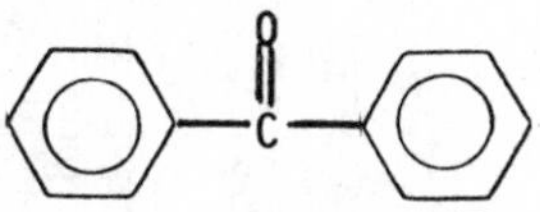

This is excited from the singlet to the triplet state[6,7] when exposed to UV light. The triplet state benzophenone abstracts a hydrogen from an adjacent donor molecule, e.g. amine synergist (Photomers 4116 and 4182) leading to radical formation and hence polymerisation.

The efficiency of the photoinitiators used will be dependent on their UV absorption spectra and molar extinction coefficients.

1.5   __Radiation Curing Equipment__

1.5.1 __UV Curing__

The main type of lamps that are employed in UV curing are:

- Medium pressure mercury lamps

- Electrodeless lamps

The former is far lower in price than the elecro- deless type of lamps. Both emit a strong 366 nm line in the UV region. The spectral output of a number of different type of lamps are shown in Table VIII.

Table VIII—Evaluation of Ultraviolet light Sources

Item \ Lamp	Line source			
	Low pressure mercury vapour lamp	High pressure mercury vapour lamp	Metal halide lamp	Capillary super-high pressure mercury lamp
Input density (W/cm)	0.5 ~ 6	30 ~ 160	60 ~ 250	300 ~ 400
Power (W)	4 ~ 1,000	100 ~ 20,000	800 ~ 20,000	600 ~ 8,000
Arc length (cm)	10 ~ 200	4 ~ 200	10 ~ 160	1 ~ 23
Lamp cooling	Air	Air	Air	Water
Average life	long <------------------------------------> short			
Characteristic wave length range (nm)	184.9 253.7	200 ~ 300	300 ~ 400	200 ~ 400
Features as a light source	(1) Smaller heat radiation (2) Higher efficiency (3) Free shape (straight tube U type, folded coil, etc.)	(1) Abundant combinations of input densities and arc lengths.	(1) Same as left.	(1) Sharp line source (width 2mm) (2) Limited arc length (25cm or less)
Major application	(1) Surface treatment (2) Sterilization.	(1) UV curing	(1) UV curing	(1) UV curing (2) Lithography

1.6   New Application Areas in U.V.

New applications area include:

- leather finishes

- UV laminating adhesives of o.p.p./paper and board

- U.V. or E.B.C. Pressure Sensitive Adhesive

- U.V. blisterpack adhesives

- U.V. Photoimageable products

- U.V. Composite materials

- U.V. Printing and varnishing simulating halographic effects

- Textiles

- U.V. Jet printing of inks

1.7   Potential Application in Maintenance Products

Any prospective application in this area must be very long term in the future. E.B.C. cannot be entertained because of the costs involved as well as safety considerations.

However, portable low energy U.V. lamps are now available. These are used in Windscreen repair kits. With major development work now proceeding in external weatherability coatings it is possible that U.V. Curing might be employed in the future in the following areas:

- U.V. protective decorative varnishes for doors

- U.V. varnishes for repairing plastic components of motor cars

- U.V. repair coatings for roof tiles

- U.V. repair and protective coatings for skis, golf clubs, cricket bats etc.

All these type of coatings could be brush applied and then cured by either strong sunlight or using portable UV lamps, within a few seconds.

In addition, it is possible in future, that equipment
suppliers will develop portable machinery that incorporates
both application rollers, or small curtain coaters, as well
as a small UV lamp which applies the protective or decorative coating to, for example, PVC flooring in office
blocks, supermarket floors and cures the products
immediately after applying.

The advantages of the above coating will be not only that
they will be fast curing but contain 100% solids and,
therefore, avoiding pollution.

1.8 Market Volume in Europe

The European market has grown steadily since 1972.[8] Fig.1.
shows the rate of growth for the past nine years. The
market volume has grown at an annual rate of about 20-22%
and is expected to grow at an annual rate of 8-15% based on
volume of 1989.

The estimated market distribution for the raw materials
consumed in both U.V. and EBC coatings in 1989 is shown in
Table IX.[8]

Table IX

	Tonnes
Monomers	9,500
Photoinitiators	1,000
Epoxyacrylates	3,500
Urethane acrylates	1,500
Polyester acrylates	2,500
	18,000

The application areas in which these raw materials are used are shown in Table X.

Table X

Application	Tonnes
Lithographic Inks and Varnishes (UV and EB)	6,000
UV silk screen Inks	1,000
UV Overprinting Varnish	6,000
Wood, cork and plastic	4,000
EBC Coatings	1,000
	18,000

From Table X it can be seen that the graphic area accounts for 70-75% of the total volume. The coatings area is taking an increasing share of the market.

1.9.   References

1.   "UV Curing: Science and Technology" - Ed. S. Peter Pappas, Technology Marketing, 642 Westover Road, Stamford, Connecticut 06902, USA.

2.   H.er. Heine, Tetrahedron Letters, 4755 (1972).

3.   S.P. Pappas and A.K. Chattopadhyay, J. Amer. Chem. Soc., 95, 6484 (1973).

4.   A. Ledwith, P.J. Russell and L.H. Sutcliffe, J. Chem. Soc., Perkin 11, 1925 (1972).

5.   L.H. Carlblom and S.P. Pappas, J. Polym. Sci,. Polym. Chem. Ed., 15, 1381 (1977).

6.   S.G. Cohen, A. Pavola and G.H. Parsons, Jr., Chem. Rev., 73, 141 (1973).

7.   R.F. Bartholomew, R.S. Davidson, P.F. Lambreth, J.F. McKellar and P.H. Turner, J. Chem. Soc., Perkin II, 577 (1972).

8.   P Dufour, "Overview of European Rad Cure Market", Rad Cure Conference, Florence. 9-11, 1989.

Electron Beam Curing (EBC)

John Sutcliffe

CROWN INDUSTRIAL PROJECTS, P.O. BOX 37, CROWN HOUSE, HOLLINS ROAD, DARWIN, LANCS. BB3 0BG, UK

The purpose of this paper is to discuss the principles of electron beam curing and to cover the various available technologies used to produce high speed electrons.

The paper will also discuss the basic chemistry mechanisms and terminology involved with electron beam curing, since this may be unfamiliar technology to some members of the audience.

Finally some of the current end uses and projected end uses for electron beam curing will be reviewed.

WHAT IS EBC ?

EBC is a method of promoting a rapid curing reaction within a resin system by the action of high speed penetrating electrons. Unlike U.V. curing, which depends upon a photoinitiator to produce free radicals, EBC acts directly upon the reactive resins in the coating or film.

1.0   EQUIPMENT

The faster or more energetic an electron is, the deeper it will penetrate into a given substrate or target, but first we must have a source of electrons. This is usually a filament or cathode, (as it is more generally termed). The cathode is generally heated within a high vacuum chamber and emits electrons as its temperature is raised. The cathode heating and subsequent electron acceleration must take place inside a vacuum (typically 10^{-7} Torr), otherwise the electrons produced would scatter rapidly on collision with molecules in the air, resulting in no effective increase in their velocity.

Various types of equipment are available to produce high speed electrons and these are described as follows:-

1.1 Scanned Beam type (Fig. 1)

The earliest work on EBC was undertaken using this type of equipment.

In very simple terms the equipment operates in a similar way to a television cathode ray tube.

The electrons are emitted from a heated spiral point cathode with a diameter of about 5 mm which is contained within a high vacuum. The focused beam is scanned across the web width using an electromagnetic deflection system with a frequency between 50 - 200 Hz.

Delivered dose is determined by the dwell time of the beam on the substrate. The acceleration voltage is between 150 and 280 kV, depending upon the application involved.

Beam current is controlled by the temperature of the cathode and because this is non linear in relationship a very sophisticated regulation device is required to bring about a rapid response linear relationship, for positive beam current control.

Because these units use a scanning horn, which in general is required to be 1 - 2 times the processing width they are usually positioned horizontally whenever the application permits.

The beam produced passes into the process zone through a 12 - 15 micron thin titanium foil window which is supported and cooled from the vacuum side. [1]

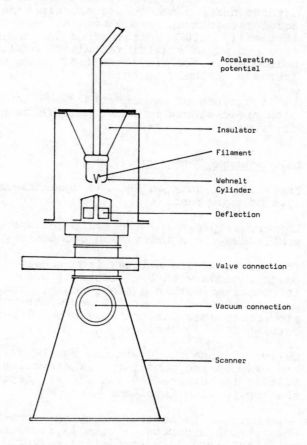

Figure 1 SCANNER TYPE ELECTRON PROCESSOR.

1.2 <u>Linear cathode type</u> (Fig. 2)

This design of accelerator is also known as the Electrocurtain^R design. It is this design of accelerator which has probably seen the widest adoption in industry, certainly in the area of curing surface coatings.

The linear cathode system works in the following way:

Electrons are drawn from the cathode (C) which is a little longer than the width of the process zone, by the application of a relatively low voltage (a few hundred volts only) to the grid (G).

This grid current is used to control the number of electrons per second extracted from the filament, this is termed the beam current.

The strip or curtain of electrons (B) is then accelerated by the electric field which rises from the large negative potential on the housing terminal (T) to the outer wall of the vacuum chamber (W). This voltage can be 100 kV to 300 kV. All the possible energy is imparted into the electrons in this space.

The beam of accelerated electrons exits the high vacuum chamber through the thin titanium foil window (F) (which is supported on the vacuum side and cooled continuously) into the process zone to contact the product (P) which is to be irradiated.

The electrons are moving at virtually light speed as they strike the window and consequently only lose about 10 - 15% of their energy in the foil.

As the beam of electrons moves into the 'air' path (i.e. outside the high vacuum area) they tend to scatter with the beam becoming wider, the further it travels.

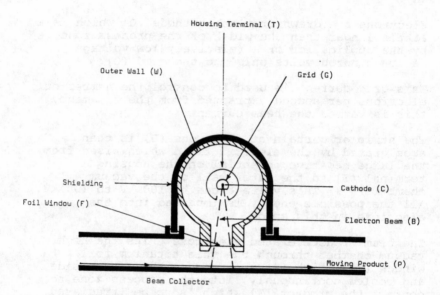

Figure 2   ELECTROCURTAIN TYPE ELECTRON PROCESSOR.

For this reason the product to be processed is placed as close as possible to the window (within a few inches) so as to reduce energy losses to negligible levels.

As processing speeds and consequently dose capacity requirements have increased many EBC installations now utilise multiple filament linear cathode accelerators.

Instead of one linear cathode traversing the web two are used, so as to proportionately increase the dose capability of the accelerator. [2]

Another type of linear cathode system is the Broadbeam[R] design.

1.3 Broadbeam[R] type (Fig. 3)

For the sake of simplicity this type of accelerator generates its electrons in a similar fashion to the previously described linear cathode assembly. However the significant difference is that the Broadbeam[R] system employs multiple linear cathodes 300 mm long sited in the web direction, arranged in modules across the web width. These cathode modules utilise a common extraction grid mounted between the filament-grid modules and the output window. The dose rate capacity is thus increased by a longer dwell time of product under the electron beam. [3]

1.4 Wire-Ion-Plasma (WIP[R]) System (Fig. 4)

The latest addition to the field of electron beam processors is the WIP[R] process, which utilises plasma electron beam technology. This technology emerged from the development of military high pulsed power, electron beam pumped CO_2 lasers.

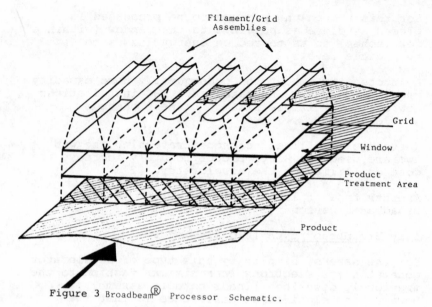

Figure 3 Broadbeam® Processor Schematic.

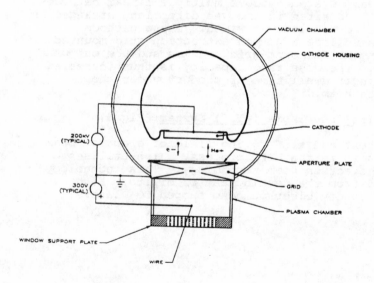

Figure 4   Schematic Illustration of WIP® Electron Beam Source

In theory the plasma technique has always offered some significant operating advantages, this theory has now been commercialised and WIP[R] processors are now available for some users. [4]

One of the very significant differences with the WIP[R] process is that it uses a cold cathode which extracts ions from a controlled gas plasma and accelerates them to the cathode or electron emitter surface.

The processor operates in the following way:-

A chamber is divided into two volumes by a grid structure; one volume is the "gun" or accelerator and the other is the plasma chamber.

At the wall of the plasma chamber, opposite the grid is a water-cooled copper plate supporting a thin metallic electron beam window foil, as in conventional technology electron guns. Inside the plasma chamber, an electron discharge is maintained between a wire anode and the walls of the chamber, which at ground potential, function as the discharge cathode.

Inside the "gun" volume is a molybdenum plate surrounded by electric field shaping electrodes. This assembly, which is the "gun" cathode structure, is supported by a high voltage feed-through bushing. A small turbo molecular vacuum pump is used to initially reduce the pressure in both chambers to below 10^{-6} Torr.

A continuous flow of helium gas is then introduced into the chambers and the pressure maintained in the range of 10 to 30 mTorr by a servo valve which is driven by an automatic pressure controller.

A positive voltage of several hundred volts is applied to the plasma chamber anode wire and the plasma discharge current maintained by the plasma power supply. A negative high voltage of several hundred kV is applied to the cathode structure of the gun.

A portion of the helium ions produced in the plasma
chamber drift towards the grid, moving under the
action of the discharge electric field. Ions which
drift into the openings of the grid structure are
accelerated towards the gun cathode by the much
stronger electric field in the gun.

Each ion gains the full kinetic energy
corresponding to the cathode potential (eg 200 keV)
and upon impact with the cathode creates about 15
secondary electrons.

The secondary electrons are then accelerated
towards the grid by the gun's electric field and
they also gain kinetic energy and momentum
corresponding to the cathode potential. The
atomic cross sections for collisions between the
high energy electrons and the helium atoms or ions
are very small; the electrons, therefore, pass
through the plasma chamber, window foil, and into
the process zone of the WIP^R processor.
Proposed advantages for the WIP^R processor over hot
cathode systems are as follows:-

(i)   Less complex, both mechanically and
      electrically

(ii) Smaller, more compact processors are possible

(iii) Potential for reduced capital investment

(iv) The design is adaptable to three dimensional
      curing

(v)   Less complex vacuum requirements

(vi) Less complicated high voltage requirements.

It is also fair to say that at present it is
difficult to achieve a stable plasma in a processor
with dimensions greater than 660 mm process width.

However, possibly a modular approach to processor
design could overcome this in the short term.

Longer term, development is ongoing to produce
WIP^R processors capable of use in wider width
applications as a single module.

2.0 Terminology

Before we go too far into this discussion it may be worthwhile to explain some of the terms associated with electron beam curing. Often newcomers to the technology find some of the terms confusing or over scientific. I have known many cases where people are warned off electron beam with the line "You need to be a PhD to use it!" I assure you this is not the case, electron beam is probably one of the most suitable curing technologies for line slaving there is. Most production facilities virtually run themselves, in that they self select curing parameters to suit prevailing line speed etc., automatically. [2]

2.1 Beam Current

The beam current, determines the number of electrons per second received by the product and hence determines the treatment rate or dose rate. Current is typically expressed in milliamperes (mA).

2.2 Accelerating Voltage (Fig. 5)

This is the potential applied to accelerate the electrons once extracted from the cathode and as their speed increases (i.e. the acceleration voltage increases) so does their ability to penetrate deeper into a substrate.

The penetration capacity of electrons is finitely determined by the acceleration voltage and their depth of penetration can not be increased by longer exposure to the beam

Acceleration voltage is measured in kilovolts (kV). (Fig. 5) shows some typical depth dose profiles.

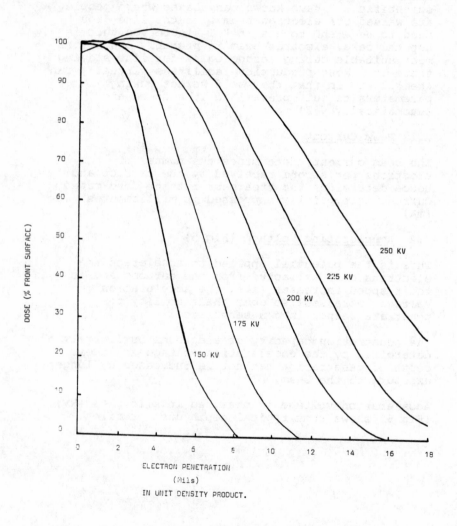

Figure 5   ELECTROCURTAIN PROCESSOR DEPTH — DOSE PROFILES

2.3 Product Speed

This will determine the exposure under the electron
beam and therefore determine the dose received.

As beam current determines the dose capability and
is measured in electrical current, it is simple to
feedback this current and slave dose rate to line
speed. As mentioned earlier this is a very
desirable facility with electron beam and results
in fully automatic running capabilities.

2.4 Megarad

This is the unit of dose most commonly quoted when
discussing cure requirements of electron beam
coatings.
It is the measure of energy absorbed per unit mass
of the product to be treated and is equivalent to
2.4 calories/gram.

Typical dose to cure values are in the region of 2-
3 Megarads for a typical paper coating.

2.5 Delivered Dose

With the relationship between beam current and
speed of product one can calculate the exposure or
dose received by the coating.

All processors have a 'yield factor' which is a
measure of the particular machines performance
characteristics. This is constant for a machine
and determined by the individual machine geometry.
This constant is termed **K**.

When looking at machine operating conditions one
must firstly look at the density of the product and
the thickness of coating to be applied, for
determination of the accelerating voltage or
penetrating power. This would remain constant for
that coating and thickness. To work out the
machine constants required to achieve the correct
dose to cure for a given coating, values are
substituted in the following equation:

$$\text{Dose (Megarads)} = \frac{K \times \text{beam current (mA)}}{\text{line speed (m/min)}}$$

eg. $\text{Dose} = \dfrac{30 \times 3}{30} = \dfrac{90}{30} = 3$

If line speed is a deciding factor then one can
calculate other line constants to give the
required dose, or conversely if Megarad dose is
critical one can calculate a maximum line speed
achievable.

Machine manufacturers often quote their machines
capacity as line speed achievable at 1 Megarad dose
e.g. 1 Megarad at 600 metres/min. Returning to the
earlier point, it does not take a PhD to work out
that if your coating requires 3 Megarads then your
maximum opeating speed would be 200 metres/min in
this case.

2.6 Machine Safety

Another criticism often levelled at electron beam
is "They produce harmful **x-rays** and you need concrete
vaults", for safety.

It is true that electron-beam processors produce x-
rays, but with modern processors operating up to
300 kV accelerating voltage, these x-rays can be
stopped by lead foil applied to all the "hot" areas
within the machine. These are termed self sheild
processors and give undetectable amounts of stray
x-rays outside the process zone. In fact these
units are so well interlocked that they can not be
operated if excess x-rays are detected outside the
process zone.

The accelerators which do require concrete vaults are those operating above 300 kV, mainly for thick film cross-linking etc. However, even this can sometimes be done with self shield systems by beams firing on each side of the web operating at 300 kV each, this gives the penetration equivalent of 600 kV which would normally require vaulting.

2.7 Inert Atmosphere

Most electron beam chemistry requires to be done in an inert atmosphere of nitrogen. This is usually from a liquid nitrogen source, as most chemistry requires oxygen levels less than 200 ppm to avoid the scavenging of radicals and to prevent ozone formation, which has a destructive effect on the titanium foil window. For this reason it is unlikely any other form of nitrogen will be adequate for inerting unless this level of purity can be continuously achieved. In the case of silicone chemistry oxygen levels of 50 ppm or less are required, thus nitrogen of the highest purity is essential.

Inerting is an expense which must be carefully considered when assessing the cost of running an electron beam operation. However, it is also true to say that such operations as film lamination with electron beam cured adhesives would not require inerting, as oxygen is physically eliminated from the reaction site when two or more films are brought together.

2.8 Ozone Production

Under nitrogen inerted conditions ozone is not significantly produced as oxygen is not present in any great proportion, however the previously mentioned non-inert laminating operation could require appropriate ozone extraction to avoid ozone damage to the titanium window.

3.0 CHEMISTRY

3.1 Initiation by UV light

The mechanism for U.V. curing has already been
described in some detail in previous papers as the
action of U.V. light on a photo initiator to cause
it to split into two radicals.

One of these radicals attacks any neighbouring
unsaturation to give a second radical in which the
initiating species is chemically bonded to a carbon
atom which was originally unsaturated. This
results in a chain reaction where the radical
generated at the initiation step attacks more
unsaturation. When two radicals combine
termination occurs.

3.2 U.V. Curing – Radical Mechanism (Fig. 6) [6]

        (PI = photoinitiator)

(i) **Initiation**

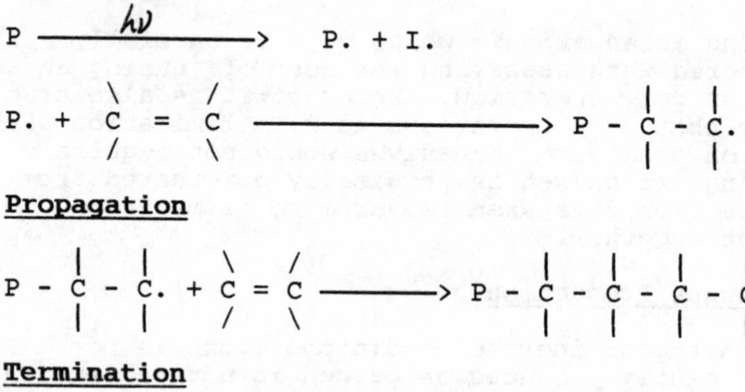

$$P \xrightarrow{\quad h\nu \quad} P\cdot + I\cdot$$

$$P\cdot + \overset{\displaystyle \diagdown \quad \diagup}{\underset{\displaystyle \diagup \quad \diagdown}{C = C}} \longrightarrow P - \overset{|}{\underset{|}{C}} - \overset{|}{\underset{|}{C}}\cdot$$

(ii) **Propagation**

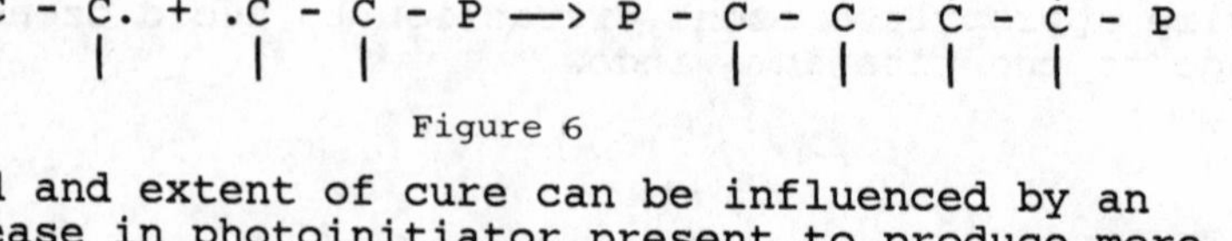

$$P - \overset{|}{\underset{|}{C}} - \overset{|}{\underset{|}{C}}\cdot + \overset{\displaystyle \diagdown \quad \diagup}{\underset{\displaystyle \diagup \quad \diagdown}{C = C}} \longrightarrow P - \overset{|}{\underset{|}{C}} - \overset{|}{\underset{|}{C}} - \overset{|}{\underset{|}{C}} - \overset{|}{\underset{|}{C}}\cdot$$

(iii) **Termination**

$$P - \overset{|}{\underset{|}{C}} - \overset{|}{\underset{|}{C}}\cdot + \cdot\overset{|}{\underset{|}{C}} - \overset{|}{\underset{|}{C}} - P \longrightarrow P - \overset{|}{\underset{|}{C}} - \overset{|}{\underset{|}{C}} - \overset{|}{\underset{|}{C}} - \overset{|}{\underset{|}{C}} - P$$

Figure 6

Speed and extent of cure can be influenced by an
increase in photoinitiator present to produce more
radicals per given UV dose or to increase the UV
dose.

3.3 U. V. Curing - Cationic Mechanism (Fig. 7)

Classic cationic polymerisation is initiated by acids and has been used to polymerise isobutylene. Another functional group which is sensitive to acids and widely available to the surface coatings industry is the epoxide group which is the foundation for this type of radiation curing.

Cationic chemistry relies upon a catalyst to produce an acid upon exposure to UV light.

Initially UV light interacts with the photoinitiator, which breaks up into a hydrogen ion, a counter ion and a neutral residue. The hydrogen ion attacks any neighbouring epoxide functionality, giving rise to ring opening which results in a hydroxyl group attached to an organic cation. This cation attacks further epoxide groups giving rise to a polymer with ether linkages.

If the organic cation comes into contact with anions, or even a neutral electron donating molecule, chain termination occurs.

(i) **Initiation** (PI = Photoinitiator)

$$\text{PI} + \text{RH} \xrightarrow{\ h\nu\ } \text{P'} + \text{P''} + \overset{\oplus}{\text{H}} + \overset{\ominus}{\text{I}} + \text{R.}$$

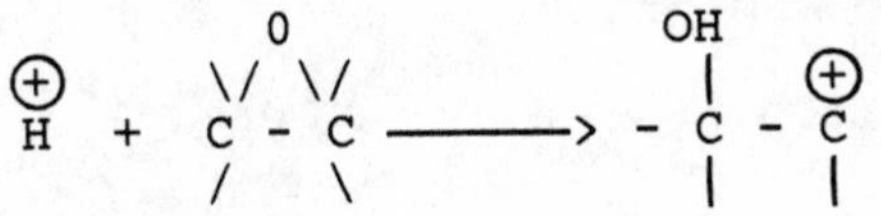

$$\overset{\oplus}{\text{H}} + \overset{\quad O \quad}{\underset{/\ \ \ \backslash}{\text{C} - \text{C}}} \longrightarrow - \overset{OH}{\underset{|}{\text{C}}} - \overset{\oplus}{\underset{|}{\text{C}}}$$

(ii) **Propagation**

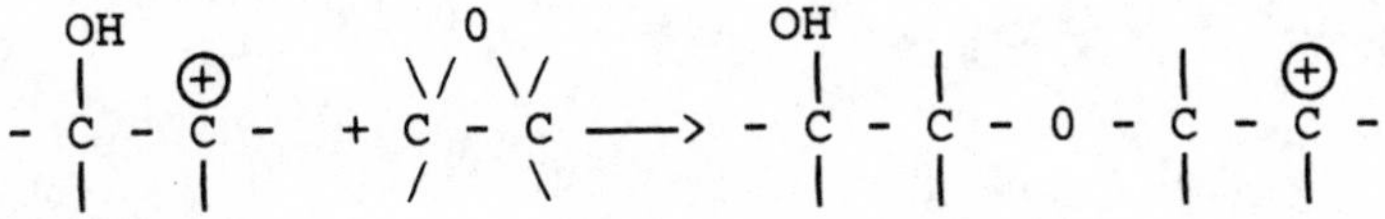

$$- \overset{OH}{\underset{|}{\text{C}}} - \overset{\oplus}{\underset{|}{\text{C}}} - \ + \ \overset{\quad O \quad}{\underset{/\ \ \ \backslash}{\text{C} - \text{C}}} \longrightarrow - \overset{OH}{\underset{|}{\text{C}}} - \overset{|}{\underset{|}{\text{C}}} - \text{O} - \overset{|}{\underset{|}{\text{C}}} - \overset{\oplus}{\underset{|}{\text{C}}} -$$

(iii) **Termination**

$$- \overset{OH}{\underset{|}{\text{C}}} - \overset{\oplus}{\underset{|}{\text{C}}} - \ + \ \overset{\ominus}{\text{I}} \longrightarrow - \overset{OH}{\underset{|}{\text{C}}} - \overset{|}{\underset{|}{\text{C}}} - \text{I}$$

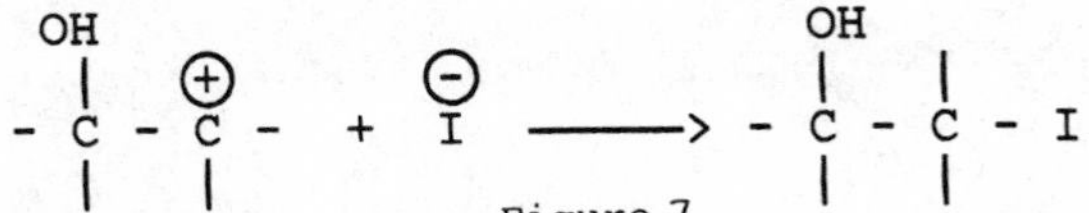

Figure 7

3.4   Initiation by Electron Beam

When considering the radical mechanism, the main difference between UV and EBC is the absence of a photoinitiator in the latter.

Radicals are generated when the coating absorbs energy from the high speed electrons via inelastic collision.

3.5   EB Curing - Radical Mechanism (Acrylate chemistry)
(Fig. 8)

(i)   Primary Process

$$\text{Oligomer (AB)} + e^{\ominus}_k \longrightarrow AB^{\oplus} + e^{\ominus}_{th} + e^{\ominus}_k$$

$$AB^* + e^{\ominus}_k$$

Figure 8

This schematic shows how the energy is absorbed.

The kinetic electron (e^-k) can either ionise the resin and also produce a thermal electron (e^-th) (slow electron) OR it can simply promote an electron to a higher level and thus leave the resin in an excited state.

(ii) <u>Secondary Process</u> (Fig. 9)

Secondary processes now occur as depicted by the following possibilities:

$$\overset{\oplus}{AB} \longrightarrow \overset{\oplus}{A}. + B.$$

$$AB* \longrightarrow \overset{\oplus}{A}. + B.$$

$$AB + \overset{\ominus}{e}_{th} \longrightarrow \overset{\ominus}{AB}.$$

$$\overset{\ominus}{AB} \longrightarrow A. + \overset{\ominus}{B}.$$

$$\overset{\oplus}{A}. + \overset{\ominus}{B}. \longrightarrow AB*$$

Figure 9

The most important of these, where acrylate chemistry is concerned is the capture of a slow moving thermal electron as follows:

$$AB + \overset{\ominus}{e}_{th} \longrightarrow \overset{\ominus}{AB}.$$

Figure 10

Fig. 11 describes how in the presence of extractable hydrogen this results in polymerisation.

AB is acrylate functionality $(\text{—} CO_2R)$

SH is a source of extractable hydrogen.

$$\text{—}CO_2R + \overset{\ominus}{e}_{th} \longrightarrow (\text{—} CO_2R).^{\ominus}$$

$$(\text{—}CO_2R).^{\ominus} + SH \longrightarrow CH_3 \overset{.}{C}HCO_2R + \overset{\ominus}{S}$$

$$CH_3 \overset{.}{C}H CO_2R = (\text{—}CO_2R)_n \longrightarrow \text{Polymer}$$

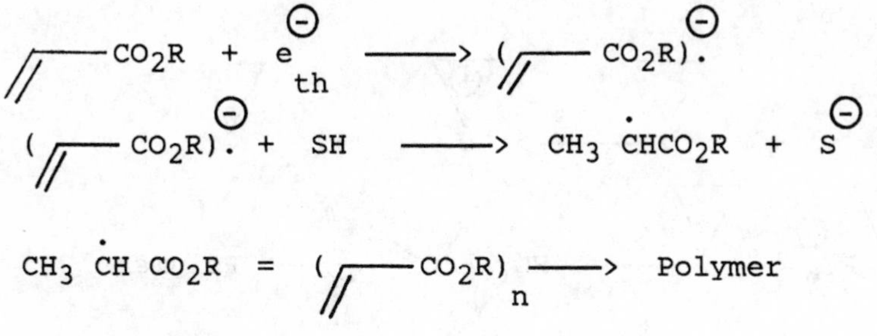

Figure 11

3.5 <u>EB Curing - Cationic Mechanism</u> (Epoxide chemistry)[7]

In the case of the cationic mechanism, the process is slightly more complex. As epoxides are unaffected by radicals and low dosages of high energy electrons, a catalyst which generates hydrogen ions is still required. Available cationic photoinitiators such as sulphonium salts can be used as this catalyst.

As before, the coating absorbs energy from high speed electrons and ionisation occurs to produce slow electrons.

These slow electrons will interact with any neighbouring photoinitiator molecule causing dissociative electron capture.

Meanwhile, the ionised resin can abstract a reactive hydrogen radical from adjacent molecules, thus producing a cation, stabilised by the PF_6^- anion. This cation can attack epoxide functionality causing polymerisation as described below.

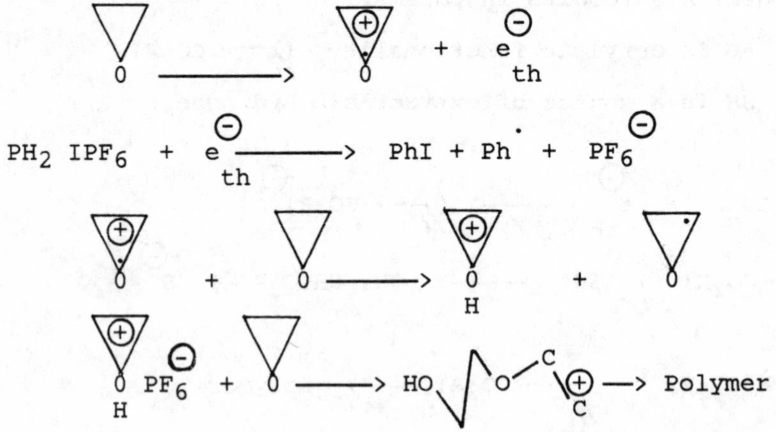

Figure 12

4.0   Comparison of U.V. to electron beam as methods of curing

4.1   <u>U.V. Light</u>

(i)   U.V. light can effect cure of coatings in an oxygen atmosphere providing that the correct chemistry is employed.

(ii)   There is a good selection of equipment available to suit most types of curing applications.

(iii)   UV technology is used mainly for the curing of clear lacquers and inks. Clear lacquers are generally 4 - 20 microns thick, however thicker films can be cured at the expense of line speed. Inks are pigmented systems but are relatively transparent at the low film thicknesses used.

(iv)   There has been progress in the curing of pigmented UV systems using special lamps and new photoinitiators.

(v)   Faster line speeds can generally be achieved by employing more UV lamps or substituting more powerful lamps, to maintain the same amount of incident luminous energy per second. This obviously requires more space and generally increases the heating effect on the substrate.

4.2 Electron Beam

(i)   Electron beam technology normally requires an
      inert atmosphere in the processing zone, as
      the components employed in UV curable
      formulations are not always effective under
      electron beam conditions.

(ii) The equipment incorporates comparatively
      expensive high vacuum and high voltage
      technology. Thus electron beam generally
      requires comparatively high capital outlay.
      For this reason it is more often employed in
      applications where high volume throughput is
      envisaged or where electron beam gives a
      distinct technical advantage over alternative
      cure methods.

(iii) Electron beam is favoured for high build or
      heavily pigmented systems and for very fast
      line speeds. Films up to 150 microns thick
      can be cured using acceleration voltages of
      175kV and line speeds can be over 200
      metres/min.

      Where thicker films are envisaged a
      proportional increase in acceleration voltage
      is required.

(iv) In order to achieve faster line speeds beam
      current can be increased at the same
      acceleration voltage, or where machine design
      will allow the use of multiple cathodes or
      double side firing can be employed with
      negligible effect on substrate temperature.

(v)   Electron beam is favoured for sensitive cure
      applications due to the higher cross-link
      density achieved.

4.3 Current applications for UV and EB curing

The scope of applications for UV and EB curing systems is continuously expanding as equipment and technology become available to overcome previous drawbacks. Also, as the constant pressure to reduce environmental pollution from volatile solvent emissions gathers pace, the benefits of solventless radiation curing techniques become increasingly attractive.

Presently the largest users of UV curable and EB curable systems are the graphic arts companies with those involved in wood and plastic finishing following closely behind.

According to information from one major resin manufacturer the split for acrylate based chemistry is as follows:- [8]

Overprint Varnishes (36%)

All low viscosity overprint varnishes used in the graphic arts industries UV cured.

Offset Printing (32%)

Inks and varnishes applied by litho/letterpress offset machines both UV and EB cured.

Wood, Plastic and Cork (20%)

Coatings for wood, furniture doors and floorcoverings, PVC doors, foil, mainly UV cured.

Silk Screen (7%)

Inks and varnishes applied by silk screen machines UV cured.

EB Coatings (5%)

Coatings for wood doors and furniture, decorative paper foils, metalisation base coats and silicone release coatings.

As we can clearly see Electron beam curing has a
relatively small share of the potential radcure
market.

The adoption of EBC has no doubt suffered from its
capital cost when comparing with UV equipment,
hence where UV could do a job, not necessarily the
best job, UV was often chosen on cost grounds.
However, there are applications where EB has been
chosen in preference to UV. Sensitive food
packaging application for example, where no risk of
residual monomer can be tolerated or for high
specification furniture foil laminates and highly
pigmented systems.

EB will see future growth, perhaps not as a
competitor to UV but in areas where EB offers
distinct technical or commercial advantages. Metal
coil coatings for example is an area where EB can
offer some real positive advantages over present
thermal drying systems and also can give very
positive economic advantages in terms of 'new for
new' capital outlay. If the polymer technology
was commercially available for EB cured coil
coatings, the cost of curing equipment would
probably be significantly less with EB than for an
equivalent thermal system.

The introduction of commercially available more
compact processors such as WIP^R will move EBC into
areas previously excluded by physical size and
cost.

Fig. 13   shows recent statistics on EBC
installations worldwide. (300 kV) in 1988 [9].

	Japan	Europe	USA	Pacific	Total
R & D	50	18	40	0	108
Pilot	12	35	45	2	94
Production	12	30	65	5	112
	74	83	150	7	314

Figure 13

A high proportion of the Worlds machines are serving an R & D function which would seem to indicate a serious general interest in the potential for EBC.

Japan has the highest number of R & D machines and also has very significant commercial interests in both major EB equipment manufacturers.

These two factors alone lead me to believe that we will witness, over perhaps the next five years, very significant growth in the use of EB technology, as companies strive to meet increasingly stringent ecological constraints and to improve their own profitability, through the application of technology.

<u>References</u>

1. Dr Peter Holl - paper presented at Radtech Florence 1989.

2. D J Maynard et Ac - Radcure '86

3. W J Ramler/A M Rodrigures - Conference on Radiation Curing, Tokyo 1986.

4. D A Meskan - RPC Industries

5. W J Ramler - Radtech N.A. 1988

6. Dr C Lowe - Radtech Seminar Copenhagen 1989

7. R J Batten etal - PPCJ March 11 1989

8. P Dufour - paper presented Radtech NA 1988

9. Tabata - Radtech NA 1988

Photoinitiators and Photocatalysts for Various Polymerisation and Crosslinking Processes

Hendrik J. Hageman

AKZO CORPORATE RESEARCH LABORATORIES, ARLA, VELPERWEG 76,
POSTBUS 9300, 6800 SB ARNHEM, THE NETHERLANDS

Introduction

Crosslinking of multifunctional oligomers and monomers to form
three dimensional networks is a rapidly expanding field on account
of the growing number of commercially important applications

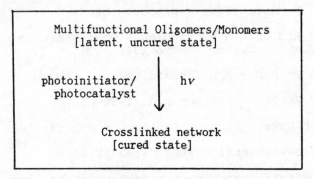

Figure 1

The concept of latency is introduced through the use of
photoinitiators and photocatalysts which, on exposure to light of
a specific wavelength, generate the necessary initiating species,
thus triggering the crosslinking process at any desirable time.

The oligomers and monomers available today stem from a variety of
chemical classes, including acrylates, unsaturated polyester
(U.P.) resins, melamine resins, epoxides and vinyl ethers.

The underlying chemistry to form crosslinked networks will as a
consequence be different, requiring a different type of
photoinitiator/photocatalyst in all these cases.

Multifunctional System	Crosslinking process	Remarks
I Acrylates, methacrylates ·	hν, radical/very fast	O$_2$.Inhibition
U.P. resins/styrene	hν, radical/slow	O$_2$.Inhibition
II Melamine resins	hν+Δ, H$^+$.catalyst/slow	T~ 100°C, No O$_2$.Inhibition
III Epoxides	hν(+Δ), cationic/fairly slow	(T ~ ?), No O$_2$.Inhibition
IV Vinyl ethers	hν, cationic/very fast	No O$_2$.Inhibition

Figure 2

In this chapter photoinitiators particularly for free-radical processes will briefly be reviewed, and some special aspects such as the configuration of the excited state involved, oxygen-inhibition and the role of tert-amines, will be elaborated.

Photoinitiators for free-radical processes [1]

The majority of commercially significant processes is still based on free-radical chemistry. This certainly is the main reason for the large number of radical-generating photoinitiators available today.

Most if not all of these photoinitiators belong to the category of aromatic carbonyl compounds. These compounds have in common that on absorption of light of the proper wavelength an excited singlet state is produced which undergoes rapid and efficient intersystem crossing (i.s.c.) to the lowest excited triplet state (n.π*,π.π*) from which the actual photochemical reactions take place (see Fig. 3).

Two major chemical deactivation pathways are available to the excited triplet state of these compounds depending on the substituents: (i) unimolecular fragmentation, and (ii) bimolecular hydrogen abstraction from a suitable hydrogen donor.

On the basis of these two major processes photoinitiators of this category have been divided into two major types as is illustrated in Fig. 4.

Major processes   **Competing processes**

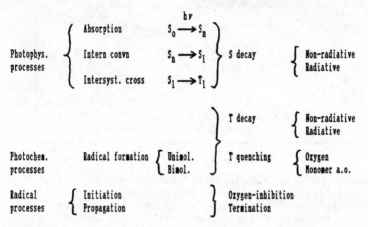

Figure 3   Important processes in photoinitiated free radical polymerisation

The Type-I photoinitiators known today have been subdivided into several different groups on account of their molecular structure, as is shown in Fig. 5. They have in common that they all produce benzoyl (or substituted benzoyl) radicals, which are known to be very reactive towards the olefinic double bond of vinyl monomers. The other (unlike) radical produced in the primary fragmentation reaction will also contribute to the initiation process, undergo further (thermal or photochemical) fragmentation, or take part in termination depending on its structure and the conditions used.

Representatives of Type-II photoinitiators are shown in Fig. 5. These compounds will only become effective in the presence of a suitable H-donor, usually tert-amines such as the various amino alcohols and N,N-dialkylamino aromatic compounds also shown in Fig. 6.

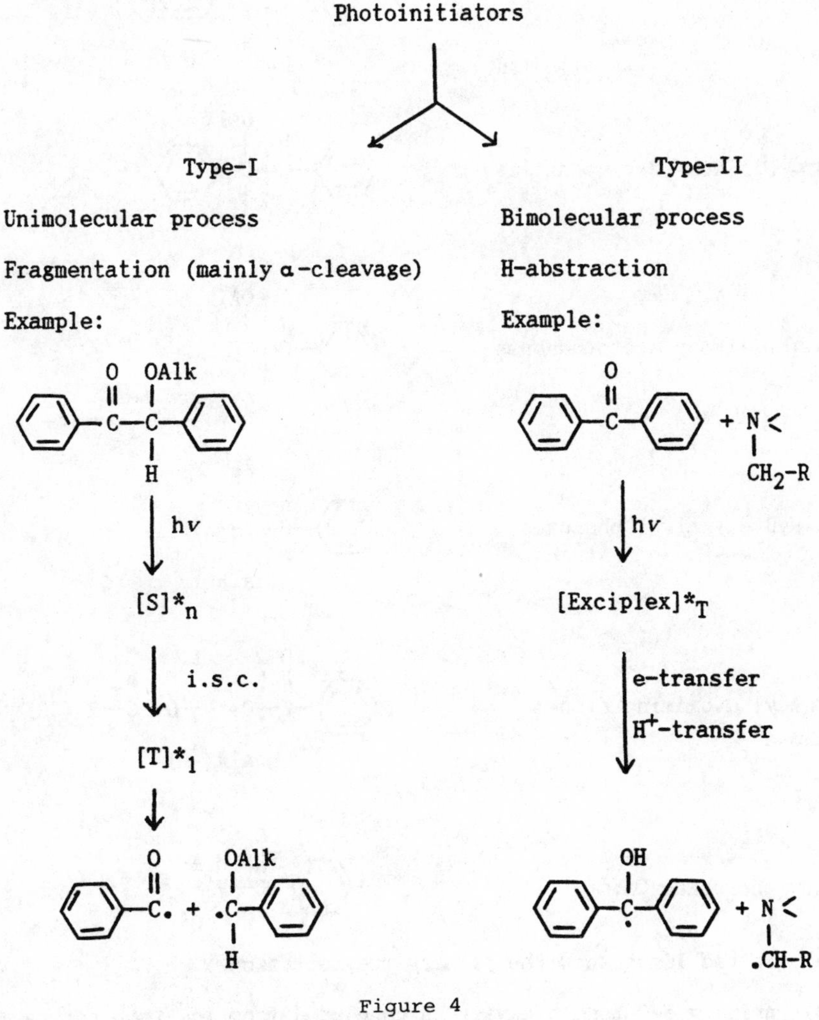
Photoinitiators
Type-I
Type-II
Unimolecular process
Fragmentation (mainly α-cleavage)
Example:
Bimolecular process
H-abstraction
Example:
O
OAlk
C
C
H
hν
[S]*n
i.s.c.
[T]*1
O
OAlk
C.
+
.C
H
O
C
+ N<
CH2-R
hν
[Exciplex]*T
e-transfer
H+-transfer
OH
C
+ N<
.CH-R

Figure 4

Type-I Photoinitiators [a]

α-Alkoxy deoxybenzoins

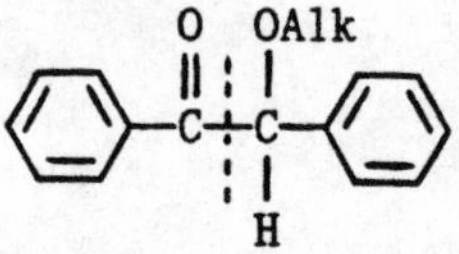

α,α-Dialkoxy deoxybenzoins

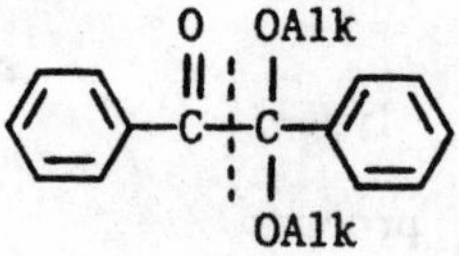

α,α-Dialkoxy acetophenones

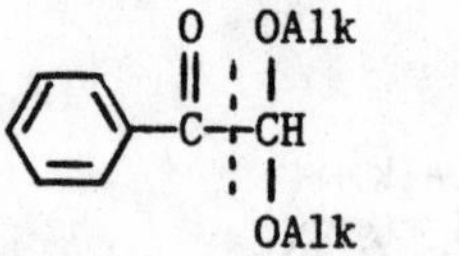

α-Hydroxy alkyl phenones

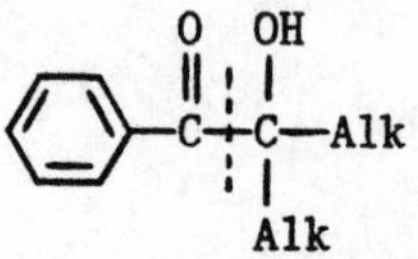

O-Acyl α-oximinoketones

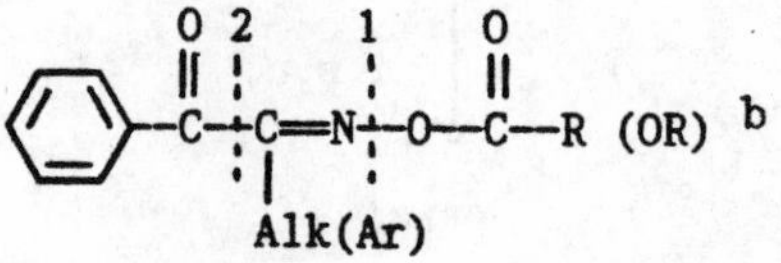

Acylphosphine Oxides

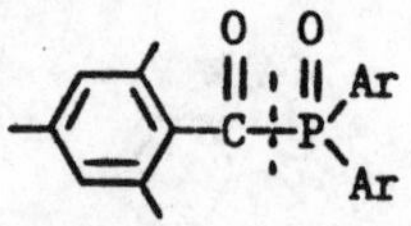

a)   dotted lines show the primary fragmentation

b)   primary N–O bond cleavage as concluded from low-temperature
     ESR spectroscopy

Figure 5

Type-II Photoinitiators

Benzophenone(s)

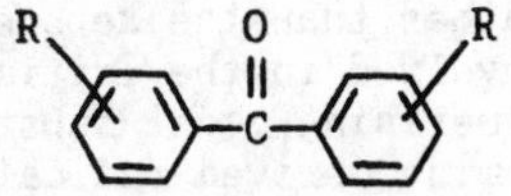

Benzophenone structure with R and X substituents on both benzene rings and a central C=O group.

(Thio)xanthone(s)

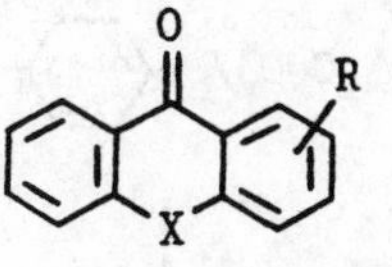

X = 0, S

α-Keto coumarins

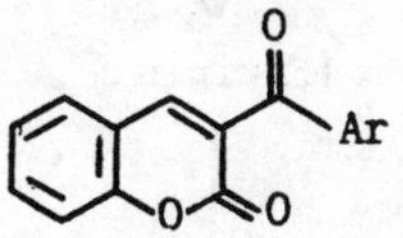

Benzils

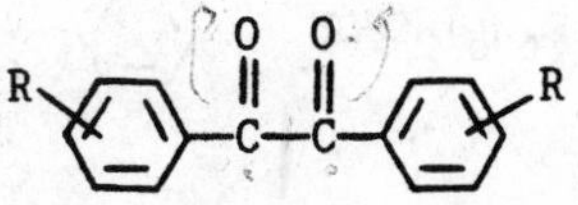

Benzoyldialkyphosphonates

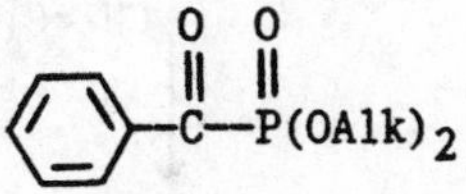

$$C-P(OAlk)_2$$

tert-Amines (H-donors)

Amino alcohols

$N(CH_2CH_2.OH)_3$

$CH_3-N(CH_2CH_2.OH)_2$

N,N-Dialklamino benzene derivatives $(CH_3)_2 N-$

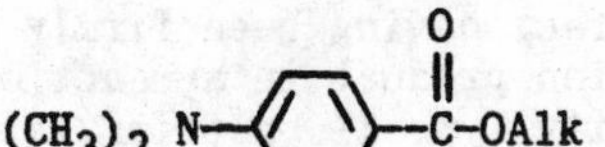

$(CH_3)_2 N-$ $C-OAlk$

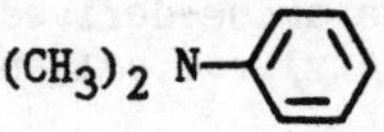

Figure 6

These Type-II photoinitiating systems have the amine-derived
radical in common, and it is generally accepted that it is this
radical rather than the ketone-derived radical (ketyl radical)
that is involved in the initiation step. Recent experiments using
a non-polymerising model substrate have now provided evidence that
the tert-amine derived radical is indeed the initiating radical in
Type-II systems.

e.g.

$$(C_6H_5)_2C{=}O \ / \ \langle\bigcirc\rangle{-}N(CH_3)_2$$

$$\downarrow h\nu$$

[Exciplex]

$$\downarrow$$

$$(C_6H_5)_2\overset{\cdot}{C}{-}OH \ + \ \langle\bigcirc\rangle{-}\underset{CH_3}{\overset{|}{N}}{-}CH_2\cdot$$

$$\downarrow CH_2{=}C(\langle\bigcirc\rangle{-}CH_3)_2$$

$$\langle\bigcirc\rangle{-}\underset{CH_3}{\overset{|}{N}}{-}CH_2{-}CH_2{-}CH(\langle\bigcirc\rangle{-}CH_3)_2 \ + \ (C_6H_5)_2\overset{\overset{OH}{|}}{C}{-}\overset{\overset{OH}{|}}{C}(C_6H_5)_2 \quad \text{a.o}$$

The major product turned out to be the product resulting from
addition of the amine-derived radical to the model substrate, the
benzophenone-ketyl radical ending up as the pinacol.

This fact having been firmly established, the occurrence of this
addition product in a reaction mixture may even be used to
identify Type-II photoinitiating systems, provided of course it is
not formed from Type-I photoinitiator/tert-amine combinations.

Photoinitiator / ⟨◯⟩-N(CH₃)₂ $\xrightarrow[\text{CH}_2\text{=C}\left(\langle\bigcirc\rangle\text{-CH}_3\right)_2]{h\nu}$ ⟨◯⟩-N-CH₂-CH₂-CH$\left(\langle\bigcirc\rangle\text{-CH}_3\right)_2$
│
CH₃

⟨◯⟩-C(=O)-⟨◯⟩ ·   +   (major product)

[thioxanthone structure] ·   +   ·

⟨◯⟩-C(=O)-C(=O)-⟨◯⟩ ·   +   ·

⟨◯⟩-C(=O)-P(=O)(OAlk)₂ ·   +

[2,4,6-trimethylbenzoyl diphenylphosphine oxide structure] ·   −

⟨◯⟩-C(=O)-C(OCH₃)(OCH₃)-⟨◯⟩ ·   −

The results shown in the table clearly indicate that the amine-
derived radical, if formed at all, does not contribute to the
initiation in combination with Type-I photoinitiators.

In the following, some attention will be given to two important
aspects of photoinitiated (free-radical) polymerisation processes.

1.   The configuration of the (triplet) excited state of the photoinitiator

Most Type-I photoinitiators possess absorption maxima λ_{max} around 330 nm with modest extinction coefficients ε_{max} between 200 and 250. For use in highly pigmented (often TiO_2) systems λ_{max} should preferably be shifted towards longer wavelengths and ε_{max} should be much higher in order to overcome the absorption of the pigment.

Attempts to achieve this have so far concentrated on substitution in the para-position of the benzoyl moiety, which is the common chromophoric group. The question can be raised, however, whether one can simply do so without adversely affecting the photoreactivity of the photoinitiator.

Comparative tests made on a series of hydroxyphenones substituted in the para-position have led to the conclusions [2] :

(i)   introduction of alkyl groups has a negligible effect if any

(ii)   introduction of alkoxy groups slightly reduces the photoreactivity, and

(iii) introduction of alkylthio and dialkylamino groups strongly reduce the photoreactivity (see Fig. 7).

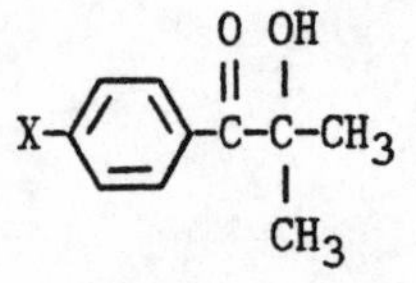

$$X-\!\!\left\langle\bigcirc\right\rangle\!\!-\overset{\overset{\displaystyle O}{\|}}{C}-\overset{\overset{\displaystyle OH}{|}}{\underset{\underset{\displaystyle CH_3}{|}}{C}}-CH_3$$ [C] = 3.10^{-1} in an acrylate system

X	Curing speed (m.min^{-1}) to obtain "Persoz" 170 s
H-	41
$(CH_3)_2CH-$	37.5
CH_3O-	27.5
$HO-CH_2CH_2-O-$	39
CH_3S-	< 5
$HO-CH_2CH_2-S-$	< 5
$(CH_3)_2N-$	< 5

Figure 7

Corroborating evidence is provided by
the crosslinking of trimethylolpropane
triacrylate (TMPTA) in solution
monitored by laser-nephelometry as
developed by Decker[3]

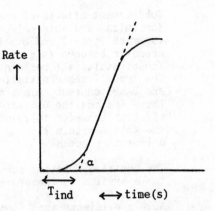

I	$T^{\ast}$-state	R = -CH$_3$		R = -(CH$_2$)$_{11}$CH$_3$		R = -CH$_3$	
		T_{ind}	tqα	T_{ind}	tqα	T_{ind}	tqα
H-	n-π*	25,3	11,2	24,9	11,0	25,2	11,0
CH$_3$O-	n-π*	n.d.[a]		n.d.[a]		n.d.[a]	
CH$_3$S-	π-π*	64,3	3,2	41,2	2,4	insoluble	
C$_6$H$_5$-	π-π*	n.d.[a]		n.d.[a]		n.d.[a]	
(CH$_3$)$_2$N-	CT	no polymerisation		no polymerisation		insoluble	

[a] not determined yet

The effect of substituents on the photoreactivity of these Type-I
(α-cleavage) photoinitiators can be explained on the basis of the
configuration of the lowest excited triplet state:

$$T_{n-\pi\ast} > T_{\pi-\pi\ast} \gg T_{CT}. \quad 4$$

Substituent effects of a similar nature are also known to occur
for intra- and intermolecular hydrogen-abstraction reactions
(Norrish Type-II and bimolecular photoreduction respectively) of
aromatic ketones (e.g. valerophenones and benzophenones). [5]
These results all pertain to <u>hydrogen-atom</u> abstraction reactions.
The Type-II photoinitiating systems used in practice, however, all
are based on tert-amines as the hydrogen-donating component. In
these systems the ultimate hydrogen-transfer is the result of an
electron-transfer followed by a proton-transfer from the amine to
the ketone within the exciplex formed, which is a fundamentally
different process.

The question, whether substituent effects are to be expected for
these systems is therefore fully justified.

Some preliminary experiments on the curing of a standard acrylate
resin then showed a clear "substituent"-effect which might well be
the result of the different nature of the lowest excited triplet
state, the order of increasing photoreactivity being

$T_{n-\pi^*} < T_{\pi-\pi^*}$ however. (See Fig. 8)

2.   <u>Oxygen-inhibition</u> and the role of tert-amines

A well known phenomenon in connection with free-radical
polymerisation processes is the adverse effect of oxygen from the
surrounding atmosphere commonly known as oxygen-inhibition,
leading to:

   (i)   an induction period

  (ii)   a reduced rate of polymerisation, and

 (iii)   an incomplete conversion of unsaturated functionalities,

which of course is most pronounced in the curing of thin coatings
owing to an unfavourable surface-to-volume ratio.

From a scrutiny of the photophysical and photochemical processes
taking place as shown in Fig. 3, oxygen can be seen to possibly
interfere in two different ways at two different stages: (i)
quenching of the excited triplet state of the photoinitiator and
(ii) scavenging of primary initiating radicals and/or propagating
macro radicals.

We will now try and estimate the actual contribution of each of
these processes to the overall oxygen-inhibition observed.

Curing speed (m.min^{-1}) in an acrylate resin [a]

Ketone/NMDEA (1:2)	T*-state	1	3	5 Wt%
(benzophenone structure)	n-$\underline{\pi}$*		6	
(benzofuran phenyl ketone structure)	n-$\underline{\pi}$*		8	
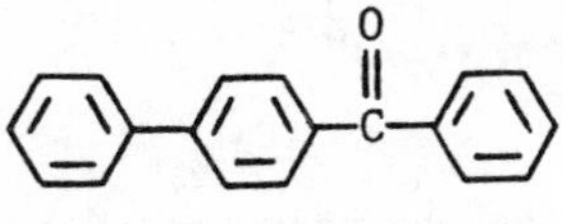 (biphenyl phenyl ketone structure)	π-π*	2	26	46
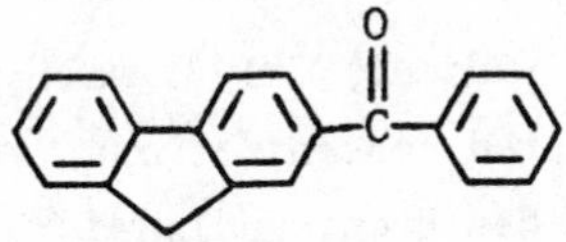 (fluorenyl phenyl ketone structure)	π-π*		28	

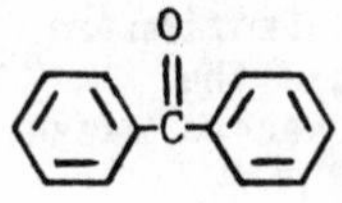

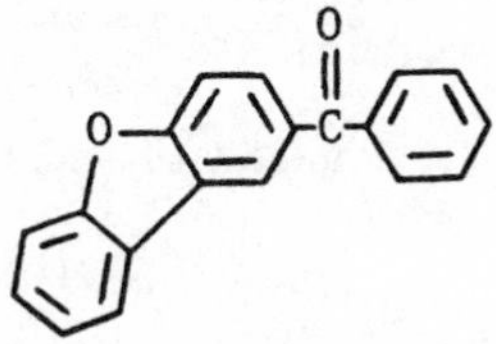

a) light-source: Philips HOK-6 lamp (80W/cm); film thickness 20μm

Figure 8

(i) <u>Oxygen-quenching</u>:

$$[\text{R-R}']^*_T \xrightarrow{k_\alpha} \text{R}\cdot + \cdot\text{R}' \qquad \text{fragmentation}$$

excited photoinitiator   primary radicals

$$[\text{R-R}']^*_T + \text{O}_2 \xrightarrow{k_q} \text{R-R}'_{\text{So}} + {}^1\text{O}_2^* \qquad \text{quenching}$$

excited photoinitiator   ground-state photoinitiator

The ratio of the two competing processes may be represented by the quenching equation:

$$(\emptyset_{R.})_{N_2} / (\emptyset_{R.})_{O_2} = 1 + k_q . \tau . [O_2]$$

in which ϕ_R. is the quantum yield of radical formation, kq is the rate constant for quenching, and τ ($=^1/_{k_q}$) is the lifetime of the excited triplet state of the photoinitiator. Inserting now the generally accepted value of $5.10^9 1.\text{mol}^{-1}.\text{S}^{-1}$ for kq and an estimated value of 10^{-3} $\text{mol}.1^{-1}$ for the oxygen concentration in the surface layer of the thin film it can be concluded that quenching is negligible for a lifetime τ in the nanosecond range.

(ii)   Oxygen-scavenging:

$$R. + M \xrightarrow{k_a} R\text{-M}. \qquad\qquad \text{initiation}$$

primary radical

$$R. + O_2 \xrightarrow{k_{O_2}} R\text{-OO}. \qquad\qquad \text{scavenging}$$

primary radical   inactive peroxy radical

The ratio of these two competing processes may be represented by the following equation: $\text{Rate}_{init}/\text{Rate}_{scav} = k_a[M]/k_{O_2}[O_2]$

From the known values for k_a (e.g. benzoyl radical) $\sim 10^5 1.\text{mol.}^{-1} S^{-1}$ and $k_{O_2} \sim 10^9 1.\text{mol.}^{-1} S^{-1}$ and the estimated concentrations of unsaturated functionalities and oxygen, it has been concluded [6] that the $[O_2]$ has to drop by more than two orders of magnitude before polymerisation can start. It will be clear then that all actions to diminish oxygen-inhibition should aim at depletion of oxygen.

Possible in principle are:

(i)   The exclusion of oxygen altogether by using an inert atmosphere (nitrogen-blanketing).

(ii)   The use of paraffin waxes, which migrate to the surface area to prevent oxygen diffusion into the coating.

(iii) Depletion of oxygen by (a) providing a high concentration of primary radicals, and (b) by the use of trapping agents such as thiols, phosphorus (III), derivatives, and tert-amines, which are all readily oxidised.

The beneficial effect of the addition of tert-amines on the
performance of common Type-I photoinitiators (which do not need
tert-amines to generate primary radicals) is well known, and has
been attributed to the reaction sequence shown.

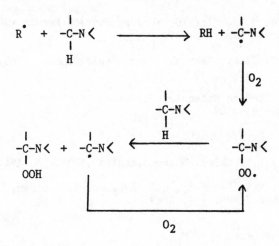

Tert-amines scavenge the oxygen present, and thus bring the $[O_2]$
down to a level that allows the initiation to become competitive
with the reaction between primary initiator radicals and oxygen.
Experiments with a Type-I photoinitiator in the absence and
presence of a tert-amine under various conditions (absence and
presence of O_2 respectively) [7] have so far not been able to
indicate any involvement of tert-amine derived radicals in the
initiation step, whereas in Type II systems tert-amine derived
radicals are in all likelihood the only initiating radicals. This
observation might be explained by higher rates of addition of the
primary radicals generated from Type-I photoinitiators as compared
with the rate of addition of the various α- amino alkylradicals.

References

1) For a recent review cf. H.J. Hageman in "Photopolymerisation and Photoimaging Science and Technology" (Ed. N.S. Allen) Ch.1. Elsevier Appl. Science, 1989.

2) J. Ohngemach, M. Koehler, G. Werner, Radtech Europe 1989 Florence, Conf. Papers p.639.

3) C. Decker, M. Fizet, Makromol.Chem. Rapid Commun. 1980, 1, 637.

4) N.J. Turro, "Modern Molecular Photochemistry" Ch.13. Benjamin/Cummings 1978.

5) Ref. 4. Ch. 10.

6) C. Decker, A.D. Jenkins, Macromolecules, 1985, 18, 1241.

7) J.E. Baxter, R.S. Davidson, H.J. Hageman, T. Overeem, Makromol. Chem. 1988, 189, 2769.

Trends in UV Equipment

R.E. Knight

SPECTRAL TECHNOLOGY LTD., 667 AJAX AVENUE, SLOUGH, BERKS.
SL1 4DB, UK

1 INTRODUCTION

One of the first production UV dryers in Europe was
installed in Glasgow in 1972 for printing on foil.

In reviewing the current situation, I think it is
probably worthwhile to examine the differences between
this first system and a typical unit as would now be
supplied for a similar application.

Firstly, the current reflector unit would be considerably smaller. The 1972 unit was 200mm wide x 150mm
deep. The current unit could be as little as 100mm wide
x 125mm deep.

Secondly, the new system would be more powerful.
In 1972, 80 watts/cm lamps only were available. Now 120-
160 watts/cm lamps are supplied as standard and in some
circumstances powers up to 240 watts/cm are available.

Thirdly, the current system would probably be water-
cooled as against air-cooling in 1972, this development
being mainly in response to the effect of higher power
in a reduced volume.

Finally, the construction in 1972 was basically
fabricated sheet metal, whereas today it would almost
certainly be of extruded aluminium.

In summary, the most fundamental change in UV equipment over the last 18 years has been in obtaining about
twice the power from a unit occupying about half the
longitudinal space.

What is perhaps surprising, is that in the vast
majority of applications the 1990 model would still contain the standard medium pressure mercury arc lamp of the
type used in the 1972 model.

The main pressures for higher power and smaller size
have been:

1.   Printing presses in particular have little space
     available for UV dryers and, although new presses
     have been developed, very few of these have signif-
     icantly increased the available space. Also, the
     increase in inter-colour and inter-coater drying
     requirements has made necessary the fitting of UV
     systems in even more restricted areas of the press.

2.   Fast drying has inevitably led to faster running
     speeds which have, in turn, increased demand for
     more power, often in situations where additional
     space for extra lamps is not available.

3.   Competition pricing meant that it was cheaper to
     increase lamp power than to provide additional ref-
     lector/lamp units.

Although higher power lamps should, in theory, be more
efficient due to the increased UV sensitivity, this
possibility has not been a serious consideration in the
development of smaller, more powerful systems.

Cooling

Small, high powered systems have one major disadvantage which can be summed up in one word - HEAT. Because of their size, all aspects of cooling control
become more critical. These can be divided into four
main areas:

(a) Lamp cooling.
(b) Reflector cooling.
(c) Substrate cooling.
(d) Cooling of exposed machine parts.

Reflector cooling has been improved by the move towards water cooling and extruded aluminium reflectors,
and this has also improved lamp cooling to some extent.

The cooling of substrate materials and adjacent machinery has become the focus of much design activity, not
just because of the higher concentration of IR energy,

but also because of parallel application trends. These are:

(a) Increased web coating of very thin heat-sensitive materials.

(b) Multi-colour and in-line coating situations where substrates can be exposed to numerous lamps in a single pass.

(c) The siting of reflector systems in the press, close to or even against cylinders, is often required as pre-coating, inter-colour and perfecting applications become more widely used.

Methods of cooling exposed materials fall into two categories. The first is to positively cool the material either during or after exposure, and the second method is to reduce the IR radiation reaching the substrate.

Positive cooling methods include:

(a) Water-cooled cylinders under the substrate. (Figure 1)
(b) Water-cooled platens against the substrate. (Figure 1)
(c) Air knife cooling.

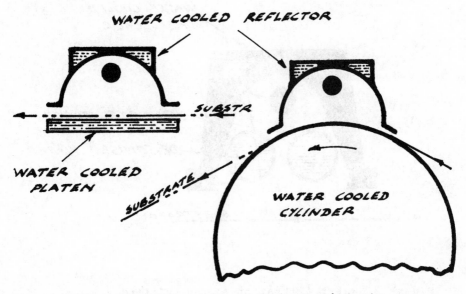

<u>Figure 1</u> Water-cooled surfaces against substrate

Water-cooled cylinders are very effective, making possible the UV coating of extremely thin heat-sensitive materials such as P.V.C. However, they are expensive, space consuming, often difficult to install, and require to be driven in synchronisation with the coater or press.

Water platens can be effective, depending on the closeness of contact with the substrate, and they have the advantage of being compact and easy to install.

Air knife cooling is of limited practical use in fast running applications, particularly if space is limited, but has uses in localised heating problems and in slow running applications.

IR reduction methods include such techniques as:

(a) Water filtration. (Figure 2).

(b) Dicroic reflectors. (Figure 3).

Both of these methods are successful in reducing direct IR radiation but they also have the disadvantage of reducing short wave UV radiations.

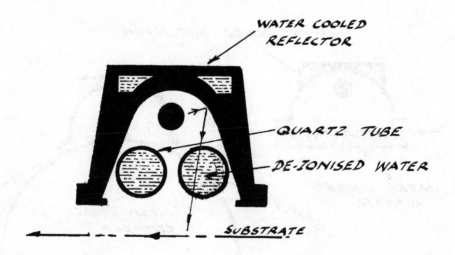

Figure 2   Water filtration system (Colordry Ltd.)

Water filtration has been widely used and is an attractive approach where the loss of short wave radiations is less important than the cooling advantages gained. Long term, it is arguably still the most practical and effective method available for the future.

Dicroic reflectors, which reflect the UV but not the IR, are an old concept which has attracted new interest recently. However, they still have the disadvantage of being difficult and expensive to manufacture. Also, experience is still required to eliminate concerns about their practical life in the hot and possibly contaminated atmosphere in which they must operate. Nevertheless, this is an an interesting area for further experimentation and development.

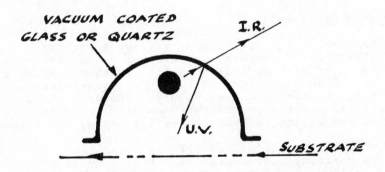

Figure 3 Diocroic reflector

Lamps

Apart from power level increases, the lamp itself has not changed significantly. As has already been said, the vast majority of equipment is still fitted with the standard medium pressure mercury arc lamp. Other lamp types do not match the medium pressure lamp in flexibility, operating costs, or even radiation efficiency.

The electrodeless lamp is the only other type which has made any significant impression on the UV market, and even here it has been mainly in specialised application areas where efficient marketing has perhaps played a more important role than any actual technical advantage.

However, in the application of metal halide or additive lamps, the electrodeless lamp has led the attack.
The fact that the metal halide electrodeless lamp is more
easily manufactured and can be interchanged with conventional lamps, has made the manufacturer, Fusion Inc., seek
out specialist applications for these lamps with some
degree of success.

Similarly, there has been some increase in metal halide medium pressure lamps for such applications, but for
the vast majority of UV users, the metal halide lamp still
has more disadvantages than advantages. These are:

(a) Only short or lower powered lamps are available.

(b) It is more expensive and has a shorter effective life.

(c) There is no universally applicable improvement in
     curing efficiency.

Control Improvements

The recent introduction by Wallace Knight Ltd. of
their "UV-Tronic" computerised control system has changed
what was a conventional large multi-switched control into
a compact high technology unit in keeping with current
trends in printing and coating press design. It is particularly relevant at a time when multi-lamp, multi-station,
combination drying is becoming an increasing requirement.

Computer control gives all the usual user advantages
of such systems, i.e. more user friendly, more reliable,
dramatically smaller control panels, and programmed warning and safety indicators. Also, it makes possible many
additional features. Typical examples are :

(a) Automatic lamp power and age monitoring.

(b) Programmed pre-selection of lamp power.

(c) Temperature monitoring.

(d) Cooling water and air flow monitoring with automatic
     correction if necessary.

(e) Early fault warning.

In the future, it is planned to build-in UV radiation
level monitoring, with the possibility of automatic up-
rating of power to compensate for lamp ageing or other
factors.

Designs for Special Applications

The continuing growth in the application of UV curable materials has led to the development of a wide range of conventional systems, together with many specialized designs for specific applications.

Nitrogen purged systems are now more generally available, and with the growth in the UV silicone coating market, these systems are becoming more varied and sophisticated (Figure 4).

Flameproof systems are now available for applications where this is required. UV systems with heat recovery arrangements have been available for some time but are now gaining in popularity as installations become larger and more energy-consuming.

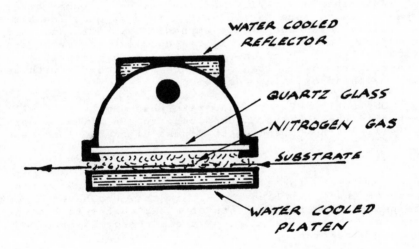

Figure 4 Nitrogen purged system

Ozone removal from exhausted air has always been technically possible, but is now becoming a more regular requirement as fears concerning air pollution gain momentum.

Alternative Technologies

In considering the future of UV it is worthwhile to compare the alternative competitive technologies.

Electron beam (E.B.) is the main competitor, but it
is unlikely that this could replace UV for the wide and
increasingly varied range of applications. Although E.B.
has advantages for thick pigmented coatings and laminating
applications, these are, and are likely to remain, a very
small proportion of the total equipment market. For thin
pigmented and clear coatings, E.B. offers few unresolvable
advantages over UV and has several disadvantages, the
main ones being the high capital cost and the large physical size of the equipment.

Water based alternatives do not have the range of
physical advantages of many UV curable materials. They
are cheaper, more flexible, and more compatible with conventional materials, but these advantages may well disappear in time as new UV materials and formulations become
available.

Water based materials are perceived in some market
areas as being more "green" , but this is questionable
and likely to change if the emphasis swings towards air
pollution problems. In this respect, ozone generation
by UV equipment is often exaggerated. The quantities
produced are low and can easily be filtered off with relatively cheap existing technology.

Obviously, UV, E.B. and water based materials will
each find their market niches but, in the writer's opinion,
UV is likely to be the most successful in the long term.

The Future

The future for UV equipment is, as always, in the
hands of the "formulators". It is they who have to meet
the performance requirements of their customers and the
end user. Naturally, most of their energy is expended
in this direction.

Similarly, equipment designers spend a great deal
of their time trying to get a quart into a pint pot, often
to the detriment of curing efficiency.

Further improvement in curing efficiency requires
a joint effort, a mutual exchange of information, and
more co-ordinated test procedures. Some form of standard-
ised reactivity measurement might now be possible with
the improved cure test equipment now available. This
would be useful to both sides.

Information on the relationship between cure rate
and UV intensity is another cloudy area and this knowledge
is essential if the most effective reflector/lamp configuration is to be provided.

Of course, increased reactivity would always be useful, but perhaps a more possible target is improved reactivity to long wave UV. This would encourage the further
development of metal halide lamps, and make cooling techniques, such as water filtration, a more universally practical tool. In addition, the better penetration of long
wave radiations, combined with improved cooling, would
make the curing of thick pigmented films more possible
as longer high-intensity exposure, without heat problems,
would be practical.

These may appear to be modest targets, but in fore-
casting the future it seems more helpful to state the possible rather than the highly speculative.

Applications

UV Curable Media for Decal Production in the Ceramics Industry

P.R. Jackson

WHITEWARES DIVISION, CERAM RESEARCH, QUEENS ROAD, PENKHULL, STOKE-ON-TRENT, STAFFS. ST4 7LQ, UK

1. INTRODUCTION

A variety of methods are employed in the Ceramic Industry to decorate tableware; these include direct screening of solvent-based inks and offset printing processes like the total transfer system, where silicon pads transfer thermoplastic inks to plates.[1] However, the only way to decorate complex-shaped ware with sophisticated multi-coloured designs is to use decalcomanias or "decals" (figure 1).

Decals are transfers and comprise a special base paper, ink layers and a clear covercoat. Depending on the nature of the base paper, either heat or water is used to effect release of the decal from the backing paper.[2,3] The decal is then applied gum (or adhesive) side down onto the ware. Subsequent firing in a kiln burns off the organics at 200°C - 400°C, leaving the inorganic colour to fuse to the ware at temperatures of 750°C +. Decals are frequently used at the expensive end of the market for the on-glaze decoration of bone china and porcelain ornaments, dinner services, etc.

Decal Types

Decals can be divided into two types according to their method of production:-

Lithographic. These decals are prepared by coating a thin varnish layer ($\sim 2\mu$m) onto decal paper and dusting the colour (pigment and flux) on top. Up to 20 or more colours can be applied in this way at a rate of < 1 colour per day. The decal is completed by screening a

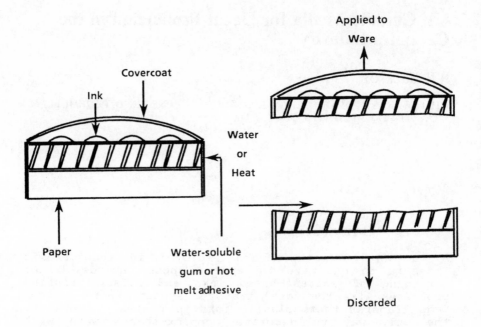

<u>Figure 1</u>   Decal composition

solvent-based covercoat (comprising polymer, solvents, dye and plasticizer) over the entire design; this too requires a day to dry.

    <u>Mitographic (or Silk-Screened).</u>   In this case the colour is incorporated into a solvent-based medium to render it screenable.   Ink layers of around 30μm are typically screened onto the paper and require drying times of up to 8 hours, even in forced air conditions with the sheets racked vertically. The next colour is usually screened at the same shift the next day.   As for lithographic decals, the final design (of perhaps 10 colours) is covercoated.

    The two types currently enjoy a 50/50 share in the decal market although improvements in screen mesh and the ease of setting up a screening unit in a factory mean mitographic decals are tending to take over. Lithographic decals, however, offer superior tonal range.   The total U.K. Ceramic decal market for 1987 was £16 million.

REFERENCES

1. B.C.R.A. U.K. PATENT APPLICATION GB 2118 900A
2. R. Gater, <u>Ind. Ceram.</u>, 1985, <u>792</u>, 197
3. J. J. Svec, <u>Ceram Ind</u>., 1980, <u>115(5)</u>, 49

2. DRAWBACKS WITH CURRENT SILK-SCREEN DECAL PRODUCTION

By replacing the usual solvent-based inks and covercoats in silk-screen decals with ultra-violet (UV) equivalents, phenomenal reductions in production time should eventually be realised: production could literally take minutes rather than days.

The drawbacks with the existing method of production are as follows:-

*   The lengthy drying times of inks and covercoat.

*   The toxicity arising from solvent emission.

*   The vast areas of wasted factory space used to rack drying decal sheets.

*   The need for humidity control to stop paper curl during ink/covercoat drying at elevated temperatures.

In addition, there are problems associated with the final product: Firstly, residual solvent in the covercoat causes decals stacked on top of one another to stick; this is termed 'blocking' and is only alleviated by interleaving with paraffin wax paper. Secondly, the covercoat needs to possess substantial flexibility to allow the decal to be moulded to the contours of the ware; although plasticizers help to achieve this, residual solvent plays a part too and because the latter evaporates with time, decals quickly become too brittle for application without strict storage conditions (e.g. 60% relative humidity, 60°F temperature, vertical sheet stacking).

Comparisons between solvent-based and UV coatings systems in general are well documented.[4] Regardless of the system used, the function of the organics in a decal will always be to place the inorganic colour in the right position on the base paper and then on the ware. Once there, organics need to be eliminated at as low a temperature during firing as possible.

Polymethacrylates are currently used as the binding polymer in solvent-based inks and covercoats, since they readily revert to the monomer or heating. Fortunately, of course, acrylates and methacrylates lend themselves very well to UV-initiated free-radical polymerisation and a UV cure system would reduce or eliminate many of the aforementioned problems like length of production time, solvent emission and variable covercoat stress/strain properties.

REFERENCE

4.   R. Holman and P. Oldring (Eds.), 'UV and EB Curing formulations for Printing Inks, Coatings and Paints', SITA Ltd., London, 1988. Chapter 1. P.16.

3.   U.V. COVERCOAT DEVELOPMENT AT CERAM RESEARCH

Work on UV decals commenced in 1986 with the UV covercoat aspect being tackled first. Covercoats must possess good flexibility but also adequate strength for handling and some plastic extension. Instron tests on fresh conventional solvent-based covercoats indicated the range of properties a UV covercoat would need to possess: A stress of > 2 Nmm^2, a strain of $> 50\%$ and a percentage permanent deformation of $>10\%$ (crosshead speed = 100mm min^{-1}). Trials over the past four years at local factories have indicated that for hand applied decals, subjective qualities like handleability and applicability are also important.

Commercially available UV media and precursors were trialled initially, but none gave coatings with all the desired properties. Flexibility in particular was consistantly absent.

Small Scale Preparation

Ceram Research have developed a suitable covercoat using a two stage UV cure process. By early 1987, promising samples had been prepared on a small (15g) scale as follows: Two methacrylate monomers were hand stirred and irradiated (cure time CT$_1$) in the presence of an intramolecular cleavage type photoinitiator using low pressure mercury lamps; on reaching a viscosity of $>100,000$ mPa.s. secondary components (including methacrylates and acrylates of varying functionality and more photoinitiators) were then mixed in with the 'prepolymer' to give the covercoat. After silk-screening, a second cure by medium pressure (80 Wcm^{-1}) lamp (cure time CT$_2$) gave the cross-linked covercoat. CT$_2$ values of > 10s were initially obtained.

Unlike most UV formulations then, the aforementioned UV covercoat contains no epoxy, urethane, polyester, polyether, etc., (meth)acrylate as reactive oligomer. Instead, a prepolymer (comprising 80% inert polymethacrylate and 20% unreacted monomer/photoinitiator) is used. The unique flexibility of the cured covercoat is believed to arise from the secondary components cross-linking amongst themselves and largely around the (pre)polymer during CT$_2$. (Some links between the network and the polymer are likely to arise by e.g, a growing polymer radical abstracting a H$\cdot$ species from the polymethacrylate to leave a radical capable of further reaction with secondary components).

Pilot-Scale Preparation.

Prepolymer can now be prepared on a 600g scale (ultimately giving ~ 1kg of covercoat) using the apparatus depicted in Figure 2. The monomer/photoinitiator mix is pumped around the right hard circuit, being irradiated as it passes up through the quartz tube between the quartz and the brass. The left hand circuit represents the temperature regulatory system, with the water passing through the central brass tube. After around 2 hours circulation/irradiation, (CT$_1$), the homogeneous prepolymer (η = 1000-3000 mPa.s) must be removed to prevent apparatus blockage. However, subsequent storage under nitrogen at 45°C allows the cure to continue (free from oxygen inhibition) and the ideal prepolymer viscosity of about 200,000 mPa.s is attained after 4 hours.

Plots of prepolymer viscosity vs. storage time under nitrogen show a slow viscosity rise followed by a sudden increase due to the Trommsdorf or gel effect. Figure 3 illustrates this and shows the importance of storage temperature; the plot closest to the x - axis indicates the effects of storage under nitrogen at 25°C with a 2 minutes exposure to the atmosphere every hour. Immediate addition of secondary components (which include an inhibitor) to a <200,000 mPa.s. prepolymer produce a covercoat with a shelf-life of > 6 months at room temperature in the dark.

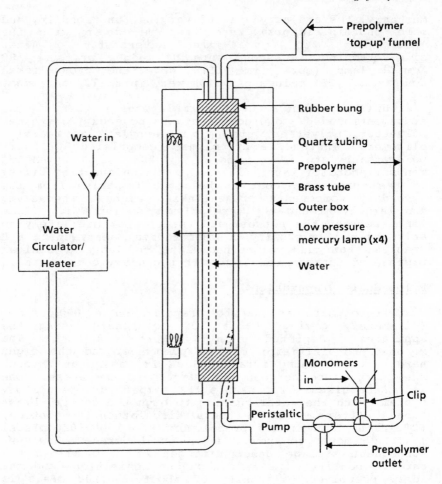

Figure 2   Pilot scale apparatus diagram

Figures 4 and 5 illustrate a typical pilot-scale covercoat preparation along with a specific formulation. Typical properties for this formulation are:

Stress = 2.0 Nmm^{-2},
Strain = 55% (at break),
% permanent deformation = 10%, at CT_2 = 3s.

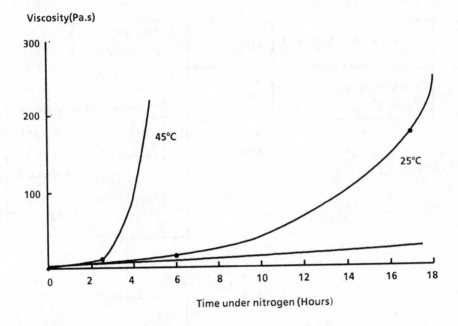

Viscosity(Pa.s)

Time under nitrogen (Hours)

Figure 3   Prepolymer viscosity vs. storage time under nitrogen (after CT_1)

4. PREPOLYMER CHARACTERISATION BY GEL PERMEATION CHROMATOGRAPHY

Gel permeation chromatography (GPC) is a form of liquid chromatography where separation is based on effective molecular size in solution.[5,6]

Work has shown that molecular weight distributions (MWD) for the polymer in prepolymers are fairly consistant from run to run and that there is only a slight shift (to higher molecular weight) as time under nitrogen at 45°C increases during any given run.

By separating the prepolymer using 2 x 100 Å columns, instead of the usual single mixed bed column, information has been obtained on the low molecular weight species in the prepolymer. Prepolymers have been shown to contain predominantly unreacted monomer/photoinitiator in addition to polymethacrylate, although other species (caused possibly by premature termination of $RM_1^{\bullet}$, $RM_2^{\bullet}$, $RM_3^{\bullet}$, etc.) occur in very small

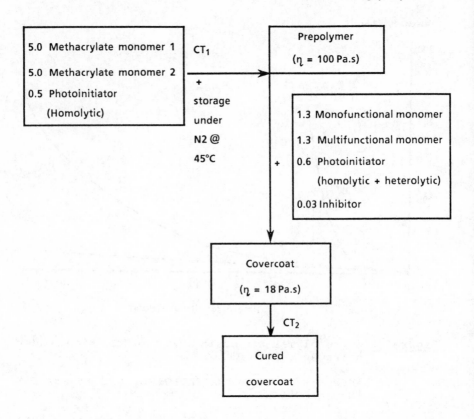

<u>Figure 4</u>   Pilot scale covercoat preparation

concentrations (M = monomer and R• = a radical from Irgacure 651); interestingly, similar species occur in unirradiated monomer/photoinitiator mixtures, although in even smaller concentrations.

The molecular weights calculated at Ceram Research are not accurate, being based on polystyrene standards and Mark-Houwink constants for methyl methacrylate.[6]

REFERENCES

5.   B. J. Hunt <u>Chrom. Anal.</u> 1990, <u>11</u>, 9
6.   J. M. Evans, <u>Polym.Engng.Sci.</u>, 1973, <u>13(6)</u>,401

COVERCOAT

10.5g prepolymer	+ 1.30g	Ethoxyethyl acrylate	(EOEA)
(η = 200,000 mPa.s)	1.00g	Ethylene glycol dimethacrylate	(EDMA)
	1.00g	Ebecryl 140	(E.140)
	0.20g	Photomer 4182	(P.4182)
	0.40g	Quantacure DMB	(Q.DMB)
	0.20g	Quantacure ITX	(Q.ITX)
	0.03	Hydroquinone	

INK MEDIUM

5.25g prepolymer	+ 3.25g	EOEA
(η = 200,000mPa.s)	2.50g	EDMA
	2.50g	E.140
	0.50g	P.4182
	0.40g	Q.DMB
	0.20g	Q.ITX
	0.03	Hydroquinone

Figure 5   A comparison of UV ink and UV covercoat compositions

5.   PROBLEMS ASSOCIATED WITH THE EXISTING U.V. COVERCOATS

Before the UV covercoat can be commercially exploited, attention needs to be turned to the following dilemmas:

(a)   The covercoat cure time, CT_2, is probably still too long at 3s.

(b)   The burn-out of UV covercoats during firing with certain colours (especially UV Inks) is poor.

(c)   Scale up of prepolymer production to a tonnage scale is required.

Regarding point (a), it is anticipated that formulation changes, the use of nitrogen blanketing and novel lamp technology will reduce CT_2. A further related problem is that conventional UV lamps emit infra-red radiation which helps CT_2, but brings about unacceptable decal paper curl.

On the subject of burn-out, UV covercoats give good results with lithographic decals, where very thin varnish layers represent the only organics below the covercoat. However, both solvent-based and UV curable inks (see Chapter 6) give disastrous burn-out, due to the dense cross-linked nature of the covercoat preventing ink "volatiles" from escaping to the kiln atmosphere during the early stages of firing. This is discussed further in Chapter 6.

Liaison with a chemical company (who are currently attempting to mimic the MWD and composition of the prepolymer, using a chemical [peroxide] route) will hopefully lead to prepolymer production on a ten tonne scale.

6.   U.V. INK DEVELOPMENT AT CERAM RESEARCH

Development of a compatible UV Ink for silk-screen decals commenced at the start of this year. The challenge is an awesome one, since depth of cure is required in 20μm screened ink layers with a colour to medium ratio of 2:1 by weight.

Initially the UV covercoat was used as the ink medium. However, since a medium of 600 mPa.s. is required and the U.V. covercoat generally has a viscosity in excess of 10,000 mPa.s. colour to medium ratios of only 1.1:1.0 were obtainable; such ratios give an inadequate depth of colour in fired decorated ware. By reducing the level of prepolymer and at the same time retaining the ratio amongst the secondary components, a satisfactory medium was produced (figure 5). 2:1 colour to medium ratios were obtained and the resulting ink cured (to thumb twist free) in < 2s under an 80 Wcm^{-1} lamp (figure 5).

Burn-out studies indicated that the UV Inks gave very poor on-glaze burn-out with UV covercoats under standard enamel firing conditions of 100°C/hr to 850°C, 1 hour soak. Interestingly, completing the same UV Ink decal with a non-cross-linked solvent-based covercoat resulted in a decal with excellent burn-out. Clearly, a cross-linked covercoat means delayed burn-out and the prevention of ink volatile escape; additionally, though, lack of through cure in the ink will increase the level of volatiles at any given temperature and aggravate the situation - support for the last hypothesis comes from the observation that red and black UV Inks (where the colour competes strongly with the photoinitiators for the available UV) produce much inferior burn-out with UV covercoats than blue inks (where the Ceramic colour has >30% reflectance at 350 - 520nm).

Some success in improving UV Ink/UV covercoat burn-out has been achieved by incorporating flux into the covercoat. Gradient firing trials have indicated that adding the flux lowers the covercoat burn-out temperature from 380° to 360°C; this is still higher than the figure of 330°C obtained for conventional solvent-based covercoats.

In addition, the problem of lack of through cure in the ink is being tackled through (a) the use of different photoinitiator/photosensitizer combinations in conjuction with ceramic colour UV spectral information and (b) the use of peroxides plus photoinitiators (thus using the heat as well as UV from lamps to promote ink polymerization).

7. THE FUTURE

The use of cationic UV, Cationic/free radical UV, or electron beam systems for inks and covercoats represent fall-back directions for further research, should a UV medium based on the UV covercoat prove unsuccessful. In the shorter term, successful large scale prepolymer production should lead the way to the commercial exploitation of UV covercoats in lithographic decals and of "UV Ink + solvent-based covercoat" decals.

The permission of the Chief Executive of Ceram Research, Dr. D.W.F. James, to deliver and print this talk is gratefully acknowledge. In addition, James Downes, Mark Crooks, Mark Hobbs and Sally Gay are all thanked for their various contributions to this work over the past four years.

The Use of Dynamic Mechanical Analysis (DMA) in the Assessment of Radiation Curable Optical Fibre Coatings

J. Mahone

HARCROS CHEMICALS UK LTD., LANKRO HOUSE, P.O. BOX 1, ECCLES, MANCHESTER M30 0BH, UK

1. INTRODUCTION

The advantages optical fibres have over conventional copper wiring (1,2,3,4) have seen them established as the preferred telecommunication medium. As their use has become more widespread, the attention given to the properties of optical fibre coatings have also become more intense. The need to fully assess coating properties is therefore of paramount importance. A vast range of physical property data are commonly used to assess optical fibre coatings and the aim of this paper is to highlight the important contribution Dynamic Mechanical Analysis has to offer. A whole host of references are available which cover optical fibre technology (1,3,11,12,13) and optical fibre coating properties (5,6,7,8,9,10) and therefore these aspects will not be referred to in this paper. This paper will however demonstrate the variety of ways in which DMA can be used, namely in (1) aiding resin design, (2) establishing degree of cure in coating films, (3) investigating post-cure effects and on a more general note to emphasise the appreciation of obtaining property information other than that obtained at room temperature.

2. DYNAMIC MECHANICAL ANALYSIS

DMA is a Thermal Analytical technique which measures the deformation of a material in response to vibrational forces over a temperature range. From these measurements, values are obtained of Storage Modulus - the "stiffness" of a system or a polymer's ability to store energy, and Loss Modulus - the ability of a polymer to dissipate energy in the form of heat.

A polymer's relative ability to store or dissipate energy is a direct measure of its viscoelastic properties and consequently DMA is one of the most sensitive methods available for measuring physical changes in a polymer (14,15).

DMA gives the important parameter of Glass Transition Temperature (Tg) which is normally allocated to the maximum in the Loss Modulus curve. This value gives a reference point which can be monitored to study any changes which may occur in a system during any investigations eg curing, post-curing, ageing etc.

DMA also gives the equally informative Storage Modulus/ Temperature profile. From this the temperature range over which the glass transition occurs can be studied and this paper will illustrate the benefit of the Tg range data in preference to Tg values.

3. OPTICAL FIBRE COATINGS

Acrylate based systems have been the preferred route for coating optical fibres for a number of years (1,3,5). The rapid ultra-violet curing of these polymers gives a neat, clean, in-line process which allows a coating to be applied to the glass fibre as it is manufactured and the coating is quickly cured.

Polyurethane acrylates have been the preferred chemical type of acrylate oligomer and the versatility of this class of polymer enables coatings to be made with a diverse range of physical properties. Polyurethane acrylate resin design differs for each coating type ie. primary coat, secondary coat or single coat and DMA offers a "picture" of how a material performs over a wide temperature range enabling judgements to be made on the suitability of that material for a coating type. Similar information can be obtained when the polyurethane acrylate oligomer is subsequently formulated with acrylate diluents. The choice and combination of reactive diluents must complement the design of the oligomer.

Figure 1 illustrates these points by considering the DMA thermogram of a potential primary coat system. The maximum in the Loss Modulus curve indicates a Tg of -60°C. It is the Storage Modulus curve which yields information on coating suitability. A primary coat system can be designed so as to have a constant change in modulus over

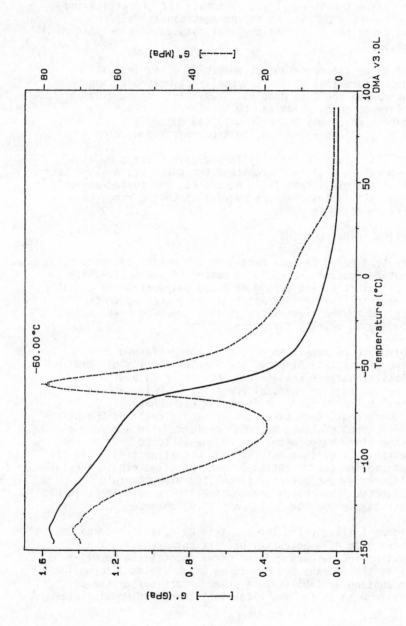

FIGURE 1   DMA   THERMOGRAM   OF   POTENTIAL   PRIMARY   COAT   SYSTEM

the temperature range so that if a fibre is subject to a drop in temperature there will not be a dramatic increase in modulus of the coating.

Alternatively, a coating can be designed to have low modulus values at low temperatures, as long as the Tg is outside the envisaged working temperatues of the optical fibre. DMA enables these types of considerations to be compared and potential performance of the coating established.

3.1. Assessing Degree of Cure

When investigating film properties of any ultra-violet (UV) curable composition it is important that a fully cured system is the starting point. It is also important to establish what "full cure" represents. A generally accepted view is that full cure is achieved when the physical properties of a film do not change on further exposure to the radiation source, (16,17). Consequently the methods employed to measure degree of cure are both important and numerous, and can vary from measuring hardness, soxhlet extraction, solvent resistance, abrasion resistance, mechanical properties, infrared spectroscopy to the relatively new photo DSC (18,19,20,21).

The following investigations show how DMA can be used to assess degree of cure, as accurately as any other available procedure and also to further demonstrate this powerful technique by revealing post-curing affects which can remain undetected by conventional property assessments.

4. RESULTS AND DISCUSSION

Of the three coating types available, i.e. primary, secondary or single, the following results   have been obtained when investigating potential single coat systems. In general, single coat systems have physical property data which is intermediate between the very soft, flexible low Tg primary coat and the high modulus, tough, high Tg secondary coat.

Table 1 shows the effect of increasing u.v. dose on a variety of properties which can be used to assess degree of cure of single coat system A. Soxhlet extraction gives the % of insoluble material in a cured film. Tensile Strength and Elongation at Break values are obtained from tensile testing measurements performed on an

TABLE 1   EFFECT OF U.V. DOSE ON PHYSICAL PROPERTIES*

U.V. dose (J/cm^2)	Soxhlet (% Gel)	Tensile Strength (MPa)	Elongation (%)	Tg (G", °C)
0.3	92	6.4	51	5
0.5	94	7.3	52	10
0.8	96	7.6	49	11
1.2	96	7.6	48	11
1.6	96	7.6	49	11

(* Single coat system(A), 150 micron unsupported film)

HARCROS
CHEMICAL GROUP

Instron machine. These types of measurements have been found to be very sensitive and very reproducible and are an excellent method for monitoring changes which may occur in film properties. The final column gives Tg data as measured by DMA where Tg is represented by the peak in the Loss Modulus curve. All results show clearly that full cure is obtained at the u.v. dose of 0.8 Joules cm^{-2}, since no further significant change in properties occur above this dose. These methods of assessing cure measure the bulk properties of the film and do not give information on the amount of conversion of acrylate groups. Fourier Transform Infra Red (FTIR) using an ATR technique does measure degree of cure of film by measuring the disappearance of the acrylate double bond. Similar results using FTIR$_2$ also indicate full cure is reached at the 0.8 Joules cm^{-2} dose. However, the limitation of this technique is that it is only the first 5-10 microns of the film which is assessed. As these results have been obtained on 150 micron thick films then FTIR data can not totally represent the remainder of the film thickness. Attempts have been made to assess cure, by FTIR, on the bottom surface of the cured film, however accurate, reproducible data is difficult to obtain.

The film thickness of 150 micron is fairly typical for assessing optical fibre coating properties and from Table 1 it is demonstrated that DMA is as sensitive a method for establishing degree of cure as any of the established methods.

Single coat System A was observed to show further changes in film properties on exposure to daylight. This effect seemed to be only observed with System A and not in any other single coat, primary coat or secondary coat systems. This warranted further investigation into this "post-curing" effect and Table 2 illustrates these effects by showing changes in film properties as measured by Instron tensile testing and DMA.

The first condition represents the optimum cure of system A previously obtained as described above. From the results it can be seen that heat treatment does not affect normal film properties but continued exposure to daylight increases the changes in film properties. The Daylight Simulator is a piece of apparatus containing a 500 watt mercury-tungsten lamp which emits a spectrum which represents/simulates natural daylight (22). The results show signifiant changes in properties from a film which was demonstrated to be fully cured according to the

TABLE 2   **EFFECT OF POST CURING ON PHYSICAL PROPERTIES OF SINGLE COAT SYSTEM (A)**

Condition	Tensile Strength (MPa)	Elongation (%)	T_g (G″, °C)
Dark, 25°C, 24 hrs	7	49	11
Dark, 80°C, 24 hrs.	7	49	11
Light, 25°C, 8 hrs.	9	53	17
Light, 25°C, 3 days	11	52	22
Light, 25°C, 6 days	12	48	25
Daylight Simulator	18	43	37

HARCROS
CHEMICAL GROUP

TABLE _3 EFFECT OF POST CURING ON PHYSICAL
PROPERTIES OF SINGLE COAT SYSTEM (B)

Condition	Tensile Strength (MPa)	Elongation (%)	Tg (G'', °C)
Dark, 25°C, 24 hrs.	8	50	-30
Light, 25°C, 3 days	9	52	-24
Daylight Simulator	8	42	-16

HARCROS
CHEMICAL GROUP

various methods indicated on Table 1. These results are
very reproducible and again illustrate the sensitivity of
both tensile testing and DMA as both can monitor/detect
film property changes.

To investigate these effects further, similar work was
done on an alternative single coat - System B - which was
based on a completely different design of resin. Table 3
shows Instron and DMA data for optimum cure, daylight
exposure and Daylight Simulator investigations. Instron
results do not indicate any significant changes occurring
in the film and using only Instron results it would appear
that changes only occur in System A. Indeed Instron
testing of a variety of primary and secondary coat systems
would also indicate no change in film properties on
exposure to daylight. However, it is the DMA information
which shows these observations are not as straightforward
as they appear. Monitoring the Tg of these systems does
indicate that "post-curing" is taking place, in
contradiction to the Instron data. The question,
therefore, arises as to why DMA data show changes in both
systems, and yet Instron data does not? To clarify this
further, again DMA investigations can suggest why these
differences occur, however instead of just using Tg data
it is the Storage Modulus/Temperature profiles which need
to be considered.

Figures 2 and 3 show the DMA thermograms of Systems A and
B respectively. As can be seen, System A not only has a
higher Tg but shows a much sharper drop in Storage Modulus
in the Tg region with System B having a much shallower
profile in the Tg region. To identify the significance of
these differences the comparison of Storage Modulus/
Temperature profiles for optimum curve and Daylight
Simulator exposure needs to be made. Figure 4 shows this
comparison for System B. The Simulator exposed material
in general shows a higher modulus value for a given
temperature than the optimum cured material and
consequently DMA will indicate a higher Tg value. However,
Instron testing is carried out at room temperature, or in a
temperature controlled room (typically 25°C), and if we
look at the 25°C region there is not much difference in
modulus for both systems. Figure 5, on the other hand,
reveals a significant difference in modulus for the System
A comparison. Consequently, Instron testing will indicate
changes in film properties for System A and not for System
B.

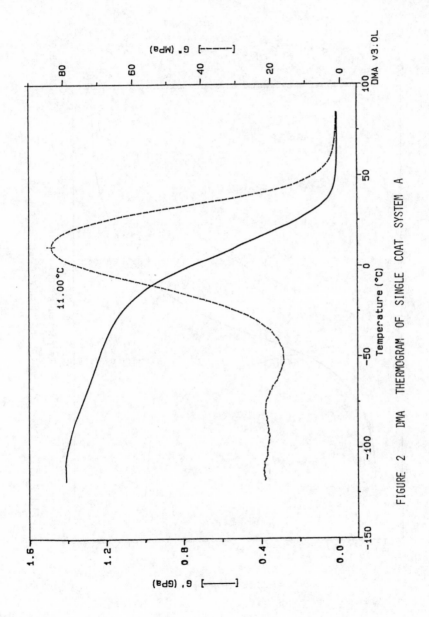

FIGURE 2   DMA THERMOGRAM OF SINGLE COAT SYSTEM A

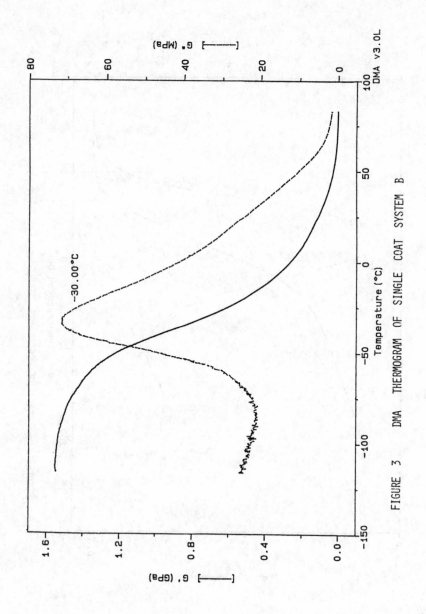

FIGURE 3   DMA   THERMOGRAM OF SINGLE COAT SYSTEM B

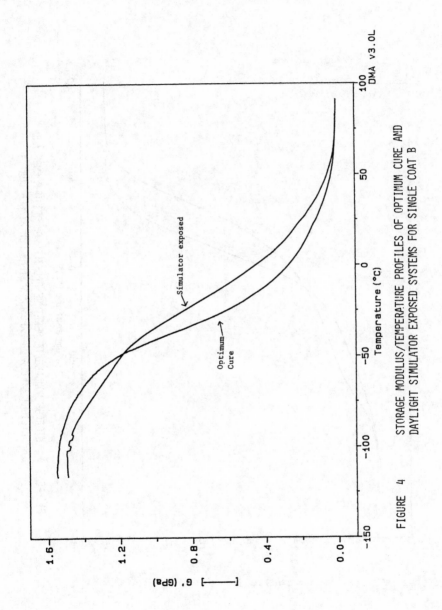

DMA v3.0L

FIGURE 4   STORAGE MODULUS/TEMPERATURE PROFILES OF OPTIMUM CURE AND
DAYLIGHT SIMULATOR EXPOSED SYSTEMS FOR SINGLE COAT B

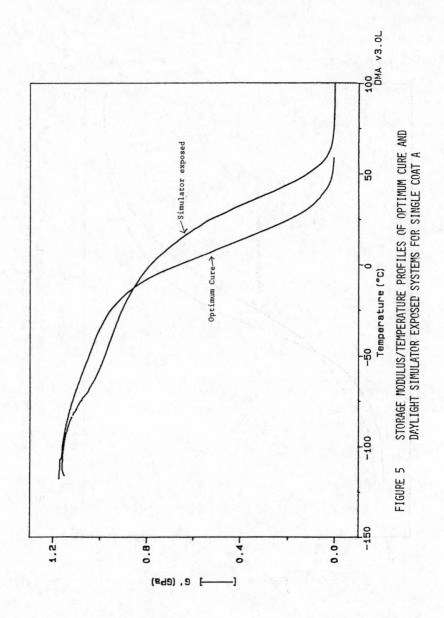

FIGURE 5   STORAGE MODULUS/TEMPERATURE PROFILES OF OPTIMUM CURE AND
             DAYLIGHT SIMULATOR EXPOSED SYSTEMS FOR SINGLE COAT A

These investigations clearly indicate the powerful nature of the DMA technique and also suggest that (i) physical property assessments should be made by more than one technique, and (ii) physical property assessments are normally carried out at room temperature and possibly other types of testing should be carried out which encompass a temperature range.

The work carried out also asks the question as to why these changes can occur so readily in films which were thought to be fully cured.

At this stage no mechanistic solution is offered however, perhaps it is possible that the longer wavelengths of light present in daylight can penetrate the 150 micron thick films and bring about further cure. To further test this general idea, two separate investigations were carried out.

Firstly, the single coat System A was taken with no photoinitiators and subsequently Electron Beam cured (EBC). Table 4 shows the physical property data for the optimum EBC system, daylight exposed and Daylight Simulator exposed systems. Immediately we see that the optimum EBC system has a Tg value of 30°C whereas for optimum u.v. cure the Tg has only 11°C. Knowing that there are mechanistic differences in u.v. and EBC systems there is still the strong suggestion that the higher energy EB penetrates the 150 micron film more readily and consequently the higher Tg reflects increased acrylate conversion. The data in Table 4 also shows minimal changes on further exposure to daylight, again suggesting minimal potential post-curing effects due to the initial increased acrylate conversion.

Secondly, and more obviously, single coat System A was taken and physical property data obtained on 75 micron thick films. Films of this thickness are also commonly used when assessing optical fibre coatings. Again the data in Table 5 surprisingly still show the significant post-curing affects as observed with the 150 micron thick films. It is also quite surprising that the magnitude of change from optimum cured system to Daylight Simulator exposed material has not diminished, even though the Tg values of the 75 micron thick systems are slightly higher than the 150 micron thick film analogues.

In view of the results obtained which show that significant changes occur in u.v. cured films on exposure

TABLE 4   ELECTRON BEAM CURED COATING

Condition	Tensile Strength (MPa)	Elongation (%)	Tg (G", °C)
5 MRAD	18	46	30
Light, 7 days	18	46	30
Daylight Simulator	20	50	33

HARCROS
CHEMICAL GROUP

TABLE 5 **PROPERTIES OF 75 MICRON THICK FILMS**

Condition	Tensile Strength (MPa)	Elongation (%)	Tg (G", °C)
Dark, 25°C, 24 hrs.	8	47	17
Light, 25°C, 24 hrs.	14	50	22
Daylight Simulator	15	45	42

HARCROS
CHEMICAL GROUP

to daylight whether they are 150 micron or 75 micron thick, it is felt that this is an area which warrants further investigation on two accounts: firstly, 75 micron thick films are close to the limit of film thickness which can be accurately and routinely handled for physical property data acquisition. However, optical fibre coatings are used in the thickness range of 20-60 microns. Since 75 micron films undergo significant post-curing, film investigation cannot reveal what, if any, affect coating changes have on optical fibre performances.

Secondly, numerous investigations have been reported in the literature which study various physical properties of U.V. cured systems (23,24,25,26,27,28). The point to consider is that all these reports have studied unsupported films in the thickness range of 100-300 microns. It could be of interest to see if post-curing can occur readily in these systems and then to see if these changes affect the initial data.

5. <u>CONCLUSION</u>

With the ever increasing usage of optical fibres, attention is being focused more on the optical fibre coating properties and from the work reported in this paper the very versatile nature of the DMA technique for assessing optical fibre coatings in demonstrated. DMA is a powerful tool in aiding the design of polyurethane acrylate resins and their derived formulations and can help in the assessment of the suitability of a system for a particular optical fibre coating type.

This particular Thermal technique has been shown to be as accurate and reproducible as any other conventional method used to establish the degree of cure of radiation curable coatings. It has also been demonstrated that DMA has the advantage of conveniently assessing systems over a temperature range and consequently can highlight changes which would otherwise remain undetected by conventional testing at room temperature - this aspect has been specifically, illustrated when investigating post curing effects.

Future developments in the field of radiation curable coatings will undoubtedly continue to benefit from the use of Dynamic Mechanical Analysis both as a research tool and a Quality Control technique.

ACKNOWLEDGEMENTS

The author would like to thank the Board of Harcros Chemicals for permission to publish this paper. In addition the author wishes to acknowledge Dr W D Davies for his continued guidance throughout this work.

REFERENCES

1. G Pasternack, Plast. Telecomm.Int.Conf. 4th Meet, 1986, 13/1 - 13/11

2. N Levy, Polym. Eng. Sci., 1981, 21, 14, 978 - 982

3. K R Lawson, Radcure 1983 Tech-Paper 1983, FC83 - 264

4. K Lawson and O R Cutler, Jr., J.Rad.Cure, 1982, April, 4-10

5. W E Dennis, Polym. Sci. Technol (Plenum), 37(Adhes, Sealants, Coat. Space Harsh Envirom) 1988, 445 - 454

6. Levy and Taylor, Wiley, New York, 1987, Vol. 7, Chapter F, 1 - 15

7. U C Paek Aiche Symp. Ser. 83 (258 Fiber Opt. Eng. Process Appl.), 1987, 83, 258, 38 - 41

8. R W Stowe Proc. Int. Wire Cable Symp. 36th, 1987, 212 - 216

9: C W Deneka et al Europ. Pat. 0261 772, 1988

10. B Joverton and C R Taylor Europ. Pat. 0314 174, 1989

11. E Miller et al. "Optical Fiber Telecommunications", Academic, New York, 1979, Chapter 10, 299 - 341

12. C J Aloisio et al. Plast. Telecomm. In, Conf. 4th Meet, 1986, 12/1 - 12/16

13. T S Wei Polym. Sci. Technol (Plenum), 37(Adhes., Sealants, Coat, Space Harsh Environ.), 1988, 455 - 466

14. T Murayama Mat. Sci. Monographs, Elsevier, Scientific Publishing Company, 1978, Vol. 1, 1 - 96

15. E A Turi "Thermal Characterization of Polymeric Materials", Academic Press, Orlando, 1981

16. B L Brann J Rad. Cur., 1985, July, 4 - 10

17. J M Julian and A M Millon J Coat. Technol, 1988, 60, 765,
     October, 89 - 95

18. C J Kallendorf and H E Pansing J Rad.Cur., 1985, April,
     18 - 28

19. C Li et al Polym. Eng.Sci., 1986, 26, 20, 1442 - 1450

20. E G Larson et al Radiat. Phys. Chem., 1987, 30, 1,
     11 - 15

21. G M Allen and K F Drain Polym. Mat. Sci. Eng., 1989, 60,
     337 - 342

22. BLMRA Journal, 1971, September, 256 - 260

23. N Levy and P E Massey Polym. Eng. Sci., 1981, 21, 7,
     406 - 414

24. C J Kallendorf and R T Woodruff Rad. Cur. '86 Conf. Proc.
     10th Meet, 1986, 9/1 - 9/19

25. M G Chan et al Plast. Telecomm. Int. Conf. 4th Meet,
     1986, 14/1 - 14/9

26. S Nakazato et al J Appl. Polym. Sci., 1989, 38, 4,
     627 - 643

27. T Kokubun et al J Lightwave Technol, 1989, 7, 5, 824 - 8

28. D A Simoff et al Polym. Eng. Sci., 1989, 29, 17,
     1177 - 81.

The Use of UV Technology to Produce Thick Pigmented Coatings

Peter G. Garratt

REICHOLD CHEMIE GESMBH, BREITENLEERSTR. 97–99, 1222 VIENNA, AUSTRIA

1 INTRODUCTION

Quality demands, whilst taking into consideration environmental pollution and energy saving, have led to an increasing use of the ultra-violet technology to cure coatings. The many advantages of UV curing are well known and must only in this case be quickly summarised

- one component paint systems
- extremely short curing times (increased throughput)
- non-polluting (solvent free or low solvent content)
- low energy consumption
- short plant lines and saving in factory space and
- high coating quality.

As a disadvantage it is often quoted, however, that the UV radiation curing process is limited in its application due to pigmentation problems. As the UV radiation is strongly absorbed by the pigments employed in the first microns of the films it is generally true that up to recent years the UV-technology has been mainly used for the curing of colourless lacquers or printing inks. The result of the pigment absorption is that in thick coatings there is retarded through cure with a complete cure only in the surface of the film.

The problem of pigmentation has always been considered by the industry as a major restraint to the UV curing process as for instance the trend in the furniture finishing industry is towards smaller production series with various colour shades. The idea of curing

pigmented coatings is not new and has been studied by
the raw material and paint manufacturing industries for
many years. The solution to the problem can be found,
either in the development of new pigmented paint systems
which can be cured using thick films at practical finishing
line speeds, or the development of new application
technological processes which produce the effect of a
pigmented coatings system.

This paper analyses the problems involved in the formulation of suitable paint systems for the production of
thick pigmented coatings and the necessary application of
technological processes. Hereby application technologies
will be described which present a specific solution to the
problem of producing thick coating systems, for instance
for kitchen and living room furniture which are manufactured in small production batches.

2   THE INFLUENCE OF PIGMENTS ON THE ULTRA VIOLET RADIATION CURING OF PAINTS

There are two fundamental mechanisms for ultra violet
radiation curing: - free radical and cationic. The principles of the more commonly used UV radiation curable coatings, which are cured by a free radical polymerisation
mechanism are well established and documented. A prerequisite for an adequate through cure is that the absorption curve of the photoinitiator used coincides with the
emission spectrum of the UV lamps employed. In the ideal
case the maximum absorption of the photoinitiator is
identical with the maximum emission of the UV source.
For pigmented systems it's absolutely necessary to ensure that
the absorption range of the photoinitiators are not only
selected to absorb energy in the spectral range of the UV
lamps used, but are also synchronized with the absorption
range of the pigments employed. In the case of the majority
of commercially available pigments there is an overlapping
of the photoinitiator absorption with the absorption curves
of the pigments. By the use of such pigments in the high
concentrations necessary to achieve suitable opacity very
little light reaches the photoinitiator in the lower layers
of the coating. Practically the total amount of the UV
radiation required for the efficient utilization of the
photoinitiator will be absorbed in the first few microns of
the coating, which results in an insufficient through cure
with the appearance of the so-called rivelling effect.

3   PREREQUISITE FOR THE USE OF UV-PIGMENTED PAINT SYSTEMS

A current major consideration for the processing of pigmented paint systems, at least in the furniture finishing industry, is that the paint systems can be processed on existing plant i.e. using medium pressure mercury vapour discharge lamps (hereafter called medium pressure Hg lamps) with predominant emission at 436, 405, 366, 313, and 302 nm. The emission spectrum of a commercial medium pressure Hg lamp is illustrated in figure 1. Working widths of 100-2000 cm are normal according to application and it is customary to use medium pressure Hg lamps with an output of 80-120 W/cm.

The problems involved in the curing of UV radiation curable pigmented paint systems using conventional photo-initiators and medium pressure Hg lamps can be simply explained using the example of white pigmented paint systems. A most important pigment for providing opacity in the paint industry is titanium dioxide. This pigment is the most efficient of the commercially available white pigments because it has the highest refractive index in the visible range of the electromagnetic spectrum. A look at the absorption characteristics of titanium dioxide-rutile, as well as anatase, (figure 2) - shows that from the theoretical point of view it would be virtually impossible to use titanium dioxide pigments to obtain a complete cure of thick UV radiation curable pigment coatings. It can be seen from this figure that the absorption is a function of wavelength and crystal type. Both the rutile and anatase forms of titanium dioxide absorb UV radiation in the same spectral regions as the majority of the industrially used photoinitiators. The rutile form absorbs markedly even at 390-400 nm. Likewise anatase absorbs UV radiation at 370-380 nm. The result is that very little light reaches the photoinitiator and practically the total amount of the UV light is absorbed in the first microns of the film.

4   INDUSTRIAL SOLUTIONS TO THE PROBLEMS OF PIGMENTATION

The challenge of curing thick pigmented coatings at practical finishing line speeds was accepted by the paint and raw material industries who developed new products for this purpose. The new developments were concerned with the manufacture of more reactive binder systems, as well as new

Rel. spectral radiation density (%)

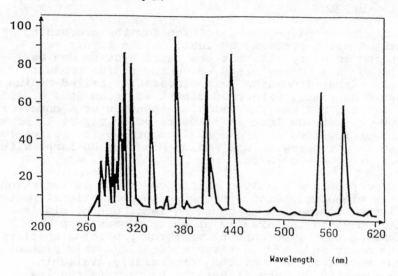

Figure 1 Emission spectrum of a medium pressure mercury vapour lar

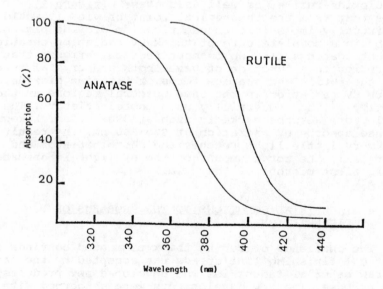

Figure 2 Absorption spectra of titanium dioxide

pigments which have a lower absorption in the UV region between 300 and 400 nm. In addition new photoinitiators were developed whose absorption ranges were shifted towards the long-wave end of the UV spectrum. A further development, which has not yet in practice been fully utilized, is the use of doped Hg vapour lamps.

The first industrial solution to the problem was found at the beginning of the Eighties when white pigmented filler and paint systems based on magnesium titanate were introduced on to the market. Figure 3 shows that magnesium titanate absorbs in comparison to the titanium dioxide pigments practically no UV radiation in the region of the predominant emission of medium pressure Hg lamps. Zinc sulphide pigments were also partially employed, because these pigments also only absorb small amounts of UV light in the near UV region. Using these pigments one component white paints were developed which could be fully cured at film weights of up to 110 g/m² at practical line speeds using standard medium pressure Hg lamps with output 30-100 W/cm.

Attempts to utilize the considerable amount of infrared radiation which is emitted by Hg vapour lamps, in addition to the UV radiation, to decompose conventional peroxide catalysts were only partially successful. The use of the so-called "Intermix" process, where the catalysts respectively the accelerators are mixed into the paint shortly before processing produced practical problems, such as a short life, when industrial line speeds were to be realised. In addition in the presence of photoinitiators and peroxides together it is difficult to achieve an uniform, respectively homogenous cure, of thick films.

Further progress was made with the introduction of the "Double or Dual-Cure" process to cure UV pigmented paints. The "Dual-Cure" process is in principle a clever kinetic combination of systems, respectively a self-regulating curing system, which produces virtually fully cured films of pigmented paints. The first stage of the process comprises of applying an active ground coat containing conventional peroxide catalysts. The second stage of the process is the application of UV radiation curable pigmented paints. For example 140-200 g/m² of a pigmented paint are first flashed off for 4 min. at 40 °C and then pregelled for 4 min. using low pressure Hg lamps (Type TL 05). The spectral energy distribution of the TL 05 lamp is shown in figure 4. The surface curing of the

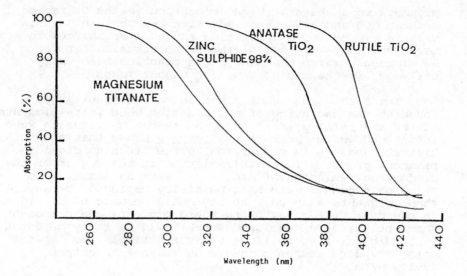

Figure 3 Absorption spectra of various white pigments

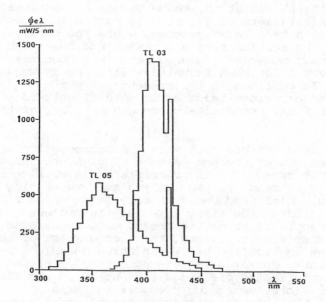

Figure 4 Spectral energy distribution of TL 05 and TL 03 fluorescent
        lamps

film is then achieved using standard Hg vapour lamps.
Generally speaking all desired colour shades can be
achieved as the UV absorption spectrum of the pigments
is not alone the determining factor for the curing
of the thick films. In the selection of suitable pigments,
however, the following factors apart from UV absorption of
the pigments must be considered: peroxide stability, light
fastness and the storage stability in the binder system
to be used.

The major break-through for the "mono-cure" process,
respectively for one component UV pigmented paint systems,
occurred with the market availability of a new class of
photoinitiators with absorption in the long-wave end of
the UV spectrum and which are suitable for use, both in
unsaturated polyester and acrylic resin systems. The
absorption characteristics of one of the representative
photoinitiators belonging to this class are shown in
<u>figure 5.</u> The scope of the use of this photoinitiator
class - the acylphosphine oxides - can be extended by
their use in paints cured on UV lines equipped with doped
Hg vapour discharge lamps. In industrial practice gallium
doped lamps are mainly employed. These doped lamps have
an improved output in their spectral range at 403 and
417 nm. The spectral emission of an industrial gallium
doped lamp is illustrated in <u>figure 6.</u> Previous experience
with photoinitiators with absorption bands above
380-390 nm has shown that practically without exception
the use of these photoinitiators leads to a yellowing
effect. The advantage of using the new class of photoinitiators lies in the production of films exhibiting an
extremely low yellowness i.e. production of practically
pure white films. The new photoinitiators combine the
advantage of absorption in the long wave-end of the UV
spectrum, high light fastness and good storage stability.

As a result of the absorption spectrum of these photoinitiators paints produced with these materials can also
be processed on plant equipped with old well known TL 03
fluorescent lamps, together with, or instead of, the
previously employed TL 05 lamps. Without doubt the new
photoinitiators used in combination with various lamp
systems have extended the scope of use of UV pigmented
radiation curable paints.

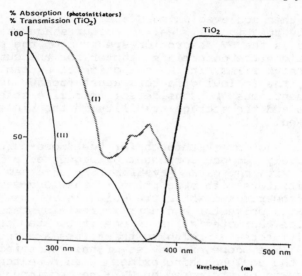

(I) 2,4,6-Trimethyl benzoyl diphenylphosphine oxide
(II) 2-Hydroxy-2-methyl-1phenylpropane-1-one

Figure 5 UV absorption spectra of photoinitiators and transmission
spectrum of rutile TiO$_2$

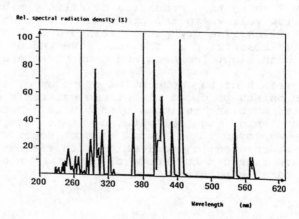

Figure 6 Emission spectrum of medium pressure gallium doped mercury
vapour lamp

5 FORMULATION AND PROCESSING PARAMETERS INFLUENCING THE CURE OF UV RADIATION CURABLE PIGMENTED COATINGS

The following experimental data are aimed at illustrating some parameters which must be considered when formulating UV pigmented paint systems.

Experimental

Unless otherwise specifically stated the experiments were performed as follows:

1. **Formulation**: Using a binder combination of a commercial monomer free unsaturated polyester acrylate and tri-methylol propane triacrylate paints were prepared containing the desired pigment concentration by grinding the pigment in the necessary quantity of the binder. After thinning with further binder and addition of photoinitiator and amine accelerator the following basic formulation was obtained:

   - commercial unsaturated polyester acrylate   70 parts
   - trimethylol propane triacrylate   20 parts
   - amine acrylate (accelerator)   7 parts
   - photoinitiator   3 parts

   Five separate pigment concentrations, namely 5, 7.5, 10, 12.5 and 15 % were studied.

   (all concentrations and parts are quoted by weight)

2. **Film application** - The films (120 μm) were applied on a polyester foil using a draw down bar.

3. **Film curing:** - Following application the films were exposed to UV radiation as follows: -

   a) **using standard medium pressure Hg lamps**

   The films were exposed to UV radiation from two medium pressure Hg lamps (CK lamps - IST Strahlen-technik, 80 W/cm, distance lamps to substrate-20 cm) by passing the films longitudinally through the UV beam. The conveyor speed was 2.5 m/min.

b) <u>using a combination of gallium doped and standard</u>
   <u>medium pressure Hg vapour lamps</u>

In this case the films were first exposed to
UV radiation from a gallium doped medium pressure
lamp (CK 1 lamp - IST Strahlentechnik, 80 W/cm,
distance lamps to substrate - 20 cm) by passing the
films longitudinally through the UV beam. The films
were then immediately exposed to UV radiation from
two standard medium pressure Hg lamps (CK lamps -
as described under a) above). The conveyor speed
employed was 3.75 m/min.

All irradiations were performed in air at room
temperature.

4. <u>Film hardness</u> - Immediately after irradiation the films
were examined for surface tackiness and their surface
hardness tested with a steel blade using a simple test
method. If the films were through cured the pendulum
hardness was measured after 5 minutes using the König
method (DIN 53.157).

5. <u>Hiding power</u> - Films for these experiments were prepared on a white/black contrast paper using the methods
previously described. (In this case curing was only performed using the curing procedure b).) The hiding power
was measured with an ACS Spectro Sensor-2 applied
colour system, using the wavelength range 400-700 nm.
A measurement value of 100 % represents full hiding
power.

<u>Results and discussion</u>

<u>UV curing of white pigmented paint systems using standard</u>
<u>medium pressure Hg lamps and a hydroxy substituted aceto-</u>
<u>phenone photoinitiator</u>

For these experiments a classical photoinitiator
which produces radicals by a Norrish 1 clevage was
selected - 2-hydroxy-2-methyl-1-phenyl-propane-1-one.
As previously shown in <u>figure 5</u> the major absorption
bands of this photoinitiator lie well below the wavelengths of the UV absorption of the titanium dioxide
pigments (< 380 nm). The results given in <u>figure 7</u>

confirm that paints containing the hydroxy substituted
acetophenone photoinitiator and the rutile form of
titanium dioxide do not completely cure even at only a
5 % pigment concentration. A slightly higher degree of
cure was obtained with the TiO_2 anatase crystal form.
As is to be expected from the absorption curves of the
pigments studied the pigments which absorb less UV
radiation exhibit a higher degree of curing at higher
pigment concentrations. The highest degree of cure is
obtained with magnesium titanate and to a slightly lesser
degree with zinc sulphide.

(Note for these experiments three different commercial
types of titanium dioxide were selected.

Type 1 is a TiO_2 prepared by the chloride process and is
stabilised with Al_2O_3.
Type 2 is an Al_2O_3 and SiO_2 coated TiO_2 pigment prepared
by the sulphate process.
Type 3 is an Al_2O_3 coated rutile pigment prepared by the
sulphate process.)

UV curing of white pigmented paint systems using standard medium pressure Hg lamps and an acylphosphine oxide photoinitiator

The positive influence of using an acylphosphine
oxide photoinitiator with a wave absorption shifted
towards visible light on the curing of white pigmented
paints is clearly illustrated in figure 8. The photoinitiator employed in these studies was 2,4,6-trimethyl
benzoyl diphenylphosphine oxide. A high degree of cure
is obtained with all investigated pigments. From the
standpoint of achieving the best cure the pigments may
be classified as follows:

magnesium titanate > zinc sulphide > TiO_2 anatase > TiO_2 rutile

No significant differences could be established between
the various rutile pigments. Thus it can be concluded
that the physical differences between the rutile grades
investigated are so small that none of them is likely to
confer any major adavantage in curing response. The
differences measured in hiding power will be discussed
later in the paper.

UV curing of white pigmented paint systems using standard med. press. lamps
Initiator: 3% Hydroxyphenyl ketone

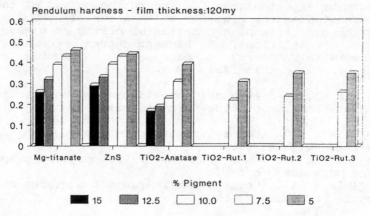

Figure 7

UV curing of white pigmented paint systems using standard med. press. lamps
Initiator: 3% Acylphosphine oxide

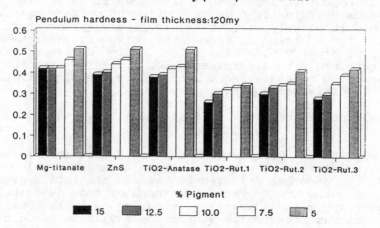

Figure 8

UV curing of white pigmented paint systems using a combination of doped and standard medium pressure Hg lamps and an acylphosphine oxide photoinitiator

A further increase in cure is observed when a combination of gallium doped and standard medium pressure Hg lamps is employed (figure 9). An increase in pendulum hardness was measured at all the concentrations of the pigment studied. The hiding power of the pigments measured using films cured with the combination of doped and standard lamps is illustrated in figure 10. Clearly from the point of hiding power the pigments may be classified in the reverse order to that for the curing characteristics. i.e.

TiO_2 rutile > TiO_2 anatase > zinc sulphide > magnesium titanate.

Obviously to obtain the best opacity with UV radiation curable white pigment paints it must be attempted to incorporate the TiO_2 rutile pigment with its high reflectivity in the visible light range into the paint in combination with a suitable reactive binder system. Also that it is necessary to employ a photoinitiator with absorption bands above 380-390 nm. Furthermore, the best cure can be obtained by using lamp combinations with at least one lamp having an improved output in the spectral range over 400 nm.

UV curing of black pigmented paint systems using a combination of doped and standard medium pressure Hg lamps and an acylphosphine oxide photoinitiator

A fashionable colour, particularly in the furniture industry in several European countries, is black. The results obtained with several commercially available black pigments used together with the previously described acylphosphine oxide photoinitiator and lamp combination are shown in figure 11. These results clearly demonstrate that little or no curing can be obtained with the films containing 5 % standard carbon black pigments. On the other hand a relatively high degree of curing can be obtained using a specially selected organic pigment which also exhibits a high hiding power (figure 12). Thus it can be concluded, that it must be possible to formulate UV radiation curable black pigmented paints which can be cured in thick films by the appropriate choice of the

UV curing of white pigmented paint systems using doped and stand. lamps 1:2
Initiator: 3% Acylphosphine oxide

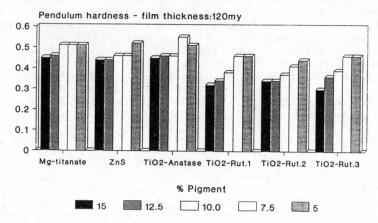

Figure 9

Hiding power and pendulum hardness
5% Pigment concentration
120 my Film thickness

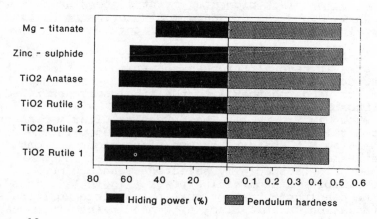

Figure 10

UV curing of black pigmented paint
systems using doped and stand. lamps 1:2
Initiator: 3% Acylphosphine oxide

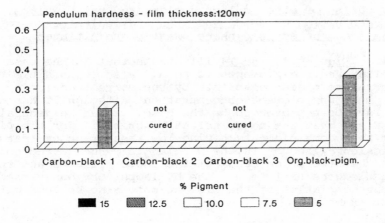

Figure 11

Hiding power and pendulum hardness
5% Pigment concentration
120 my Film thickness

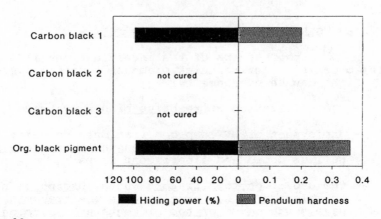

Figure 12

reactive binder together with the correct selection of
black pigment and its concentration.

UV curing of other coloured pigmented paint systems using a combination of doped and standard medium pressure Hg lamps and an acylphosphine oxide photoinitiator

Figures 13 and 14 illustrate that in the case of
yellow and red pigments it should also be possible to
formulate pigmented paints by the correct choice of
pigment and pigment concentration. Although it has not
always been proved to be the case, it can be generally
assumed that the major selection criteria for the pigments
is to choose pigments having low absorption or transmission "windows" in the wavelength range corresponding
to the absorption bands of the photoinitiator and the
wavelengths emmitted by the UV lamps. Obviously other
characteristics of the pigment must also be considered
for example

- optical properties, light scattering and packing
  effects, and chemical properties such as inhibition
  of a radical polymerisation and possible reactions
  involving chain termination. Furthermore any
  possible rheological restraints on flow must be
  taken into account.

6   GENERAL FORMULATION - KEY POINTS

As the results of the investigations illustrate
the problem of curing thick pigmented coating systems by
UV light can be overcome by:

- the utilization of reactive binder systems

- employment of UV lamp combinations including lamps
  having a spectrum markedly enhanced over 400 nm e.g.
  gallium doped and fluorescent lamps,

- the use of photoinitiators whose absorption range
  is selected to synchronize with the transmission
  wavelength bands of the pigments and emission of
  the UV lamps employed, i.e. photoinitiator mixtures

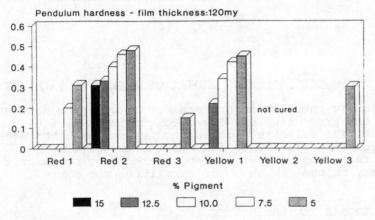

Figure 13

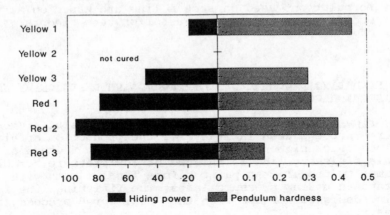

Figure 14

incorporating materials whose absorption band is
displaced in the direction of visible light,

- the selection of <u>pigments</u> having <u>low absorption</u>
  <u>over the range of wavelengths emitted by the UV</u>
  <u>lamps.</u>

7   PROCESSING OF UV RADIATION PIGMENTED PAINT SYSTEMS

Obviously there is no one standard processing method
for the production of pigmented coatings cured by UV
light. The following processing schemes are, therefore,
designed to illustrate some of the possibilities of
manufacturing UV radiation pigmented coatings particularly in the finishing of furniture components.

a) <u>Process for the production of thick UV white pigmented</u>
   <u>coatings on chipboard using standard Hg vapour lamps</u>

<u>Figure 15</u> shows a standard UV line used for the
coating of chipboard. This example illustrates the
fact that thick coatings can be processed on finishing
lines equipped only with customary lamps. Gallium
doped lamps are recommended to be used in combination
with standard lamps when faster line speeds, or the
application of thicker coatings are desired. The paint
formulations used on such a line are based on acrylic
modified unsaturated polyester/styrene binders.

<u>Alternative process for the production of thick UV white</u>
<u>pigmented coatings on chipboard</u>

<u>Figure 16</u> illustrates the various stages of a "mono-
cure" process used in Italy for the production of high
gloss thick pigmented coatings. The major feature of
this finishing line is the use of low pressure mercury
lamps to pregel the paint before final curing with
standard medium pressure lamps. The finishing line is
also designed for use with the "dual-cure" process if
desired.

RADIATION FINISHING PLANT FOR THE PRODUCTION OF THICK PIGMENTED COATINGS ON CHIPBOARD (USING STANDARD MERCURY VAPOUR LAMPS)

FOR WHITE PIGMENTED COATINGS
BINDER: ACRYLIC MODIFIED – UNSATURATED POLYESTER/STYRENE
LINE SPEED: 10 m/min
SUBSTRATE: CHIPBOARD

MULTI-BELT SANDING MACHINE   – calibration of boards

↓

DUST REMOVAL

↓

FILLER APPLICATION   –   80–100 g/m² – white pigmented filler
Application with reverse roller coater

↓

CURING OF FILLER   * 3 UV MEDIUM PRESSURE MERCURY LAMPS – 80 W/cm

↓

SANDING   – Drum sanding paper 280–400

↓

GROUND COAT APPLICATION   –   25–30 g/m² – white pigmented ground coat.
Application with twin head roller coater (second roller coater – dosing roller – reverse)

↓

CURING OF GROUND COAT   * 3 UV MEDIUM PRESSURE MERCURY LAMPS – 80 W/cm

↓

TOP COAT APPLICATION
(matt lacquer)   –   110–120 g/m² – white pigmented paint, matt.
Application with curtain coater

↓

FLASH OFF   –   90 sec at 20–30 °C

↓

CURING OF TOP COAT   * 8 UV MEDIUM PRESSURE MERCURY LAMPS – 80 W/cm

↓

COOL

↓

STACKING

---
* THE USE OF GALLIUM DOPED MERCURY VAPOUR LAMPS IN COMBINATION WITH THE STANDARD LAMPS IS RECOMMENDED. IN THIS CASE FASTER LINE SPEEDS CAN BE ACHIEVED AND/OR THICKER FILMS CURED

Figure 15

RADIATION FINISHING PLANT FOR THE PRODUCTION OF THICK PIGMENTED COATINGS ON CHIPBOARD (Italy)

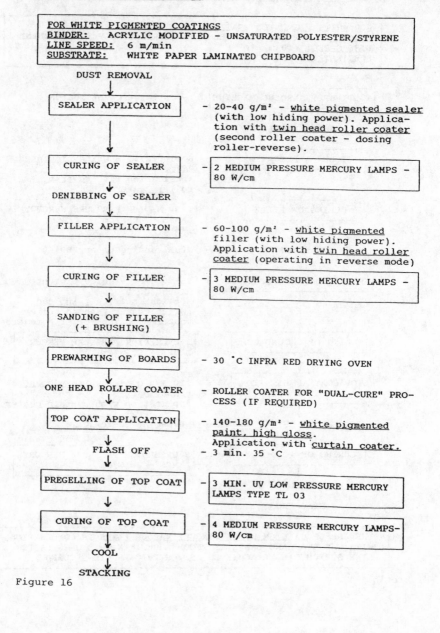

FOR WHITE PIGMENTED COATINGS
BINDER: ACRYLIC MODIFIED - UNSATURATED POLYESTER/STYRENE
LINE SPEED: 6 m/min
SUBSTRATE: WHITE PAPER LAMINATED CHIPBOARD

DUST REMOVAL

SEALER APPLICATION

- 20-40 g/m² - white pigmented sealer (with low hiding power). Application with twin head roller coater (second roller coater - dosing roller-reverse).

CURING OF SEALER

- 2 MEDIUM PRESSURE MERCURY LAMPS - 80 W/cm

DENIBBING OF SEALER

FILLER APPLICATION

- 60-100 g/m² - white pigmented filler (with low hiding power). Application with twin head roller coater (operating in reverse mode)

CURING OF FILLER

- 3 MEDIUM PRESSURE MERCURY LAMPS - 80 W/cm

SANDING OF FILLER
(+ BRUSHING)

PREWARMING OF BOARDS

- 30 °C INFRA RED DRYING OVEN

ONE HEAD ROLLER COATER

- ROLLER COATER FOR "DUAL-CURE" PROCESS (IF REQUIRED)

TOP COAT APPLICATION

- 140-180 g/m² - white pigmented paint, high gloss. Application with curtain coater.

FLASH OFF

- 3 min. 35 °C

PREGELLING OF TOP COAT

- 3 MIN. UV LOW PRESSURE MERCURY LAMPS TYPE TL 03

CURING OF TOP COAT

- 4 MEDIUM PRESSURE MERCURY LAMPS- 80 W/cm

COOL

STACKING

Figure 16

Future trends

The last two years have plainly seen great advances in the development of thick pigmented coatings which can be cured by UV light at practical line speeds. Many colour shades can already be produced with the market availability of new photoinitiators and doped mercury vapour lamps. The major drawback to manufacturing all desirable colour shades is now probably basically the lack of process pigments. It is probably fair to say that in particular no major pigment manufacturer has specifically developed pigments for UV radiation curable paint systems due to the current market volume and finance required to cover such developments. The area of UV radiation curable pigmented systems still presents major development challenges, however, not only to the pigment maker but also to resin manufacturer, the paint industry as well as the paint producer.

Photoinitiators for Litho Inks

S.J. Wilson

THE PROCESS INK COMPANY (COATINGS) LTD., CARLYON ROAD,
ATHERSTONE, WARWICKSHIRE, UK

1 INTRODUCTION

Offset lithography is a printing process which uses water to separate the image and non-image areas of the printing plate and in which the ink is transferred from the printing plate to the substrate via a rubber blanket, hence the name offset.The printed ink films are typically 2-4 microns thick.

The two important points which affect ink formulation and consequently photoinitiator selection are that water plays a fundamental part in the process and that lithography is a thin film application.

The ink formulator needs to understand both the lithographic process and the available raw materials in order to formulate an ink or coating that, firstly, performs correctly on the printing press and, secondly, has the desired properties in the cured film. We will assume that all the other ingredients of the inks have already been chosen and that it is only the photoinitiator selection that needs to be made.

I consider that a detailed knowledge of the chemistry of the various initiators is not necessary although a certain familiarity with the curing mechanisms is, the knowledge allows us to predict and thereby avoid possible problems in specific formulations.

It is convenient to list the more commonly used photoinitiators under separate headings in order to

look at their properties.They can be split into three
general groups

PHOTOINITIATORS USING HYDROGEN EXTRACTION

Benzophenone   Relatively cheap
                                Good solubility
                                Good solventcy
                                Possible odour problem
                                Absorbs at 240-340nm

Substituted Benzophenones

Alkyl   Liquids
                                Good solventcy
                                Less yellowing
                                More surface inhibition
                                Better through cure
                                Absorb down to 200nm (if this
                                is useful)

Phenyl   Very bad yellowing
                                Poor colour
                                Low odour
                                Enhanced cure
                                Poor solubility

Thiozanthones

    Of those available the solubility and colour of
the Isopropyl make it the more generally adopted.

                                Good solubility in benzophenone
                                type initiators
                                Poor colour
                                Yellowing
                                Low odour
                                Expensive
                                Absorbs at 200-400nm

    The most recent sources of this material have much
paler colour and improved solubility forming stable
solutions at high concentrations down to 6-7C.

HYDROGEN DONORS - AMINE SYNERGISTS

Tertiary   Cheap
                                Water soluble
                                Yellowing
                                Reactions with some pigments

                              May cause surface effects with
                              atmospheric moisture

Acrylated Amines   Little migration from the cured
                             film
                             Yellowing
                             Partially water soluble
                             Liquid

Aminobenzoates   Mostly insoluble in water
                             Some liquid
                             Yellowing
                             Absorb strongly at 300-330nm

PHOTOINITIATORS THAT GENERATE FREE RADICALS

Acetophenone derivatives
                             Good colour
                             Liquid or good solubility
                             Odour problems associated with
                             some (benzaldehyde formation)
                             Absorb at 250-350nm

Benzoin and Benzoin Ethers
                             Poor dark stability in highly
                             pigmented systems
                             Yellowing
                             Insoluble in water
                             Absorb at 225-350nm

Benzil ketals   Insoluble in water
                             Good solubility in benzophenone
                             type initiators
                             Some yellowing
                             The unpredictable formation of
                             methyl benzoate can give odour
                             problems

CHARACTERISTICS REQUIRED FROM THE PHOTOINITIATOR
COMBINATION

                             Cure under all practical
                             applications
                             Stable in highly pigmented
                             systems
                             Efficient from cost/performance
                             Water insoluble
                             Low odour in most applications
                             Suitable colour

Should not adversely affect the
printing machine (by the
formation of HCL for example)

SELECTION

From the list of available photoinitiators that
have practical application it can easily be seen that
the entire UV spectrum of a medium pressure mercury
vapour lamp can be covered by selecting and combining
photoinitiators (see fig.1). And it is also possible
to include initiators of different curing mechanisms.
It is necessary to do this because, if we consider the
list of desirable characteristics mentioned earlier and
consider one of the major print uses of UV curing inks
(high quality packaging for the food, cosmetic and
pharmaceutical industries), then it is obvious that the
requirement for low taint and odour is essential.This
restricts the use of many initiators and, if we also
include storage stability as desirable, then many of
the benzoin ethers and acetophenone derivatives can be
eliminated together with specific materials from other
groups which I shall mention in a moment.

One of the primary requirements of any ink is that
it must be suitable for printing. To this end the use
of tertiary amines is detrimental, being water soluble
they neutralize the fount solution and result in higher
water levels being needed in order to keep the
non-image area of the printing plate ink free. This in
turn causes other printing problems, some of which may
persist even after the ink has been replaced.There may
also be a direct reaction between the amine and some
pigments.

So there is a strong limitation on the choice of
suitable photoinitiators and this limitation is not
only based on our own technical knowledge but on the
demands of major print buyers. Both the grade of
benzophenone to be used and limitations on the use of
one of the major benzil ketal initiators have been
specified.The costs involved if taint problems are
encountered can be enormous and with the sophisticated
analysis equipment now in common use it would be a
brave person that ignored these recommendations
irrespective of personal views, however my own
experience has led me to formulate away from both
benzophenone and benzil ketals in inks.

It is possible to make photoinitiator selections

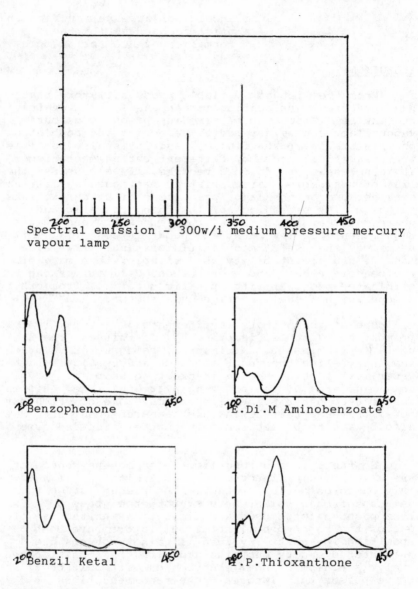

Spectral emission - 300w/i medium pressure mercury vapour lamp

All P.I. concentrations 0.01 g/l in methanol

<u>Figure 1</u>   Comparison U-V emission / absorption

based on the requirements of each ink taking into account the transmission spectra of the pigment and transparency in order to reach the most economical formulation and under some circumstances this is desirable, however litho printing is normally a multi-colour process with successive inks superimposed on each other and, while the situation is changing rapidly, this is still largely without interdeck curing.Consequently inadequate curing can result when the selection has been made in this way, and a similar effect is possible when mixing colours particularly if there is a strong increase in grey factor or opacity

In practice blends of benzophenone (or a substituted material) with aminobenzoate and a thiozanthone or alternatively, benzophenone, benzil ketal and thiozanthone are most commonly used.Both cover the spectral distribution of medium pressure lamps well and provide a simple stock solution which is suitable for all colours with a few exceptions such as black and white. In fact there is little difference economically between using one solution at varying concentrations to suit pigmentation and making the sort of selection mentioned above.From a practical point of view the minimum number of photoinitiator solutions necessary is desirable in the production area.

These typical P.I.blends are suitable for the majority of litho inks, however some have higher demands while the specification for others is lower and the photoinitiators are changed accordingly.

BLACK INKS

The absorption of U.V. light, adsorption of free radicals and the opacity of carbon black present a technical problem to the formulator. This is best overcome by limiting the carbon content of the ink as far as possible, using a photoinitiator choice which covers the entire uv spectrum with particular regard to the 350-420nm wavelengths and which contains materials that use all the different curing mechanisms. Heat from I R absorption is also an advantage with some photoinitiators, mainly those that adsorb at the longer end of the spectrum, and with experience this can also be used by the formulator.

WHITE INKS

As with black inks the opacity, a prime requirement of white inks, makes UV penetration difficult and the further requirement of non yellowing adds to the problem.

In litho filmweights there are various initiators
which can be used to cure white inks with only a
minimum of discolouration these include some of the
hydroxy ketones and BMDS with benzophenone and small
quantities of thiozanthone. The alkyl substituted
benzophenones give considerably better colour and
recent sources of ITX are also much paler with better
solubility.Used in conjunction with other techniques
available for disguising yellowing an acceptable
product can be achieved without the need for very
expensive initiators such as the acylphosphine oxides.

Other ink uses such as metal decorating, web
offset and printing on various plastic substrate do not
have the requirement for low odour that packaging inks
carry and here the benzil ketals can be used with no
reservation as can some acetophenone derivatives that
are otherwise unacceptable such
as Methyl [4-(methylthio)phenyl]-morpholino-propane
provided that there is adequate extraction for
environmental acceptance. Using these materials
economical inks can be formulated that display good
rates of cure even at the high press speeds associated
with web printing.

FUTURE
UV curing is now an important sector of the
British litho ink market and over the past 5 years it
has made strong advances. However these advances have
largely been concerned with the printing
characteristics of the inks while the photoinitiators
used have remained constant.Indications from the major
suppliers are that this is about to change with new
products showing improvements in:
>   Absorptivity
>   Use of the 380-420 wavelengths
>   Odour levels
>   Solubility
>   Colour

Recommendations are also being made by some
companies linking initiators to specific resin types.
Perhaps we shall see the development of PIs actually
designed to match the chemistry of the various acrylate
oligomer groups.

Certainly the increasing speeds of most new
printing presses together with more printing units per
press are making demands on the curing ability of inks
and will lead to a requirement for more efficient
initiators.Possibly the other wavelenghts generated by

the lamp systems and at present largely ignored or treated as a nuisance, namely the IR and visible light could also be used in the curing reaction?

With the development of photoiniators designed to absorb more in the longer UV wavelengths it is necessary to investigate the use of doped lamp systems which produce higher emissions in the 380-420 nm bands These lamps are available, though not commonly used in the printing industry, and, on a press with inter deck curing, it may be possible to use these lamps to enhance the cure of the more difficult colours.

There is also a major developement taking place with the adoption of waterless litho printing plates. These plates allow the use of water soluble initiators and synergists in acrylate formulations which should lead to cost advantages and these plates are also suit able for printing cationic inks and coatings.This makes it possible to develop products with major advantages regarding irritation to skin and respiratory systems. Waterless litho plates will also broaden the range of substrates which can be successfully used by litho printers.However the initiators presently available for cationic systems have problems of odour or colour associated with them.

There is another group of questions which are asked with increasing frequency, and failure to answer them could limit the use of U-V curing products.These are concerned with the degree of cure obtained commercially and with the reaction by-products either extracted during curing or remaining in the board or film and how will these affect the product or further processing such as recycling of board? These are very difficult problems for most ink companies to answer and they will probably be passed back to raw material suppliers or onto research organizations, however information gained from answering them should be useful in the future development programs of companies such as our hosts today.

The Use of UV Curable Prepolymers in Ink Jet Inks

A.C. Marshall, M. Sutty, N. Miller, and A.L. Hudd

DOMINO AMJET LTD., BAR HILL, CAMBRIDGE CB3 8TU, UK

Ink jet is a versatile, non contact printing technique capable of printing fixed and variable data at high speeds (up to 2100 characters per second). The fastest form of ink jet printing is continuous ink jet (CIJ), where a fine jet of ink is broken into a stream of drops which are electrically charged and then deflected. The amount of deflection together with the motion of the substrate produces a dot matrix pattern.

This paper describes some aspects of the design of a solvent based ink jet ink suitable for applications requiring resistance to post solvent treatment. It considers the chemical and physical properties required and demonstrates the need for high chemical stability during lifetime in a printer, where most of the droplets are recycled many times and exposed to oxidative, mechanical, thermal and hydrolytic processes. The approach adopted combines catioinc UV curing technology modified with free radical initiated monomers to provide maximum solvent resistance and fast cure speeds. Factors influencing the cure and chemical stability, notably the effect of ingredients such as dyes and salts have been examined using Gel Permeation Chromatography and viscometry and by conducting accelerated ageing trials and ink jet printer performance trials.

A solvent resistant UV curing ink jet ink has been succesfully developed using a cycloaliphatic epoxide cationic curing prepolymer. The incorporation of free radical initiated monomer into this system to produce a hybrid cure formulation has shown significant improvements in solvent resistance and up to 100% increase in cure speed. The stability of a hybrid UV curing ink jet formulation is affected by the choice of dyestuff and conductive salt. Chromium complex dyes have been shown to be capable of initiating the polymerisation of the cationic curing system. On the other hand, the use of Cobalt complex dyes produce inks which exhibit reliable printer performance and good ageing stability. Similarly the choice of conductive salt is important as these have been shown to initiate the cure of the free radical curing system.

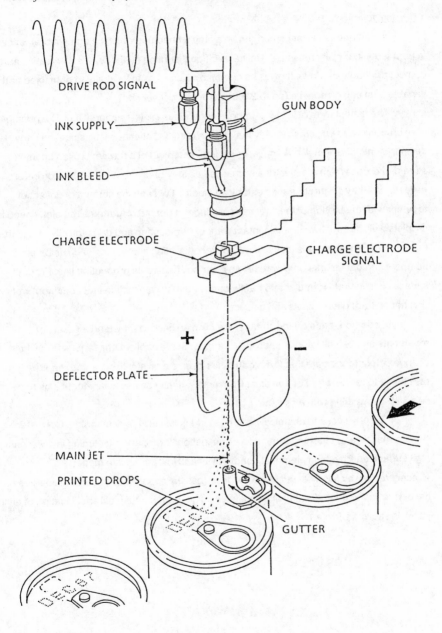

Figure 1 : Principle of Operation

1.0   Introduction

Ink jet printing is fast emerging as a high technology which has grown dramatically in popularity in the last ten years. Unlike traditional printing methods, it is a versatile, non contact printing technique. It owes its commercial success to its ability to print fixed and variable data at high speeds (up to 2100 characters per second).

The fastest form of ink jet printng is continuous ink jet printing (CIJ). The principle of operation is shown schematically in Figure 1. and has changed little since the early developments of Sweet .[1] A fine jet of ink is directed out of a nozzle, broken up into a stream of drops which are electrically charged as they break from the jet. The drops are deflected as they go through the head by an electric field and the amount of deflection together with the motion of the substrate produces a dot matrix pattern. The high speed ink droplets wet the substrate, remain intact without fragmenting and dry, typically in less than one second. The technique makes it attractive for printing information such as batch numbers and date marks onto virtually any surface on a rapid production line. Product surfaces can include irregular and fragile ones and may vary in type and condition; wet or dry; hot or cold; glass metals and plastics.

In addition, once on the surface, the ink may have to withstand a variety of treatments and exhibit various special properties. Some applications require inks which change colour to indicate that the product has been steam sterilised, some require resistance to temperature or solvents. Conversely, it may be necessary for the ink to be removed by washing with a solvent.

The physical and chemical properties of ink jet inks are demanding. Consistent drop formation and charging requires a combination of properties including low viscosity, high surface energy, high electrical conductivity and low particulate matter. A recent review by Mitchell and Hudd [2] describes the importance of these properties on the performance and reliability of an ink in an ink jet printer. A typical specification of an ink jet ink is shown in Table 1.

Ink Jet Ink at 25C

Viscosity   4 +/- 0.25 cP

Surface Tension   27 +/- 3 N/m

Conductivity   >1000 uS/cm

Particulate matter   < 3um

Table 1   Typical Specification of an Ink Jet Ink

This paper considers the criteria for the design of a solvent based ink jet ink suitable for applications requiring resistance to post solvent treatment, such as Mil Spec 202F [3]. It highlights the chemical and physical properties required and demonstrates the need for high chemical stability during service lifetime in a printer, where most of the droplets are recycled many times and exposed to oxidative, mechanical, thermal and hydrolytic processes.[4] The approach adopted combines cationic UV curing technology [5] modified with free readical initiated materials [6,7,8] to provide maximum solvent resistance and fast cure speeds. Factors influencing the cure and chemical stability, notably the effect of ingredients such as dyes and salts have been examined using Gel Permeation Chromatography and viscometry also by conducting accelerated ageing trials and ink jet printer performance trials.

2   Ink jet ink formulations

A series of model ink jet ink formulations have been examined based on a cycloaliphatic epoxide prepolymer and a triarylsulfonium photoinitiator. The chemistry of this system is understood[5,9] and is thought to proceed through two competing mechanisms.The first mechanism involves a pure cationic route and this is illustrated in Figure 2. In detail the reaction consists of three stages :

i) Photolysis of the cationic initiator yields a Bronstead acid and free radical species.

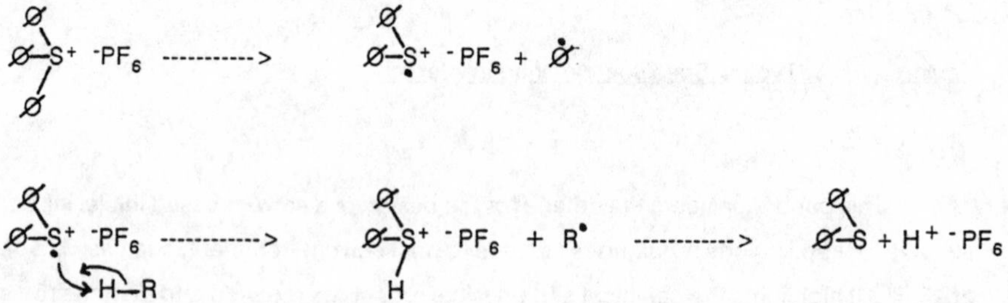

ii) A proton attacks the epoxy group forming an alcohol and a carbonium ion.

iii) The carbonium ion can then propogate the reaction by reacting with another epoxy group forming an ether and a carbonium ion.

Figure 2   Mechanism of Pure Cationic Polymerisation

An alternative mechanism requires an active hydrogen other than the original protic acid to continue the propagation. This route is illustrated in Figure 3.

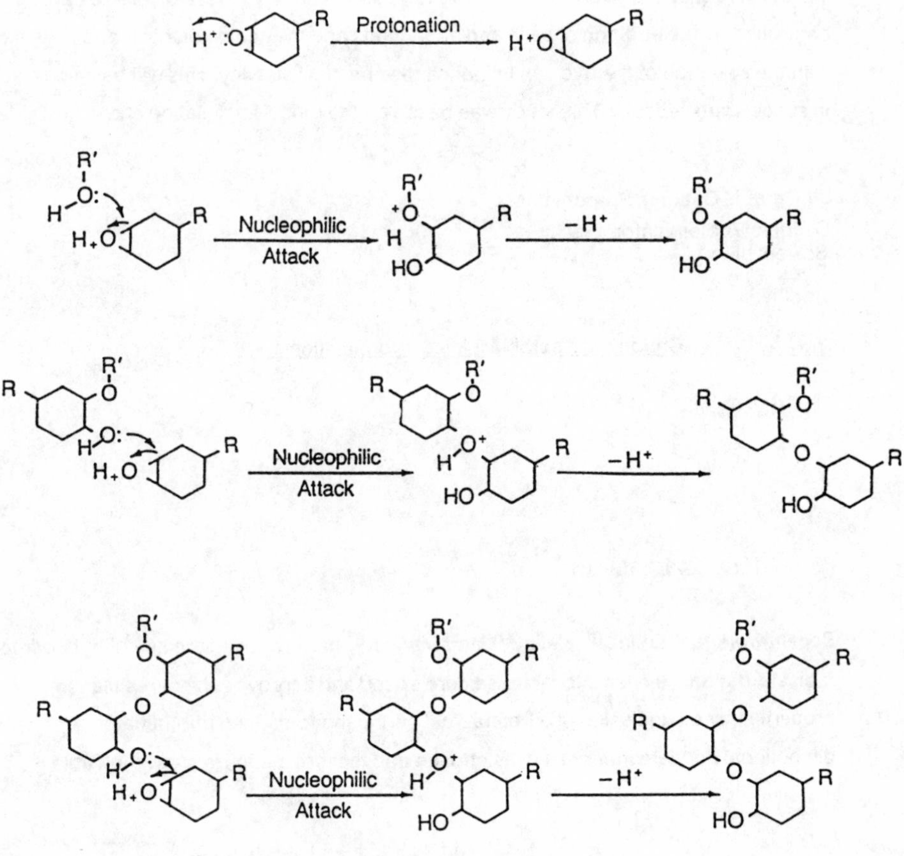

<u>Figure 3</u>   Alternative cationic polymerisation mechanism.

The cycloaliphatic epoxide prepolymer used in the model formulation was a high viscosity liquid ; approximately 350cP. On the other hand, an ink jet ink formulation requires low viscosity, typically 4cP and in this case it was achieved by incorporating high levels of volatile solvent such as methyl ethyl ketone and alcohol. In addition to viscosity an ink jet ink requires high conductivity and low surface energy. These properties are controlled by the addition of salts and surfactants and the choice of these additives along with the selection of the dye can influence the chemical stability. This will be discussed later in succeeding sections.This study was based on the model formulation shown in Table 2.

UV Curable Cationic Prepolymer	49 %
Cationic Photoinitiator	13 %
Solvent blend	35 %
Dye	2 %
Additives	1 %

Table 2.   Cationic UV curable ink jet - formulation 1.

3.   Hybrid cure studies

Recent work by Perkins [6]. Ketley[7] and Manus[8] has proposed the use of hybrid free radical / cationic systems to increase cure speed and improve solvent resistance properties by the inclusion of a free radical curing monomer and photinitiator. A difunctional acrylate monomer was chosen and the formulation is shown in Table 3.

UV Curable Cationic Prepolymer	46 %
Cationic Photoinitiator	13 %
Difunctional Free Radical Monomer	12 %
Free Radical Photinitiator System	1 %
Solvent blend	24 %
Dye	2 %
Additives	2 %

Table 3.   Hybrid UV curable ink jet : formulation 2.

Ink jet prints were cured using a Fusion F200 medium pressure mercury lamp source powered at 200 Watts per inch. The cure speed was assessed by determining the maximum speed at which the prints would pass under the lamp and become fully cured. The cure speed of the model formulation compared to the hybrid system is shown in Table 4.

Substrate	Formulation 1 Cationic System	Formulation 2 Hybrid System
Stainless Steel	6.0	9.0
Mild Steel	4.2	7.2
Glass	3.0	6.0
Polyamide Sheet	7.2	12.0
Extruded PVC	6.0	10.2

Table 4   Maximum cure speed (metres per minute) of UV curable ink jet inks.

An increase of between 50 and 100% in cure speed is provided by the hybrid curing system compared to the purely cationic formulation.

The solvent resistance properties of the two systems were compared by casting a 24 micron film of each composition onto a range of substrates and curing using a 6" UV lamp rated at 200 Watts per inch. A drop of solvent, such as alcohol or methyl ethyl ketone was applied to the surface of the films and allowed to remain in contact for 1 minute. The solvent along with any dissolved dye was then absorbed into a piece of filter paper. This technique allows a rapid visual assessment of the degree of solvent bleed, as inks with poor solvent bleed resistance colour the filter paper to a much greater extent than those with good solvent bleed resistance . The cationic system alone shows considerable dye bleed under these conditions. The hybrid system exhibits a marked improvement in solvent resistance and on most substrates no dye bleed was observed.

In addition, ink jet prints of the hybrid system were subjected to solvent resistance testing according to MIL SPEC 202F[3] . The prints passed this test on a wide range of surfaces including plastics, metals and glass.

The adhesion of the two formulations was compared using a dolly pull off technique[10] . An aluminium dolly was adhered to the surface of a cured ink film using a solvent free adhesive. The adhesive was allowed to set, the dolly removed and the pressure required to remove the ink film from the surface was recorded. The technique expresses adhesion in quantitatvie terms and allows comparison between formulations.

A comparison of adhesion between the two systems is illustrated in Table 5.

Substrate	Formulation 1 Cationic System	Formulation 2 Hybrid System
Stainless Steel	2.3	5.6
Mild Steel	1.4	9.6
Glass	2.1	11.3
Polyamide Sheet	0.4	1.8
Extruded PVC	1.5	8.0

Table 5.   Adhesion measured in Kg/cm² of UV curable ink jet inks.

It is worth noting that the cure rate, solvent resistance and adhesion properties vary considerably from one substrate to another. It appears that in addition to the surface chemistry, a combination of the surface tension, and heat capacity of the substrates plays a large role in determining the performance characteristics of these ink jet inks. For example, the hybrid system gives a cure speed of 6 metres per minute on glass but 12 metres per minute on polyamide surfaces. The adhesion is recorded as 5.6 Kg/cm² on stainless steel but 8.0 Kg/cm² on extruded PVC. The hybrid system is extremely effective in improving adhesion over a wide range of substrates. In general,these results compare favourably with the work reported by Perkins[6] , Ketley[7] and Manus[8]. In their work they suggest the inclusion of a similar carbonyl type initiator such as benzophenone results in an improvement in photo response of hybrid systems containing epoxy resins and acrylate monomers.

4   The Effect of Conductive Salts on Stability

The inclusion of difunctional acrylate monomer to the cationic ink jet formulation can be seen to offer many advantages. However one adverse effect of including this type of material is a significant reduction in the conductivity of the formulation. This causes a deterioration of the performance of the printer as the ink droplets do not receive sufficient charging thus preventing satisfactory drop deflection. It is therefore necessary to supplement the conductivity by inclusion of a conductive salt. A number of salts have been evaluated for use as conductive agents for hybrid systems. Most salts were found to show poor solubility and others showed a potential for initating the polymerisation reaction. One specific example is the effect of Ammonium Thiocyanate on a free radical curing monomer. The salt shows good solubility and provides high conductivity. However, the viscosity of solutions of Ammonium Thiocyanate increase dramatically when subjected to accelerated ageing by storage at 60C, as illustrated in Figure 4. A commercial formulation has been developed which satisfactorily overcomes this problem.

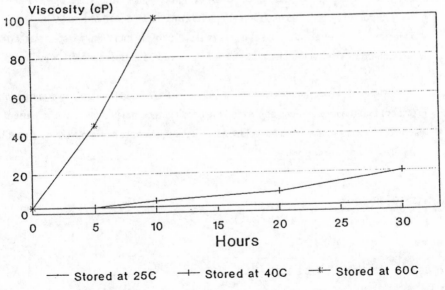

Figure 4   Stability of solutions of Ammonium Thiocyanate in Free Radical Curing Monomers

This effect is possibly due to the free radicals being produced by the monomer. A stabiliser, typically a tertiary amine is added to the monomer during production to eliminate the free radicals, however the addition of the ammonium thiocyanate conductive salt prevents the stabiliser from working thus allowing the free radicals to initiate polymerisation. The use of this combination in an ink jet formulation would lead to significant viscosity increases over a very short timescale and result in highly unreliable printer performance and ultimately to the ink system solidifying in the printer.

<u>5</u>   <u>Dye Selection</u>

The effect of dyestuff on the chemical stability was examined. Two cobalt complexes and two chromium complexes were chosen for the evaluation. The formulations were stored at 60 C . Withdrawals of ink samples were made at suitable intervals and the viscosity was measured using a Brookfield viscometer, model LV, fitted with a low viscosity adaptor. The spindle was set at 60rpm to measure values between 1 and 10 cP and at 30rpm to measure between 10 and 100cP. Molecular weight measurements were made using gel permeation chromatography (GPC). A high pressure Waters' Gel permeation chromatography system was used with columns at 1×10^4A, 1×10^3A and 500A. The mobile phase was tetrahydrofuran at a flow rate of 1.0 ml per min. Peaks were detected using a differential refractometer and a uv spectrophotometer operating at a wavelength of 254nm.

Figure 5 shows the increase in viscosity over the first 60 hours of the ageing trial. It is evident that the formulations containing the chromium complex dyes undergo a rapid viscosity increase whereas the formulations containing the cobalt complex dyes are stable.

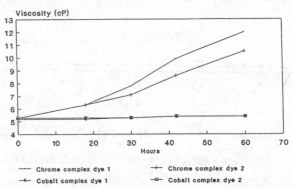

<u>Figure 5</u>   <u>Viscosity increase of model ink jet inks during accelerated ageing trial</u>

The reason for this viscosity increase is shown in the GPC traces, figures 6 and 7. Chromatograms of an ink containing a chromium dye and an ink containing a cobalt complex dye at the start of the trial are presented in Figure 6.

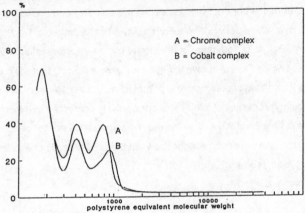

Figure 6.   GPC curves of inks at the start of the accelerated ageing trial

The same comparison carried out after 60 hours of ageing, figure 8, shows the increases in molecular weight distribution of the ink containing the chrome complex dye while over the same period the ink containing the cobalt complex dye shows little change in molecluar weight distribution.

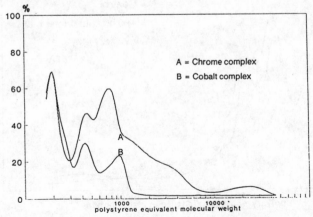

Figure 7.   GPC curves of inks after 60 hours storage at 60C

The reliability of a UV curable ink formulation containing the chrome complex dye was also tested in an ink jet printer. The performance of the ink was monitored daily. This included performing viscosity measurements using the printers in built ball fall time viscometer, measuring the conductivity, assessing the pressure and applied voltages necessary to achieve satisfactory print quality and investigating chemical changes in the formulation using GPC. After 400 hours of the printer trial a withdrawal of ink was taken from the machine. The GPC trace shown in Figure 8 shows evidence that the molecular weight of the binder in the ink taken from the printer is increasing. Recent work by Kelly [7] examining the ageing properties of ink jet inks conclusively confirmed that increases in molecular weight during the lifetime of an ink dramatically reduce printer performance and reliability. The increase in molecular weight shown in the GPC trace is characteristic of an unstable and unreliable product.

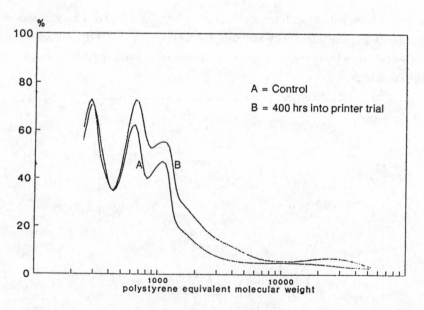

Figure 8.   GPC curves of ink containing chromium dye after 400 hours of printer trial

A similar formulation containing a cobalt complex dye was also tested in an ink jet printer. This formulation showed no change in performance throughout an extended printer trial lasting 1000 hours during which the ink was subjected to temperatures ranging from 5 to 55 C.

The change in physical properties of the ink throughout the trial is shown in Table 6.

	Start of Trial	End of Trial (1000 hrs)
Viscosity	5.58 cP	5.96
Conductivity	603 uS/cm	560
Filtration Time	29 secs	16 secs
Surface Tension	30.1dynes/cm	31.5

Table 6.   Change in Physical Properties during 1000 hour printer trial.

GPC measurements of ink samples taken before and after the trial confirm that this formulation is stable.

It is thought a change in oxidation state of the Cr^{3+} dye may provide an alternative oxidising agent capable of initiating polymerisation. On the other hand the cobalt based metal complex dyes are unlikely to show the same capability to initiate the polymerisation since complexes of Cobalt 3 + are extremely stable. This is contrary to the widely accepted view that the chromium metal is held strongly in the dye complex and is unlikely to undergo a change in oxidation state. Further work is planned to examine this effect.

Conclusions

An ink jet ink has been satisfactorily developed based on a hybrid cationic / free radical system by using cycloaliphatic epoxides and acrylate functional materials. The ink exhibits enhanced performance when compared to a purely cationic system reflected by improvement in the cure speed, solvent resistance and adhesion properties. These properties have been shown to vary depending on the substrate onto which the ink is printed.

The presence of certain transition metal complexes and conductive salts in UV curable ink jet formulations have demonstrated the ability to initiate polymerisation causing rapid increases in viscosity. This results in an unstable formulation demonstrating poor reliability in an ink jet printer and highlights the need for careful and scientific selection of raw materials when formulating ink jet printing inks.

References

[1]   Sweet, R.J., US Patent 3 596 275

[2]   Mitchell R.F. and Hudd A.L.
        Advances in ink jet inks. Dairy Industries International 1990,55 (6)

[3]   MIL-STD-202F METHOD 215A 1 April 1980

[4]   Kelly M., Gilkison I., Hudd A., Rock B.   The Ageing Properties of High Performance
        Jet Printing Inks - PRA Ninth International Conference on High Coatings

[5]   Manus P J-M.   Cationic UV-Curable coatings for metal and plastic substrates.

[6]   Perkins W.C.   Journal of Radiation Curing, Volume 6 (2),(22) 1979.

[7]   Ketley A.D. ; J. TSAO,   Journal of Radiation Curing Volume 8 (1), (16) 1981

[8]   Manus P J-M   The coating performance and formulation parameters of cationic
                       systems. Radtech Florence 1990

[9]   Hanrahan B.D., Manus P.,Eaton R.F.   Radtech 1988
                       Cationic UV Curable coatings and inks for metal substrates

[10]   BS3900 : Part E10 : 1979 , ISO 4624 - 1978 : Pull off test for adhesion.

Radiation Curable Adhesives with Auxiliary Cure Mechanisms

S. Peter Pappas and John G. Woods

LOCTITE CORPORATION, 795 NORTH MOUNTAIN ROAD, NEWINGTON, CT 06111, USA

1 INTRODUCTION

UV curable adhesives, which were virtually unknown 20 years ago, have enjoyed relatively rapid growth, particularly within the last 10 years, and are now well established in both industrial and consumer markets. Although they represent only a small fraction of the world-wide multi-billion dollar adhesives business, their share is currently growing at a faster rate than many other classes of adhesives.[1]

UV curing is based on photoinitiated polymerization of functional oligomers and monomers into a crosslinked polymer network. The process is mediated by a photoinitiator system which absorbs and converts UV-visible light into reactive intermediates, such as free radicals and radical ions, and/or long lived intermediates, such as acids or bases, which initiate the polymerization process. Recent reviews on photoinitiated polymerization,[2] photocrosslinking,[3] photoinitiators,[4] functional oligomers/monomers[5] and characterization of crosslinked polymer networks by chain growth polymerization[6] are available.

Commercially viable classes of UV curable adhesives,[7] which essentially span the gamut of adhesives applications, are provided in Figure 1.

These applications span an equally diverse range of industrial and consumer areas, including engineering (both OEM and after-market), electrical/electronic, optical, dental/medical, telecommunications and household repair.

Laminating and Structural Assemblies
Surface Mounting, Tacking and Fixturing
Sealing, Potting and Encapsulating Compositions
Pressure-Sensitive and Hot-Melt Adhesives
Conducting Adhesives
Conformal Coatings

Figure 1 Applications for UV Curable Adhesives

Attractive features and limitations of UV curable
adhesives are provided in Figure 2.

ATTRACTIVE FEATURES

Rapid Polymer Network Formation =>
High Processing Speed and Productivity

Low Heat Generation => Heat-Sensitive Substrates

One-Pack Compositions => Automatic Dispensing

Economical, ecological =>
Low Energy and Space Requirements; Low Capital Cost
Low Organic Emissions

LIMITATIONS

Pigmentation, Fillers => Opacity and Rheology

Cure Thickness => Limited by [Photoinitiator]

Geometry => "Shadow Areas," Opaque Substrates

Glass Transition Temperature (Tg) => Limited
by Ambient Cure

Exterior Durability =>
Light Stabilizers Tend to Reduce UV Cure Rate

Figure 2 General Characteristics of UV Curable
Adhesives

The important limitation of cure thickness arises
from the necessity to decrease photoinitiator
concentration with increasing film thickness in order
to achieve sufficient light penetration for through-
cure. As photoinitiator concentration decreases,
however, less light is absorbed in unit volumes
within the film resulting in slower cure rates.

These opposing factors generally result in an optimum photoinitiator concentration in UV cure.

The inter-relationships of UV cure rates, photoinitiator concentration and film thickness have been discussed in detail, as have strategies to cope with film thickness together with the related problem of air-inhibition (in UV curing by free radical polymerization).[8]

Herein, we present concepts and compositions, challenges and opportunities for UV curable adhesives with auxiliary cure mechanisms, which are designed to overcome one or more of the limitations noted in Figure 2. UV-thermal, UV-moisture, UV-aerobic, and UV-anaerobic dual cure systems will be considered for various applications, including the bonding of opaque substrates. Relatively recent overviews on UV curable coatings[9] and adhesives[10] with auxiliary cure processes are available.

2 UV-THERMAL CURE ADHESIVES

An immediate question is why UV, i.e. why not simply thermal cure? There are several answers to this question, including the following considerations. (1) UV cure provides more rapid "immobilization" or "fixturing" rates than can generally be achieved thermally. (2) The utilization of UV curable reactive diluents eliminates or reduces the need for solvent to achieve appropriate application viscosity. (3) UV/thermal cure can provide long term ambient (i.e. one-pack) stability together with a lower cure temperature than corresponding thermosets. This objective can be readily accomplished by photo-generation of a long-lived catalyst for subsequent thermal cure, which avoids the usual high temperature requirement for thermal activation of a latent catalyst or curing agent. The approach has been reported in coatings applications based on the photogeneration of strong acids, such as p-toluene sulfonic acid.[9,11]

Compositional requirements for UV-thermal cure include a photoinitiator system together with functional oligomers and monomers which are polymerized and crosslinked by the photogenerated initiators/ catalysts and heat.

Photoinitiators which produce long-lived catalysts, such as acids or bases, may not require an additional thermally-activated curing agent. Long-lived catalysts can effect both ambient and thermal cure. However, for adhesives with thick sections or "shadow" areas, this approach requires that the photogenerated catalyst diffuse into unexposed regions, which is an important consideration.

The utilization of photoinitiators for free radical polymerization generally requires the presence of thermally-activated curing agents owing to the short lifetime of free radicals. This approach introduces the challenge of achieving both a one pack composition and moderate cure temperatures. The problem has been discussed in detail within the context of kinetic limitations.[12] Approaches to obviate the limitations have also been presented, including the expedient of utilizing a phase change to effect thermal cure within narrow temperature ranges.

Aside from curing in unexposed regions, UV-thermal cure can be utilized to achieve higher glass transition temperatures (Tg) than possible by ambient UV cure. Although Tg values as high as ca. 70 $^\circ$C above UV cure temperatures have recently been reported for free radical cure of acrylated compositions, the experiments were carried in a N_2 atmosphere; and cure rates were substantially slower in the glassy state.[13]

Photoinitiators are available which produce both free radicals (for ambient cure) and acid (for thermal cure).[14] Of course, combinations of appropriate photoinitiators can also be utilized for this purpose.

Selected Applications for UV-Thermal Cure Adhesives

Several UV-thermal cure compositions have been described, which can be used as multi-layer laminating adhesives in the production of printed circuit boards. The adhesive compositions on a release-coated substrate are partially cured by UV exposure and subsequently utilized for bonding thermally cured laminate assemblies.

One composition includes a photoinitiator for free radical generation, a multifunctional thiol, such as (1), phenolic diene (2), a novolac epoxy, and a thermally latent curing agent for the epoxide-phenol

reaction.[15] An interesting feature of this
composition is the dual function of the phenolic diene
component, which undergoes photoinitiated free radical
polymerization with the thiol (via the ene) and
thermally induced condensation with the epoxy (via the
phenol). Related compositions contain acrylated resins
in place of the thiol-ene system.[16]

UV-thermal cure adhesives are also used to mount
components, such as chip capacitors, chip resistors and
packaged semiconductors, to printed circuit board
surfaces. This process is replacing the earlier
approach of mounting components by inserting leads
through holes in the circuit board. An important
function of the dual cure adhesives is to allow rapid
"fixturing" prior to soldering the component to metal
pads. As illustrated by Figure 3, the component is
positioned on an over-sized adhesive drop to form a
fillet which is UV cured. Thermal cure is required for
the shadow region below the component. The metal pads,
which are not shown, are positioned in front and behind
the figure.

(1)   (2)

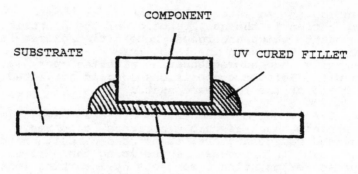

COMPONENT
SUBSTRATE   UV CURED FILLET

INDEPENDENTLY CURED BONDLINE

Figure 3 Surface Mounted Device (SMD) on a PCB

Typical dual UV-thermal adhesive compositions for
surface mounted devices include both a photoinitiator
and thermal initiator for free radical polymerization
together with (meth)acrylated oligomers/monomers. The
(meth)acrylated resins serve as curing medium for both
the photo- and thermal initiator, although ambient
stability may be compromised by thermally-labile
initiators, such as peroxidic species. An interesting
modification is the utilization of a perester, such as
(3), which absorbs UV-visible light sufficiently to

(3)

function as both the photo- and thermal initiator.[17]
Thiol-enes can also be utilized in place of
(meth)acrylated resins. The ambient stability of
thiol-ene or (meth)acrylate compositions can be
extended by using aromatic pinacols as the thermal
initiator in place of peresters.[18]

3 UV-MOISTURE CURE ADHESIVES

Compositional requirements for UV-moisture cure include
a photoinitiator together with functional oligomers and
monomers which are polymerized and crosslinked by the
photogenerated initiators/catalysts and moisture. A
potential advantage over UV-thermal cure is avoidance
of a secondary heating step. A formulation challenge
is to achieve adequate storage stability by minimizing
reaction with inadvertent water, generally associated
with oligomers, fillers and pigments.

Moisture-reactive functionality includes
isocyanates and silanes, in which silicon is bonded to
leaving groups, such as acetoxy, alkoxy and oximino
groups. Moisture cure is facilitated by high
diffusability of water, but also depends on humidity.
High conversions generally require one or more days at
ambient temperature or hours at moderately elevated
temperatures. In principle, moisture cure rates could
be enhanced by photogenerated catalysts (acids, bases).

Recent studies have been reported on photoinitiated
free radical polymerization of an acrylated oligomer/
monomer composition together with acrylate and thiol
functional trimethoxysilane coupling agents (4) and
(5), respectively.[19] The trimethoxy-silane groups
are incorporated into the crosslinked network in situ
via the acrylate and thiol functionality, followed by
(dark) moisture cure, as illustrated in Scheme 1 for
the acrylated trimethoxysilane coupling agent. The
studies included the kinetics of the alkoxysilane-
moisture reaction by FTIR at varying temperatures and
water vapor pressures. Better adhesion was obtained
with the thiol functional coupling agent (5).

Methacrylated silane coupling agents, with methoxy,
acetoxy or oximino groups, have also been prereacted
(in molar excess) with silicone hydrolysates to yield
silicone oligomers with terminal methacrylate and
moisture-reactive silane groups, as shown in Scheme 2
for the trimethoxysilane.[20] Applications include

adhesives, sealants, formed-in-place gaskets and
conformal coatings.

$$CH_2=CH-CO-O-(CH_2)_3-Si(OMe)_3 \qquad HS-(CH_2)_3-Si(OMe)_3$$

$$(4) \qquad\qquad\qquad (5)$$

IN-SITU INCORPORATION INTO ACRYLATE NETWORK:

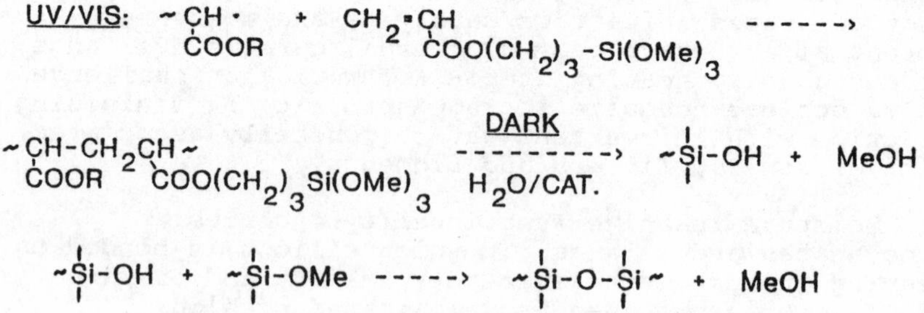

Scheme 1

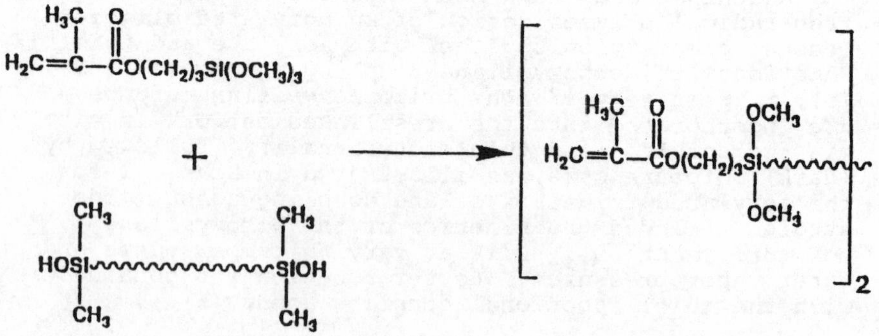

Scheme 2

4   UV-AEROBIC CURE ADHESIVES

Compositional requirements for UV-aerobic cure include a photoinitiator together with functional oligomers and monomers which polymerize and crosslink with the photogenerated initiator/catalyst and oxygen. Typically, cobalt carboxylate salts, together with auxiliary salts, such as zirconium and magnesium carboxylates, so-called "driers," are also required to effect the oxidative cure. Oxidative cure is facilitated by activated methylene groups, particularly 1,4-diene groups (6), which are present in drying oils, as well as allyl ether groups (7).[9]

$$R-CH=CH-\underline{CH_2}-CH=CH-R \qquad R-O-\underline{CH_2}-CH=CH_2$$

$$(6) \qquad\qquad\qquad (7)$$

A significant systems design challenge is achieving adequate storage stability for a one pack composition together with rapid UV cure and sufficient aerobic "shadow" cure. In this regard, high crosslink density, characteristic of UV cure, tends to reduce oxygen permeation required for aerobic cure. Dual UV-aerobic cure compositions, containing photoinitiators for free radical polymerization, (meth)acrylated resins, driers, and resins with allylic methylene groups, have been described for conformal coating applications.[21] A conformal coated PCB, which illustrates variable thickness and shadow regions, as well as adhesive requirements, is shown in Figure 4.

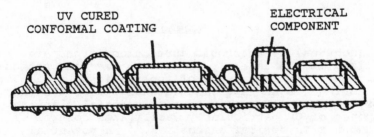

PHENOLIC/CERAMIC SUBSTRATE

<u>Figure 4</u>   Conformal Coated PCB

5   UV-ANAEROBIC CURE ADHESIVES

The marriage of UV with anaerobic cure generally
constitutes a relatively simple, compatible alliance.
For UV cure compositions based on free radical
polymerization of (meth)acrylate resins, an anaerobic
cure system can simply be added for continued dark
reaction. Alternatively, a photoinitiator for free
radical polymerization can simply be added to an
anaerobic cure composition to produce the dual cure
system. A simplified depiction of anaerobic cure
systems is provided in Scheme 3, including metal ion
catalyzed redox generation of initiating radicals from a
hydroperoxide and inhibition in air.

$$ROOH + M^{+n} \longrightarrow RO\cdot + HO^- + M^{+(n+1)}$$

$$ROOH + M^{+(n+1)} \longrightarrow ROO\cdot + H^+ + M^{+n}$$

$$RO\cdot + \text{METHACRYLATE RESIN} \longrightarrow \text{POLYMER NETWORK}$$

AIR-INHIBITION (AMBIENT STABILITY)

$$\sim \underset{\text{COOR}}{\overset{\text{Me}}{\text{C}}}\cdot + O_2 \longrightarrow \sim \underset{\text{COOR}}{\overset{\text{Me}}{\text{C}}}\text{-OO}\cdot \quad \text{(NON-PROPAGATING)}$$

Scheme 3

Such dual UV/anaerobic cure compositions are
utilized in the assembly of electronic[10] and optical
equipment. In both types of applications, rapid UV
"fixturing" is an important advantage over two-pack
epoxy or thermal cure systems. Essentially instant
response of UV particularly facilitates control of
alignments in optical assemblies.[22] A potential
limitation of such compositions is the general
requirement of a metal surface (or metal-containing
primer) for anaerobic cure.

6 UV CURABLE ADHESIVES FOR OPAQUE SUBSTRATES

UV curable adhesives for opaque substrates must be exposed prior to assembly and subsequently cure in the absence of light. Important characteristics are (1) exposure dose, (2) open time following exposure, (3) fixture (or set) time, following assembly, during which sufficient strength has developed for handling, and (4) the time for development of ultimate properties, such as strength.

Potential approaches include (1) photogeneration of long-lived catalysts (e.g. acids or bases) with cure rate controlled by subsequent dark reactions, and (2) photogeneration of free radicals which are converted into more latent or longer-lived initiators. Utilization of a volatile inhibitor could also be an attractive feature of either approach.

Interesting systems which combine photogenerated acid with anaerobic cure have been developed.[23-25] Compositions contain methacrylated oligomers and monomers; a photoinitiator for acid generation (e.g. triarylsulfonium salt); and an anaerobic cure system, including peroxide (e.g. cumene hydroperoxide) and an acid activated promoter (e.g. ferrocene).

UV exposure of adhesive on one or both substrates results in photogeneration of acid which activates the ferrocene/ peroxide system. However, polymerization by the resulting initiating radicals is air-inhibited, which provides open time. Anaerobic cure proceeds following structural assembly.

Characteristically, lap shear strengths increase to maximum values (ca. 16-20 MPa) with UV dose and subsequently decrease. With constant dose, lap shear strength remains relatively constant with open time (to ca. 2 min) and subsequently decreases slowly to ca. 11 MPa (1600 psi) after 1 hour.[24]

Optimum lap shear strengths with UV dose can be attributed to increasing photogenerated acid, which would tend to promote adhesive strength, and increasing polymerization prior to assembly (by overcoming air-inhibition), which would tend to reduce bond strength. Long open-times may also reduce bond strength by slow polymerization prior to assembly or by consuming the peroxide.

An interesting visible light activated adhesive for opaque substrates, based on photogeneration of free radicals, has also been reported.[26]

7 CONCLUSION

It is our hope that this account will impart to the reader a sense of the challenge and opportunity in this area as well as the realization that one of the most important assets is a vivid imagination.

REFERENCES

1. C. Bluestein, Adhesives Age, 1982, 25, 19.

2. S.P. Pappas in "Comprehensive Polymer Science", Vol. 4, Part II, G. Allen, J.C. Bevington, G.C. Eastmond, A. Ledwith, S. Russo and P. Sigwalt, eds., Pergamon Press, 1989, pp. 337-55.

3. S.P. Pappas in "Comprehensive Polymer Science", Vol. 6, G. Allan, J.C. Bevington, G.C. Eastmond, A. Ledwith, S. Russo and P. Sigwalt, eds., Pergamon Press, 1989, pp. 135-48.

4. S.P. Pappas in "Handbook of Organic Chemistry", Vol. 2, J.C. Scaiano, ed., CRC Press, 1989, pp. 329-39.

5. K. O'Hara in "Radiation Curing of Polymers", Special Publ. No. 64, D.R. Randell, ed., Royal Society of Chemistry, 1987, pp. 116-27.

6. J.G. Kloosterboer, Adv. Polym. Sci., 1988, 84, 1.

7. J. Woods in "Radiation Curing of Polymers", Special Publ. No. 64, D.R. Randell, ed., Royal Society of Chemistry, 1987, pp. 102-15.

8. S.P. Pappas, J. Rad. Curing, 1987, 14(3), 6.

9. S. Peeters, J.-M. Loutz and M. Philips, Polym. Paint Colour J., 1989, 179, 304; see also, "Radtech '88 - North America, Conf. Papers", Radtech International, 1988, pp. 79-83.

10. S. Grant in "Radcure Europe '87, Conf. Proceedings", Soc. Manuf. Eng., 1987, pp. 7/5-7/14.

11. G. Berner, R. Kirchmayr, G. Rist and W. Rutsch, J. Rad. Curing, 1986, 13(4), 10.

12. S.P. Pappas and H.B. Feng, Org. Coatings, 1985, 8, 139.

13. J.G. Kloosterboer and G.F.C.M. Lijten, Polymer, 1990, 31, 95.

14. J.V. Crivello, J.L. Lee and D.A. Conlin, J. Rad. Curing, 1983, 10(1), 6.

15. G.E. Green and A-C. Zahir, U.S. Patent 4,308,367 (1987) (to Ciba-Geigy).

16. J.E.Gervay, Eur. Pat. Appl. 270,945 (1987) (to DuPont).

17. N.S. Allen, S.J. Hardy, A. Jacobine, D.M. Glaser and F. Catalina, Eur. Polym. J., 1989, 25, 1219.

18. C.R. Morgan, U.S. Patent 4,288,527 (1981) (to W.R. Grace).

19. G. Gozzelino, A. Priola and F. Ferrero, Makromol. Chem. Makromol. Symp. 23, 1989, 393.

20. S. Nakos, U.S. Patent 4,699,802 (1987) (to Loctite).

21. L.A. Nativi and K. Kadziela, U.S. Patent 4,451,523 (1984) (to Loctite).

22. N. Siga, N. Seo, T. Harutake and M. Shirae, U.S. Patent 4,778,253 (1988) (to Olympus Optical).

23. J.M. Rooney, J. Woods and P. Conway, U.S. Patent 4,525,232 (1985) (to Loctite).

24. P. Conway, D.P. Melody, J. Woods, E. Casey, B.J. Bolger, and F.R. Martin, U.S. Patent 4,533,446 (1985) (to Loctite).

25. J. Woods and P. Coakley, Eur. Pat. Appl., 251,465 (1987) (to Loctite).

26. R.A.A. Gonzalez, H.C. Nicolaisen and H.P. Handwerk, U.S. Patent., 4,882,001 (1989) (to Henkel).

Initiator Chemistry

New Optimised Alpha-Aminoketone Photoinitiators

M. Koehler[1], L. Misev[1], V. Desobry[2], K. Dietliker[2], B.M. Bussian[3] and H. Karfunkel[3]

[1] CIBA-GEIGY APPLICATION LABORATORIES, ADDITIVES DIVISION, CH-4002 BASEL, SWITZERLAND
[2] CIBA-GEIGY RESEARCH CENTER, ADDITIVES DIVISION, CH-1701 FREIBURG, SWITZERLAND
[3] CIBA-GEIGY CENTRAL RESEARCH, COMPUTER CHEMISTRY, CH-4002 BASEL, SWITZERLAND

1   INTRODUCTION

Acetophenone derivatives with a certain substitution pattern on the alpha-carbon atom (compare Table 1) are efficient photoinitiators for the polymerisation of lacquers or ink formulations. A photochemical cleavage of the bond between the carbonyl group and the adjacent saturated carbon atom leads to the formation of radicals which add to the double bonds of the polymerisable components. The resulting curing process is a fast polymerisation which leads to the formation of an extended macromolecular network[1,2].The photoinitiator is a key component of the UV curable system because it consumes the radiation energy and converts it into reactivity.
To better understand the process of photoinitiation and enable the development of more optimised initiators, it is important to examine the influence of the substituents (R1,R2,R3 and X) of the acetophenones (Table 1).The presence of a heteroatom substituent X (hydroxy or dialkylamino) at the alkylated carbon atom is a common structural feature of the compounds 1-4 on Table 1. The hydroxyketones 1 and 2 are mainly employed in clear (nonpigmented) formulations. Recently it has been shown that radiation energy in the wavelength range between 340 and 350 nanometers is important for the activation of the hydroxyketone 2 [3].In pigmented formulations,

there is less light for the initiators within this wavelength range because the pigment is also a light absorbing species. The aminoketone <u>3</u>,with a sulfur atom on the benzoyl chromophore is a highly effective photoinitiator for such systems [4].The sensitization of the sulfur substituted aminoketone <u>3</u> by thioxanthones extends the area of light absorption to the longer wavelengths[5].

<u>Table 1</u>: Acetophenone type photoinitiators

Initi- ator	3 R	2 R	1 R	X
<u>1</u>	H	-(CH2)5-		OH
<u>2</u>	H	CH3	CH3	OH
<u>3</u>	CH3S	CH3	CH3	morpholino
<u>4</u>	morpholino	C2H5	CH2-C6H5	N(CH3)2

(morpholino =)

Substituent R3 on the benzoyl chromophore has a strong influence on the spectral absorption of the compound. However,the heterosubstituent X and the groups R1 and R2 seem to be more important for the formation and the properties of the substituted methyl radical which is produced by the photocleavage reaction.
The replacement of the methyl groups of <u>3</u> (R1 and R2) by an ethyl and a butyl substituent,for example, leads to a significant increase in the photoinitiator efficiency in UV curing of a white pigmented lacquer. The enhanced spectral absorption at longer wavelengths

of such aminoacetophenones having larger substituents R1 and R2 than methyl groups is responsible for this surprising result[6].

Recently the alpha-dimethylamino acetophenone 4,having ethyl and benzyl substituents was presented as a new photoinitiator [7].This derivative has a higher reactivity than the sulfur substituted compound 3 (compare Table 1).This was explained by a comparison of the absorption spectra of the two compounds. These results indicate that the substituents have an important influence on the reactivity of aminoketones in pigmented coatings. Investigations of a broader series of alpha-dimethylamino acetophenones could show structure-reactivity relationships in more detail.

2 UV CURABLE SYSTEMS

The UV curable systems which are used in this study consist of unsaturated compounds and a pigment. The photoinitiators are varied within an experimental series to obtain information about the influence of the substituents R1,R2 and R3 on the curing process. A blue printing ink and two white pigmented lacquers which are similar to industrial formulations were chosen as model systems. The blue offset printing ink is based on a polyurethane acrylate, a diacrylate and a pigment of the phthalocyanine type. The white lacquers are composed of an epoxyacrylate or polyesteracrylate which are diluted with monomers such as trimethylolpropane triacrylate, hexanediol diacrylate,or N-vinylpyrrolidone. Titanium dioxide (rutile modification) in amounts of 50 and 25 weight percent of the curable composition was used as a white pigment. The photoinitiator content is 3 weight percent in the case of the printing ink and 2 weight percent in the case of the white lacquer formulations I and II (compare Table 2).

<u>Table</u> <u>2</u>: UV Curable Systems

<u>Blue</u> <u>Offset</u> <u>Printing</u> <u>Ink</u>

Initiator Content : 3 weight percent of the printing
                    ink
Polymerisable   : polyurethaneacrylate (Setalin
Components   AP 565 from Synthese, Netherlands),
                    diacrylate (Ebecryl 150 from UCB,
                    Belgium)
Pigment   : Irgalith   Blue   GLB   (from Ciba-
                    Geigy, Switzerland)

<u>White</u> <u>Lacquers</u>

Initiator Content : 2 weight percent of the lacquer

<u>Formulation</u> <u>I</u>

Polymerisable   : epoxyacrylate (Ebecryl 608 from
Components   from UCB, Belgium),
                    trimethylolpropane triacrylate
                    (from Degussa, Germany),
                    N-vinylpyrrolidone (from Fluka,
                    Switzerland)

Pigment   : titaniumdioxide (rutile type
                    R-TC 2 from Tioxide, France),
                    50 weight percent

Formulation II

Polymerisable Components	: polyesteracrylate (Ebecryl 830 from UCB, Belgium), trimethylolpropane triacrylate (from Degussa, Germany), hexanediol diacrylate (from Roehm, Germany)
Pigment	: titaniumdioxide (rutile type R-TC 2 from Tioxide, France), 50 weight percent

3 CURING OF THE OFFSET PRINTING INK

The printing ink was applied on an aluminium coated paper with a laboratory offset printing machine (from Pruefbau, Germany). The thickness of the ink layer (1,5 micron) was determined by the weight loss of the printing cylinder upon ink transfer. The prints were cured by irradiation with one or two medium pressure mercury lamps in an ultraviolet processor (from PPG Industries, Illinois,U.S.A.).The performance of each lamp is 80 Watts per centimetre of the light bulb length. The speed of the conveyor belt is given in m/min. The reactivities of the photoinitiators are expressed in terms of the fastest cure speed which corresponds to the minimum exposure time. The results of structurally related aminoketones are given on Table 3. The cured prints were tested for the through cure by applying a rotary twisting motion on the ink film by means of a mechanical thumb.

Table **3**: Curing of the blue offset printing ink with
two 80 W/cm medium pressure mercury lamps

Initi- ator	1 R	2 R	Through Cure [given in m/min]
4	C2H5	CH2-C6H5	130
5	C2H5	CH2-CH=CH2	150
6	CH2-CH=CH2	CH2-CH=CH2	150
7	CH3	CH2-C6H5	140
8	CH3	CH2-CH=CH2	130

Formulations containing the (4-morpholino)acetophenone
derivatives **4** - **8** can be cured at very high speeds.
The reactivity does not depend significantly on the
structure of the substituents at the alpha-carbon atom
of the ketones. The influence of the different
substituents on the phenyl ring is shown in Table 4.
The light source was a 80 Watt per centimetre medium
pressure mercury lamp. The surface of the print was
cured if the color could not be transfered onto white
paper.
The morpholino derivative **4** affords the fastest cure
followed by the the dimethoxy compound **9.** The
monomethoxy compound **10** is significantly less reactive
and the aminoketones **11** and **12** do not allow cure even
at a speed of 10 m/min. The results on Table 4 show
the strong influence of the substituent R3 on the
photoinitiator activity in a printing ink system. The
reactivity corresponds well to the electron donating
ability of the substituent R3. A similar influence of
the phenyl substituent on the curing speed had
previously been found in a series of alpha-morpholino
acetophenones [6].

Table 4: Curing of the blue offset printing ink with
         one 80 W/cm medium pressure mercury lamp

$$R^3 \text{—} \underset{\text{CH}_2\text{-C}_6\text{H}_5}{\overset{\overset{\displaystyle O}{\parallel}\quad \overset{\displaystyle C_2H_5}{|}}{C\text{-}C\text{-}N\,(CH_3)_2}}$$

Initi-	3	Cure Speed [given in m/min]	
ator	R	Surface Cure	Through Cure
4	4-morpholino	140	70
9	3,4-(CH3O)2	120	60
10	4-CH3O	40	20
11	4-CH3	<10	<10
12	H	<10	<10

4 CURING OF WHITE LACQUERS

The alpha-aminoketones with different substituents R1
and R2 (compare Table 3) were used as photoinitiators
for curing of a thin (30 micron) white lacquer coating
(Table 5). The formulation I (compare Table 2) was
coated on a glass plate. The layers were cured with
two different light sources, a 80 W/cm medium pressure
mercury lamp and a Fusion D bulb (120 W/cm). The Fusion
D bulb has a broader and stronger spectral output in
the wavelength range from 350 to 400 nm than the
standard mercury lamp [8].
The influence of the substituents R1 and R2 on the
initiator activity is seen more clearly with the 80
W/cm medium pressure mercury lamp than with the Fusion
D lamp. Contrary to the results obtained with the
offset printing ink (Table 3), only the aminoketone 4
with the ethyl/benzyl substitution affords a fast
curing whereas the other substituent patterns are less
efficient. The most striking difference is found
between the compounds 4 and 7 which have a similar
chemical structure. The methyl derivative 7 is much
less efficient than its ethyl homologue 4.

Table **5**: Curing of the white lacquer formulation I
         under a 80 W/cm medium pressure mercury lamp
         and a 120 W/cm Fusion D lamp

Initi-ator	1 R	2 R	Cure Speed [given in m/min] 80 W/cm	120 W/cm
4	C2H5	CH2-C6H5	140	>200
5	C2H5	CH2-CH=CH2	60	130
6	CH2-CH=CH2	CH2-CH=CH2	70	150
7	CH3	CH2-C6H5	50	170
8	CH3	CH2-CH=CH2	20	30

Five ethyl/benzyl substituted alpha-dimethylamino
acetophenones with different substituents R3 on the
phenyl ring were investigated in the white lacquer
formulation II (Table 6). The lacquer was applied as a
200 micron layer on a coil coated aluminium and cured
with two 80 W/cm medium pressure mercury lamps.

(2,4,6-Trimethyl-benzoyl)diphenylphosphine   oxide
(TMDPO) is a commercial photoinitiator for white
lacquer applications [9].   In addition to the cure
speed the pendulum hardness of the cured coatings
provides valuable information on the thorough curing.
The hardness is measured directly after the curing
process and after 16 hours of additional exposure to
TL20/05 lamps(fluorescent low pressure mercury lamps
from Philips, Netherlands).
Two or three passes at a belt speed of 10 m/min are
generally necessary to achieve through cure of the
thick layer. Only the compounds 4 and 13 with
aminosubstituents at the phenyl ring produce
significant yellowing under these conditions. The
yellowing values obtained with the other aminoketone
derivatives 9, 10 and 12 are comparable with
that of the acylphosphine oxide TMDPO.

Table 6: Curing of the white lacquer formulation II
         (200 micron thickness on coil coated
         aluminium)

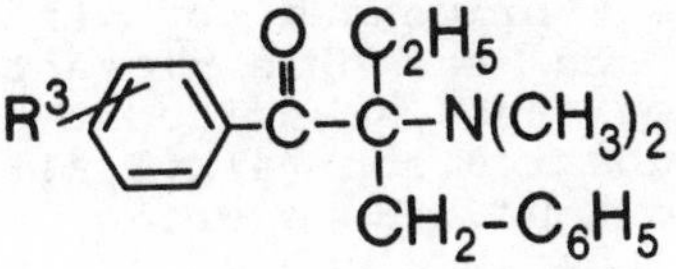

$$R^3 \underset{}{\overset{}{\bigcirc}} -\overset{O}{\overset{\|}{C}} - \overset{C_2H_5}{\underset{CH_2-C_6H_5}{C}} - N(CH_3)_2$$

R3 = 4-morpholino (4), 3,4-(CH3O)2 (9), 4-CH3O (10),
     H (12) and 4-N(CH3)2 (13)

Initiator	number of passes at 10 m/min and yellowing()	pendulum hardness [given in seconds]	
		directly after curing	add.irradiation with TL20/05 lamps

4	2 (22,0)	57	126
9	2 (3,2)	78	155
10	2 (2,3)	70	148
12	2 (2,7)	81	168
13	3 (28,0)	52	106
TMDPO	3 (1,5)	48	113

Directly after the curing process the pendulum
hardness was found below 100 seconds. Upon additional
exposure to low pressure mercury light the hardness
values nearly doubled. The aminoketones 9, 10 and 12
afford the hardest films after both exposure steps.
The compound 12 (R3 = H) is comparable in reactivity
to the other aminoketones on Table 6. Therefore, the
presence of an electron donating substituent R3 is not
absolutely necessary in the white lacquer application.

6   DISCUSSION OF THE SUBSTITUENTS EFFECTS

From the curing results it is concluded that the reactivity of aminoketones is influenced by their substituents R1,R2 and R3. In the blue pigmented printing ink the influence of the substituent R3 dominates, while the alpha-substitution (alkyl,allyl or benzyl)seems to be less important. However,in the white lacquer application the influence of the substituents R1 and R2 at the carbon atom is revealed. Maximum reactivity results from the combination of an ethyl with a benzyl group.

The substituents on the photoinitiator have a significant influence on its UV absorption. The absorption properties are especially important in pigmented systems because the pigment acts as a filter and only a part of the light is available to the initiator. The comparison of the UV absorption spectra of compounds 4,9,10 and 12 shows the influence of the different R3 substituents on the chromophore (Figure 1). The absorptions in the range between 300 and 400 nm relate well to the reactivity of the initiators in the offset printing ink (Table 4).

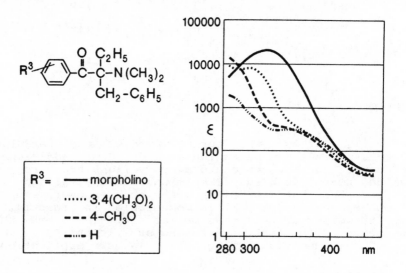

Figure 1: UV Absorption spectra of the compounds 4, 9, 10 and 12 (0,001 weight percent in methanol)

For a blue (cyan) pigment a transmission in the region around 300 nm was measured [10] which is assumed to be sufficient for the excitation of the photoinitiator in a thin layer. In the white pigmented lacquer,however,the rutile pigment is transparent for radiation only in the range above 400 nm [11]. Alpha-aminoketones on Figure 1 show a very similar absorption behaviour in this region.

By excluding the white pigment it can be determined whether the difference between the ethyl/benzyl and the methyl/benzyl derivatives is due to the filter effect of the rutile.

The aminoketones 4,7,9 and 14 were used as photoinitiators in two different nonpigmented coatings based on an epoxyacrylate and an unsaturated polyester (Table 7). The samples were cured in 100 micron layers with two 80 W/cm medium pressure mercury lamps at a belt speed of 20 m/min (epoxyacrylate formulation) or 5 x 20 m/min (unsaturated polyester formulation). The pendulum hardness of the coatings can be directly related to the photoinitiator reactivity.

Table 7: Curing of clear coatings

$$R^3 \text{—} \bigotimes \text{—} \overset{\overset{O}{\|}}{C} \text{—} \overset{\overset{R^1}{|}}{\underset{\underset{CH_2\text{-}C_6H_5}{|}}{C}} \text{—} N(CH_3)_2$$

4: R1 = ethyl, R3 = 4-morpholino
7: R1 = methyl, R3 = 4-morpholino
9: R1 = ethyl, R3 = 3,4-(CH3O)2
14: R1 = methyl, R3 = 3,4-(CH3O)2

Initiator	pendulum hardness [given in seconds]	
	epoxyacrylate	unsat.polyester
4	178	39
7	168	not cured
9	162	98
14	151	32

In both formulations the ethyl derivatives 4 and 9 produce higher hardness values than their methyl homologues 7 and 14. This difference becomes more apparent in the less reactive unsaturated polyester formulation. These findings lead to the conclusion that the higher reactivity of the ethyl derivatives is not exclusively due to the filter effect of the rutile pigment.

Complementary to the UV curing experiments, Molecular Modelling is employed to calculate the most stable conformations of the photoinitiators 9 and 14.

7 MOLECULAR MODELLING OF PHOTOINITIATORS

There are different computational techniques which are commonly used when studying molecular conformations or energies and other physical properties like charges, dipole moments or spin densities of radicals. Molecular mechanical methods give fast results for molecular conformations at ground states whereas properties related to charges are not computed. Quantum mechanical calculations allow the consideration of ground states and excited states of uncharged and charged molecules or radicals.

In this study molecular mechanical methods [12] and semiempirical quantum mechanical methods [13] were used. The modelling package was Macromodel [14] in conjunction with the MM2 force field to generate a classical mechanically optimised structure of the aminoketone photoinitiators. The theoretical investigation will also elucidate implications of the chemical structure on reactivity and stability of the initiator molecules. The aminoketones 9 and 14 were selected for a comparison because of their different reactivities in the curing experiments. Molecular modelling of the initiators 9 and 14 might predict differences in their ability of radical formation according to Figure 2.

<u>Figure</u> <u>2</u>: Photocleavage reaction of the aminoketones
<u>9</u> (R = C2H5) and <u>14</u> (R = CH3)

Our **approach is to ask whether formation of the** ethyl
or the methyl substituted radical is thermodynamically
more favourable. The calculations are based on the
photocleavage reaction (Fig. 2) where the educt
aminoketones <u>9</u> and <u>14</u> are promoted into the T1
precursor state upon photon absorption and subsequent
intersystem crossing. Thus, the model reaction
pathway covers three different electronic states, the
singlet ground state (So) for <u>9</u> and <u>14</u>, their first
excited triplet states (T1), and the resulting doublet
states (Do) of the product radicals. The first excited
singlet state (S1) is not relevant for the
thermodynamical calculations and was not considered.
This situation implicates some fundamental problems
for the computational method. Since the three states
are of different symmetry the conversion from one to
the other state implies intersystem crossing.
Therefore the three states have to be calculated
individually.
At first the educts and the products in their singlet
ground states have been computer modelled and some 50
different conformations have been generated by means
of a molecular mechanical method. The MM2 method [12]
is not parameterised for excited states and for

radicals, therefore the electronic ground states are
considered and the radicals have been substituted by
their parent molecules. From the resulting geometries
10 structures have been selected for quantum
mechanical energy minimisation at the T1 state. The
product radicals are developed by a formal hydrogen
abstraction step. The correct electronic state to be
considered is set then as a condition for the quantum
mechanical calculation. The resulting heats of
formation are finally compared for the educt molecules
9 and 14 and the product radicals. Since the
3,4-dimethoxybenzoyl radical is identical for both
aminoketones it can be omitted at this point. The
comparison between ethyl and methyl substitution shows
that the homolytic cleavage of the ethyl substituted
compound 9 is thermodynamically more favourable by 5,7
kcal/mol.

A kinetic treatment was not attempted because it is
known from the literature that the photocleavage
reaction proceeds via tunneling from T1 to the Do
state [15].

A separate quantum mechanical calculation of the
singlet ground states, from which the reaction starts,
shows drastic conformational differences between the
ethyl and the methyl compound (Fig.3).

In Figures 3, 4 and 5 the compared ethyl and methyl
substituted aminoketones are oriented with the
3,4-dimethoxy group upward.The substituents at the
alpha-carbon atom are designated with the symbols of
the adjacent atoms.

In Figure 4 the optimised geometries of the T1 states
are shown. The calculation shows that the methyl
group is located between the two aromatic ring
systems. In contrast to that the ethyl derivative 9
exhibits a bent arrangement of the aromatic rings
with the ethyl group pointing outwards.

To compare the steric effect of the small sized methyl
group to the larger ethyl group both compounds are
shown in a projection along the cleaving bond (Fig.5).
The steric interaction of the methyl substituent with
the remaining molecule is weak.It therefore fits well
between the two large planar aromatic ring systems.

The comparison of the educt geometries for the singlet
and triplet states shows that the size of the alkyl
substituent at the alpha-carbon atom determines the
conformation of the total molecule when undergoing the
homolytic cleavage.

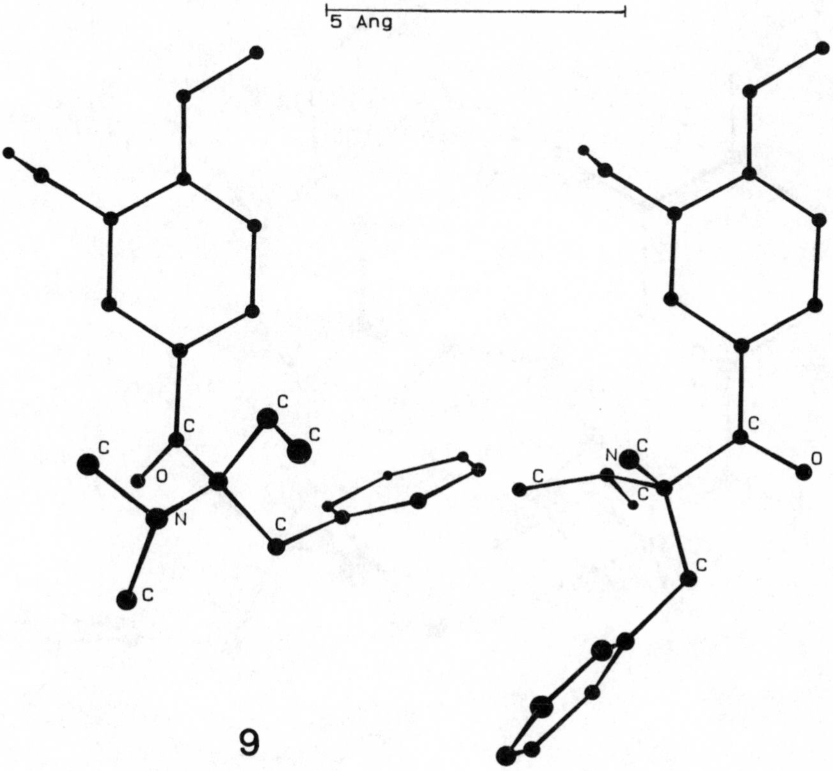

9

14

Figure 3: The most stable conformations of the ground state So of the aminoketones 9 and 14

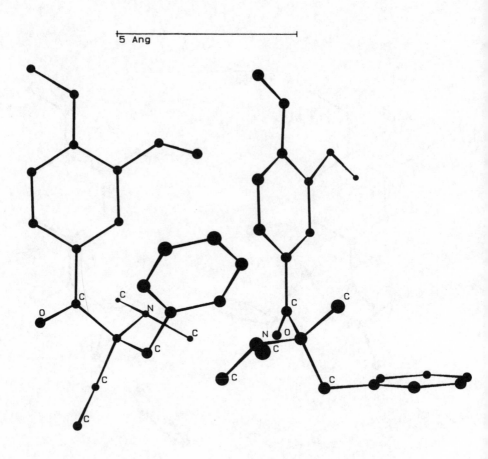

9   14

<u>Figure 4</u>: The most stable conformations of the first
          triplet state T1 of the aminoketones
          <u>9</u> and <u>14</u>

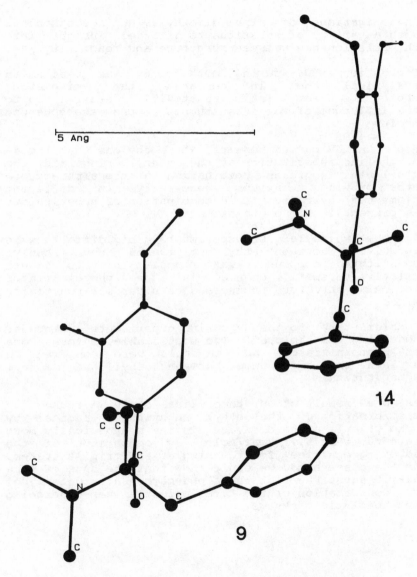

Figure 5: Projection of the aminoketones 9 and 14 in the ground state So along the cleaving bond

8   CONCLUSIONS

Investigations of alpha-dimethylamino acetophenones
with a variety of substituents afforded insight into
the relationship between structure and reactivity.

Conventional UV curing methodology was used as an
analytical tool in comparing the aminoketone
photoinitiators. Their reactivities were related to
the influence of the substituents on the acetophenone
nucleus.

From the UV curing results, it is obvious that the 4-
morpholino substitution at the phenyl ring and the
ethyl/benzyl groups combination at the alpha-carbon
atom provides maximum reactivity in different
pigmented systems. This combination of substituents
is present in the photoinitiator <u>4</u>.

In the white lacquer experiments a big difference in
reactivity between ethyl/benzyl and methyl/benzyl
substituted compounds was found. In this case a
relatively small change in the photoinitiator
molecular structure produces a considerable reactivity
effect.

Complementary to the UV curing experiments, Molecular
Modelling was employed. The most stable conformations
of photoinitiators and radicals were compared to
explain the difference between ethyl and methyl
substitution.

As one result of our theoretical study we found that
the formation of the ethyl substituted radical by
homolytical alpha-cleavage is thermodynamically more
favourable by 5,7 kcal/mol. By comparison of the
educt geometries for the singlet and triplet states,
on the other hand, we could show that the size of the
alkyl substituent at the alpha-carbon atom determines
the conformation of the total molecule when undergoing
the homolytic cleavage.

ACKNOWLEDGEMENT

The authors are grateful to Mr.W.Friedli for the absorption spectra and to the management of Ciba-Geigy AG for permission to publish these results.

REFERENCES

1. C.Decker and K.Moussa, J. Appl. Polym. Sci., 1987, 34, 1603.
2. H.M.J.Boots, J.G.Kloosterboer and G.M.M. van de Hei, British Polymer J., 1985, 17, 219.
3. W.Baeumer, Kontakte (Darmstadt), 1989, 3, 42.
4. K.Meier, M.Rembold, W.Rutsch and F.Sitek, "Radiation Curing of Polymers", Ed.: D.R. Randell, London, 1987, p.197.
5. K.Dietliker, M.W.Rembold, G.Rist, W.Rutsch and F.Sitek, Radcure Conference Papers, 1987, 3-37.
6. W.Rutsch, G.Berner, R.Kirchmayr, R.Huesler, G.Rist and N.Buehler, "Organic Coatings, Science and Technology", Eds.: G.D.Parfitt and A.V. Patsis, New York, 1986, p.175.
7. V.Desobry, K.Dietliker, R.Huesler, L.Misev, M.Rembold, G.Rist and W.Rutsch, "Radiation Curing Polymeric Materials", Eds.: C.E.Hoyle and J.F. Kinstle, Washington, 1990, p.92.
8. W.R.Schaeffer, Polymers Paint Colour J., 1989, 179, 19.
9. M.Jacobi and A.Henne, Polymers Paint Colour J., 1985, 175, 636.
10. J.W.Vanderhoff, "Ultraviolet Light Induced Reactions in Polymers", Ed.: S.Labana, Washington, 1976, p.162.
11. B.E.Hulme and J.J.Marron, Paint and Resin , 1984, 54, 31.
12. N.L.Allinger, J. Amer. Chem. Soc., 1977, 99, 8127.
13. J.J.P.Stewart, J. Comput. Chem., 1989, 10, 209 and 221.
14. F.Mohamadi, N.G.J.Richards, W.C.Guida, R.Liskamp, M.Lipton, C.Caufield, G.Chang, T.Hendrickson and W.C.Still, J. Comput. Chem., 1990, 11, 440.
15. L.G.Arnaut and S.J.Formosinho, J. Photochem., 1985, 31, 315.

Photochemistry and Photopolymerisation Activity of Novel Perester Derivatives of Fluorenone

N.S. Allen[1], S.J. Hardy[1], A. Jacobine[2], D.M. Glaser[2], F. Catalina[3], S. Navaratnam[4], and B.J. Parsons[4]

[1] DEPARTMENT OF CHEMISTRY, MANCHESTER POLYTECHNIC, CHESTER STREET, MANCHESTER M1 5GD, UK
[2] LOCTITE CORPORATION, 705 NORTH MOUNTAIN ROAD, NEWINGTON, CONNECTICUT, USA
[3] INSTITUTO PLASTICOS Y CAUCHO, CSIC, 3 JUAN DE LA CIERVA, 28006 MADRID, SPAIN
[4] SCHOOL OF NATURAL SCIENCES, KELSTERTON COLLEGE, N.E. WALES INSTITUTE OF HIGHER EDUCATION, CONNAH'S QUAY, CLWYDD, N. WALES CH5 4BR, UK

1 INTRODUCTION

The photochemistry of fluorenone and some of its derivatives is well established in the literature[1-5], particularly with regard to its ability to initiate the photopolymerisation of vinyl monomers in the presence of a tertiary amine co-synergist[6]. Fluorenone is an aromatic ketone having a lowest lying triplet excited state which is π-π^* in character and consequently photoreduction of this molecule does not occur in hydrogen atom donating solvents. However, photoreduction has been reported to occur in the presence of tertiary amines and that this process involves electron transfer through an intermediate exciplex[4-6] and is able to initiate the photopolymerisation of vinyl monomers.

Of more recent interest in the field of photopolymerisation are t-butylperester derivatives of benzophenone substituted in either the 4,4' or 3,3,',5,5' positions of the phenyl rings[7-9]. These compounds initiate photopolymerisation through homolytic scission at the peroxy link to give benzoyloxy and t-butyloxy radicals. Detailed studies on the photopolymerisation activity of compounds of this

type have shown that their efficiency depends on the
absorption characteristics of the associated chromophore[7].
In this paper we report some of our findings on the
photochemistry and photopolymerisation activities of three
novel t-butylperester derivatives of 9-fluorenone
(structures II-IV)[10]. For reference purposes similar studies have been included on 9-fluorenone itself (I).
Absorption, fluorescence and photoreduction properties of
these molecules are related to their photopolymerisation
activities in bulk methylmethacrylate together with the
effect of a tertiary amine co-synergist and their behaviour on nanosecond laser flash photolysis.

2 EXPERIMENTAL

Materials

The compounds 4-tertbutylperoxycarbonyl-7-nitro-9-
fluorenone, 2-tertbutylperoxycarbonyl-9-fluorenone and
2,7-ditertbutylperoxycarbonyl-9-fluorenone of the structures II-IV respectively were supplied by the Loctite
Corporation, Connecticut, USA.

Structures I-IV

All compounds were chromatographically pure. The 9-fluore-
none (structure I) was obtained from the Aldrich Chemical
Company, U.K and was purified on a silica column using
chloroform as the eluant followed by recrystallisation
from "Analar" ethanol. The 2-propanol, furan and chloroform ("Analar" grades) were obtained from the Aldrich
Chemical Company, U.K as was the triethylamine, methyl-

diethanolamine and the methylmethacrylate monomer which
was used as supplied.

Spectroscopic Measurements

Normal and second-order derivative absorption spectra
were obtained using a Perkin-Elmer Model Lambda 7 spectrometer. Fluorescence spectra were obtained using a Perkin-
Elmer Model LS-5 fluorimeter and quantum yield measurements were obtained by the relative method using quinine
sulphate in 0.1 molar sulphuric acid as a standard (0_f =
0.54)[11].

Photopolymerisation

Photopolymerisation efficiencies were determined gravimetrically by irradiating a quartz cell apparatus containing 10 cm^3 of methylmethacrylate, 0.25% w/w of initiator and 0.25% w/w of methyldiethanolamine. The light
source used was a Thorn 100 watt tungsten-halogen lamp set
at a distance of 10 cm and the reaction mixture was
continually bubbling with white spot nitrogen gas. After
irradiation the reaction mixture was poured into methanol
for precipitation, and this was followed by centrifuging,
filtration and drying to constant weight in a hot air oven
at 60 °C.

Photoreduction Quantum Yields

Absolute quantum yields of photoreduction (ϕ_r) for
the compounds were determined in 2-propanol at 5 x 10^{-4} M
for compounds II-V and 10^{-3} M for compound I in the
absence and presence of triethylamine at 10^{-3} M concentration. An irradiation wavelength of 365 nm was selected
from a Philips high pressure mercury lamp (HB-CS 500 W/2)
and a Kratos GM 252 monochromator. Sample cells were
thermostatted at 30°C and the solutions were deoxygenated
using white spot nitrogen gas. The absorbed light intensity was measured using an International Light Model 700
radiometer previously calibrated by Aberchrome 540 actinometry[12,13]. The photolysis of the compounds was monitored
by measuring the change in their absorption maxima.

Laser Flash Photolysis

Laser flash photolysis experiments were carried out
using a frequency tripled neodymium laser (J.K. Lasers Ltd)
which delivered 15 ns pulses of 355 nm radiation of 50-60
mJ energy. Transient absorption changes were measured by
illuminating the 1cm path length quartz reaction cell with
light from a pulsed xenon lamp. Wavelength selection was
achieved with a diffraction grating high irradiance monochromator with a 5 nm bandwidth. Kinetic changes in the light

signal at preselected wavelengths were detected and amplified using a photomultiplier (RCA IP28A) prior to collection by a storage oscilloscope.

3 RESULTS AND DISCUSSION

Spectroscopic Measurements

Ultra-violet-visible absorption data in furan, chloroform and 2-propanol solvents are shown Table 1. It is seen that the longest wavelength absorption band for 9-fluorenone is little influenced by solvent polarity probably because of its close proximity to the next lying upper $n-\pi^*$ singlet state. The lowest excited singlet state is essentially $\pi-\pi^*$ in nature with extinction coefficients of 2.41 to 2.43. Substitution of a single t-butylperester group in the 2-position blue shifts the longest wavelength absorption maximum in all three solvents and enhances the extinction coefficient to a maximum of 2.86 in chloroform. The shorter wavelength transitions are little influenced by substitution although there is an appearance of a shoulder at about 310 nm which blue shifts with increasing solvent polarity and may be characteristic of a close lying $n-\pi^*$ state. Substitution of a second t-butylperester group in the 7-position further blue shifts the longest wavelength absorption maximum but has no additional effect compared with that of mono-substitution.

The next shorter wavelength absorption bands show an interesting effect in the observation of three shoulders between 285 and 312 nm with enhanced extinction coefficients compared with those of the mono-substituted t-butylperester compound. Thus, whilst substitution of the t-butylperester groups appears to have an electron withdrawing effect on the chromophore as seen by a blue shift in the absorption maxima the extinction coefficients are enhanced which may be associated with increased molecular rigidity. In the case of the nitro substituted compound (structure II) the extinction coefficients of the longest wavelength absorption band in all three solvents are significantly enhanced together with a marked red shift in the absorption maxima of the two shorter wavelength transitions. Indeed the second group of bands at 323 and 338 nm in furan are markedly red shifted from that of 9-fluorenone at 291 nm. Thus, although the nitro group is essentially electron withdrawing in character it is complementary with that of the perester group in enhancing the charge-transfer content of the molecule. Fluorescence emission maxima and quantum yields are shown in a range of solvents in Table 2.

TABLE 1 Absorption Maxima and Extinction Coefficients for Compounds I-IV

Solvent	λ max nm	Log E	λ max nm	Log E	λ max nm	Log E
			Compound I			
Furan	245.6	4.58	291.2	3.61	376.6	2.41
Chloroform	247.2	4.55	292.8	3.56	374.8	2.43
2-Propanol	244.6	4.56	291.0	3.56	378.0	2.43
			Compound II			
Furan	278.4	4.38	323.0(s)	3.92	385.2	3.19
			338.6(s)	3.82		
Chloroform	278.6	4.36	321.2(s)	3.83	370.0	3.29
			336.6(s)	3.73		
2-Propanol	273.1	4.37	321.6(s)	3.78	372.0	3.18
			335.7(s)	3.68		
			Compound III			
Furan	242.5	4.50	292.2(s)	3.87	371.9	2.86
			309.2(s)	3.67		
Chloroform	244.0	4.44	293.6(s)	3.54	372.0	2.84
			310.4(s)	3.20		
2-Propanol	248.8	4.49	291.5(s)	3.62	372.6	2.85
			306.9(s)	3.26		
			Compound IV			
Furan	257.5	4.46	285.8(s)	4.14	368.2	2.96
			296.5(s)	4.19		
			312.7(s)	3.98		
Chloroform	255.5	4.47	288.6(s)	4.14	368.0	2.70
			292.9(s)	4.15		
			315.2(s)	3.95		
2-Propanol	254.4	4.43	295.3(s)	4.04	365.4	2.86
			310.6(s)	3.81		

It is interesting to note that all three substituted 9-fluorenones exhibit higher fluorescence quantum yields than those of 9-fluorenone especially in chloroform. This is consistent with a possible increase in structural rigidity and an increase in the charge transfer content of the lowest excited singlet state. The latter is confirmed by the marked

TABLE 2 Fluorescence Properties of Compounds I-IV

Compound	Furan		Chloroform		2-Propanol	
	λmax	$\emptyset_{f}$	λmax	$\emptyset_{f}$	λmax	$\emptyset_{f}$
I	442,575	5.2×10^{-4}	650	5.5×10^{-4}	442,555	1.6×10^{-4}
II	480	1.5×10^{-4}	508	1.8×10^{-3}	500	6.4×10^{-4}
III	475	3.0×10^{-3}	495	9.9×10^{-3}	520	2.6×10^{-3}
IV	465	1.6×10^{-3}	495	8.3×10^{-3}	502	6.6×10^{-3}

red shift in the fluorescence maxima for all three substituted molecules with increasing solvent polarity. Thus, the lowest excited singlet state of all three molecules is evidently n-π^* in nature. The lowest excited singlet state of 9-fluorenone appears to be less sensitive to solvent polarity and to some degree inconsistent with those observed in the literature[1-5]. However, this may be associated with differences in the solvents studied, the purification of the 9-fluorenone and the fact that our measurements were carried out in air equilibrated solutions. The fluorescence of the substituted molecules was found to be unaffected by the presence of oxygen whereas that of 9-fluorenone was quenched by 10 %.

Photoreduction Quantum Yields

The photoreduction quantum yields for all four compounds except those for structure II are compared in Table 3 in 2-propanol in the absence and presence of triethylamine. The photoreduction quantum yields for the 2-t-butylperester-7-nitro derivative were unobtainable due to product formation which interfered with the uv absorption analysis. Firstly, it is seen that the 9-fluorenone has no photoactivity in 2-propanol and a quantum yield of 0.034 in the presence of a tertiary amine which is consistent with the literature[5]. Secondly, both the mono and di-t-butylperester derivatives of 9-fluorenone exhibit very high photoreduction quantum yields, both of which are enhanced in the presence of the triethylamine and with a higher magnitude than that observed for fluorenone. The most interesting feature of the data is the observation that the quantum yield of photolysis for the monosubstituted derivative is half that of the di-substituted molecule and is consistent with photoreaction occurring via photolysis of the t-butylperester groups.

Photopolymerisation

Rates of photopolymerisation (R_p) for all the compounds in bulk methylmethacrylate monomer are compared in Table 4.

TABLE 3 Photolysis Quantum Yields for Compounds I-IV in
2-Propanol

Compound	No Amine ϕ_r	Triethylamine (10^{-3}M) ϕr
I	0.0	0.034
II	Not Possible to Measure by UV Absorption Analysis	
III	0.245	0.327
IV	0.479	0.698

TABLE 4 Rates of Photopolymerisation Rp (L-1 Moles s-1)
for Compounds I-IV (0.25% w/w) in Bulk Methyl
Methacrylate in the Absence and Presence of
Methyldiethanolamine (0.25% w/w)

Compound	No Amine Rp	Amine Rp
9-Fluorenone	0.00	1.00×10^{-4}
II	5.33×10^{-4}	2.08×10^{-4}
III	5.23×10^{-4}	5.50×10^{-4}
IV	5.16×10^{-4}	4.17×10^{-4}

As expected 9-fluorenone operates as a photoinitiator
under the excitation conditions used here only in the
presence of a tertiary amine (methyldiethanolamine). However, all three substituted compounds are effective photoinitiators with the mono-t-butylperester being the most
efficient. In the latter case the presence of the amine
slightly enhances the photopolymerisation rate whereas it
markedly reduces the rate for both the nitro (compound
II) and the di-t-butylperester (compound IV) derivatives.
These results are interesting in that the order in the
rates of photopolymerisation correlate closely with the
extinction coefficients of the longest wavelength absorption bands of the molecules ie. compound II > III > IV.
Furthermore, all three substituted derivatives of 9-
fluorenone exhibit much higher fluorescence quantum yields
than those of the unsubstituted chromophore although
enhanced reactivity through the lowest excited singlet
state appears unlikely. However, the photoreduction quantum yields do indicate a much higher degree of photoreactivity which associated with homolysis of the perester
groups and subsequent photoinitiation via the benzolyoxy
radicals of 9-fluorenone (scheme 1). This is confirmed by
the analysis of the end groups on solvent cast films using
second-order derivative uv spectroscopy (Figures 1 and 2).
Solutions of all the polymers, after 40 minutes irradiation, were prepared in chloroform and then subsequently
cast onto salt flats at 10 microns thickness for analysis

by uv derivative absorption spectroscopy. The spectra in Figures 1 and 2 show that polymethylmethacrylate exhibits characteristic absorption spectra for residues of the associated mono and diperester initiators respectively used for initiation. The second-derivative absorption spectrum of the mono perester compound in chloroform is shown in Figure 1 for comparison purposes and is seen to match that of the solvent cast polymer film. The interesting feature of these results is the much lower absorption for the polymer prepared using the monoperester than that for the polymer initiated with the di-perester. In the case of the latter there is a much higher degree of self-termination by the bis-benzoyloxy radicals and also possibly the formation of some in chain initiator residues.

The photopolymerisation data in Table 4 also shows the influence of the presence of a tertiary amine co-synergist, methyldiethanolamine. It is seen that for both the nitro (compound II) and bis-t-butylperester (compound IV) molecules the amine reduces the rate of photopolymerisation. In the case of the monoperester (compound II) the rate is only slightly enhanced. The reduction in the rate of photopolymerisation for compound II is evidently associated with alkylamino radicals competing with the benzoyloxy radicals in either the initiation or termination processes (scheme 1). The second-order derivative uv analysis of the films shown in Figures 1 and 2 for polymers produced using compounds III and IV respectively indicate that the latter is a more likely process. Both spectra clearly show that the presence of terminal or in chain 9-fluorenone units (in the case of the diperester) are significantly reduced by the presence of the tertiary amine. Similar results were obtained for polymer prepared using compound II. Apart from competitive termination reactions the formation of benzoyloxy radicals may be supressed by hydrogen atom abstraction from the tertiary amine by the lowest excited triplet state of the 9-fluorenone chromophore through the formation of an intermediate exciplex.

Alternatively, and in addition to the latter, the benzoyloxy radicals may be removed by abstracting hydrogen atoms from the tertiary amine (scheme 1). From the data the latter appears more likely since, in the case of the monoperester derivative (compound III), the rate of photopolymerisation is enhanced probably through initiation via the more efficient alkylamino radicals. In the case of the di-t-butylperester compound competitive termination reactions are more likely even between the radicals themselves.

The 2-t-butylperester-7-nitro derivative is anomalous

with regard to the above since the presence of the
tertiary amine reduces the rate of photopolymerisation.

<u>Scheme 1</u>

$$Bu^tOOCO-p-C_6H_4-p-CO-p-C_6H_4-p-OCOOBu^t$$
$$\downarrow$$

$$\cdot OCO-p-C_6H_4-p-CO-p-C_6H_4-p-OCO \cdot \ + \ 2Bu^tO \cdot$$
$$(\cdot A \cdot \ or \ A \cdot)$$

$$\overset{nM}{}$$
$\cdot A \cdot$ or $\cdot A + M \ ----> \ \cdot AM \cdot$ or $AM \cdot \ ----> \ \cdot MnAM \cdot$ or $AMn \cdot +1$ etc.

$\cdot A \cdot$ or $\cdot A + NR_3 \ -----> \ AH \cdot$ or $AH + NR_2R \cdot \ -----> $ Termination

$Bu^tO + NR_3 \ -----> \ Bu^tOH + NR_2R \cdot$
$$\overset{nM}{}$$
$NR_2R + M \ -----> \ NR_2RM \cdot \ ------> \ NR_2RMn \cdot +1$
$$\overset{XMi \cdot}{}$$
$NR_2R \cdot$ or Bu^tO or $AH \cdot + XMn \cdot \ ------> $ Termination

where A is 9-Fluorenone chromophore
   XMi· is any growing polymer radical
   NR₃ is a tertiary amine

However, the same effects were observed by second-order
derivative uv spectroscopy as was shown for the bis-
perester in Figure 2. Thus, it would appear that the nitro
group itself may be involved in some way in the photoinitiation process and/or reaction with the monomer radicals.
Suppression of the photopolymerisation by the tertiary
amine may well be associated with a photoinduced reaction
between the amine and the nitro group of the photoexcited
9-fluorenone chromophore.

<u>Laser Flash Photolysis</u>

  All the solutions used in this study were prepared to
give an absorbance of 1.0 at the excitation wavelength of
355 nm. For compounds II and III this was obtained at a
concentration of 10^{-3} molar. For the nitro derivative of
fluorenone the concentration used was 4×10^{-4} molar and
for fluorenone 5×10^{-3} molar. The end-of-pulse transient
absorption spectrum of fluorenone in acetonitrile is shown
in Figure 3. It has absorption maxima at 320, 425 and 630
nm which corresponds closely with those reported earlier
for the triplet of fluorenone[14]. The transient is formed
within 2.4 micro-seconds with a half-life of 1.2 microseconds and decays by first-order kinetics leaving no
significant residual absorption. Furthermore, the transient is effectively quenched by oxygen confirming the
assignment to that of the triplet. The end-of-pulse

transient absorption spectrum of the nitro, t-butylperester derivative (II) is shown in Figure 4 in acetonitrile. This strong transient is formed within 180 ns with a half-life of 220 ns and has absorption maxima at 390, 440, 470, 500 and 630 nm and decays by two first order processes. The transient is again effectively quenched by oxygen confirming the assignment to that of the triplet state. At longer time delays there is a residual transient absorption which may be associated with that of the benzoyloxy radical produced by homolysis of the t-butylperester group. Both the concentration and energy of the laser had to be reduced compared with those used for the other molecules.

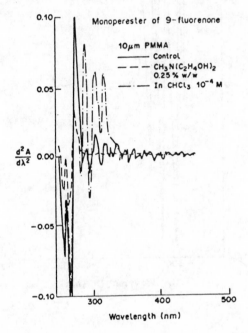

Figure 1   Second-order derivative uv absorption spectra of polymethylmethacrylate after 40 minutes irradiation in the bulk (150 watt tungsten-halogen source) in the presence of the mono-t-butylperester derivative of 9-fluorenone (0.25% w/w) (_____) alone and (-------) in the presence of methyldiethanolamine (0.25% w/w) compared with that of the initiator in chloroform solution (10^{-4}M) (-.-.-.-.).

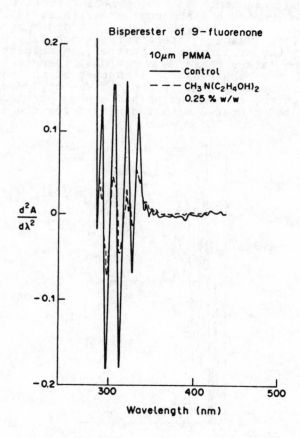

Figure 2 Second-order derivative uv absorption spectra of polymethylmethacrylate after 40 minutes irradiation in the bulk (150 watt tungsten-halogen source) in the presence of the bis-t-butylperester derivative of 9-fluorenone (0.25% w/w) (_____) alone and (-------) in the presence of methyldiethanolamine (0.25% w/w).

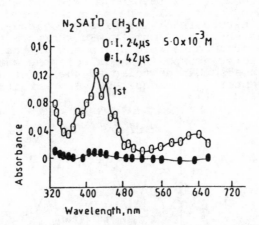

Figure 3 End-of-pulse transient absorption spectra of 9-fluorenone in nitrogen saturated acetonitrile (5 x 10-3 M) after (O) 2.4 and (●) 42 microseconds delay.

On this basis the actual corrected maximum absorbance at 470 nm should be 0.64. At this time scale the transient would not therefore be expected using the conventional method above since the radical species would have decayed within the lifetime of the xenon flash lamps (100 microseconds). Figure 5 shows the end-of-pulse transient absorption spectra for the mono-t-butylperester of fluorenone (compound III) in hexafluorobenzene initially after its formation and after 500 ns. In this case absorption maxima are observed at 350, 420 and 450 nm with the decay kinetics obeying two first order processes. The transient which has a half-life of 220 ns in acetonitrile is again assigned to that of the triplet which is effectively quenched by oxygen. A residual absorption is also observed after 500 ns which closely matches that seen in acetonitrile and is therefore assigned to the benzoyloxy radical. This is supported by the fact that in hexafluorobenzene the triplet is unable to decay by a process of hydrogen atom abstraction. The end-of-pulse transient absorption spectrum for the bis-t-butylperester of fluorenone in acetonitrile (compound IV) is shown in Figure 6 after 100 and 980 ns and 3.98 microseconds. In this case absorption maxima are observed at 330, 380, 420, 440 and 560 nm with the transient undergoing both 1st and 2nd order decay kinetics with a half-life of 110 ns. A negative absorption is observed at 500 nm which is due to the strong fluorescence emission from this molecule. This transient which is also assigned to that of the triplet is effectively quenched by oxygen and decays to give the benzoyloxy radical. Whilst all the above triplets were effectively quenched by oxygen all the perester derivatives displayed a novel feature in giving rise to a long-lived transient growth as shown in Figure 7. Furthermore the intensities of these transients were dependent on the number of t-butylperester groups in the fluorenone chromophore. Thus, transient absorbances followed the order nitro, mono-t-butylperester (A) = mono-t-butylperester(B) < di-t-butyl-perester (C).

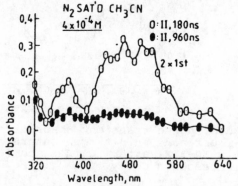

Figure 4 End-of-pulse transient absorption spectra of the nitro, mono-t-butylperester of 9-fluorenone in nitrogen saturated acetonitrile (4 x 10⁻⁴ M) after (O) 180 and (●) 960 nanoseconds delay.

Both transients are formed by a first-order process and are not observed in the case of unsubstituted fluorenone. Thus, since this transient growth appears to be associated only with the t-butylperester groups three feasible explanations are possible. These are shown in scheme 2 with the first being associated with the formation of some type of singlet or triplet complex of the perester groups with oxygen. No enhanced absorption was observed in the ground state for oxygen saturated solutions of the perester compounds using absorption spectroscopy. The second explanation is the possible formation of benzoyloxy radicals from the singlet state and the abstraction of an electron by the oxygen to give a benzoyloxy radical cation and an oxygen radical anion. The third explanation, and one which appears most likely, is the rapid dissociation of either the benzoyloxy radical or complete de-esterification to give a phenyl type radical followed by reaction with oxygen to give a long-lived aromatic peroxy radical. Further discussion on this mechanism would be speculative and requires clarification through detailed oxygen concentration studies and product analysis. Of the three explanations the latter appears more likely although further experimental work is required to clarify this point. Both oxygen complexes and radical cations are short-lived and difficult to account for the growth. In the second case it could be argued that homolysis in the lowest excited singlet state should therefore give rise to oxygen insensitivity during photopolymerisation. From our previous study this was not the case and may be accounted for on the basis of the inefficiency of the radical cation to initiate free radical polymerisation.

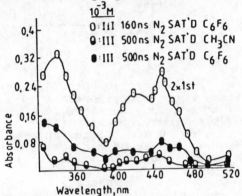

Figure 5 End-of-pulse transient absorption spectra of the mono-t-butylperester of 9-fluorenone in nitrogen saturated (◓) acetonitrile (10^{-3} M) after 500 nanoseconds delay and nitrogen saturated hexafluorobenzene after (O) 160 and (●) 500 nanoseconds delay (10^{-3} M).

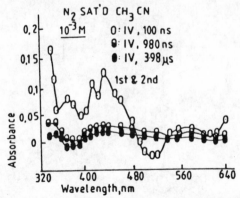

Figure 6 End-of-pulse transient absorption spectra of the bis-t-butylperester of 9-fluorenone in nitrogen saturated acetonitrile (10^{-3} M) after (O) 100 and (◓) 980 nanoseconds delay and (●) 3.98 microseconds delay.

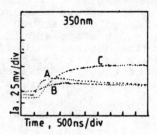

<u>Figure 7</u> Transient absorption profiles for the (A)
nitro, mono-t-butylperester (4 x 10⁻⁴ M),
(B) mono-t-butylperester (10⁻³ M) and
(C) bis-t-butylperester (10⁻³ M) of
9-fluorenone and (D) the mono-
t-butylperester of benzophenone (2.5 x 10⁻³ M)
in oxygen saturated acetonitrile.

<u>Scheme 2</u>

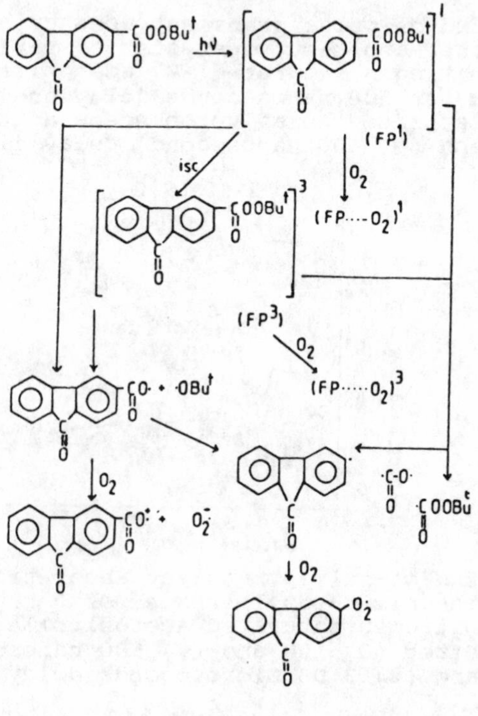

Photopolymerisation and Laser Flash Photolysis

Rates of photopolymerisation (R_p) for all the compounds in bulk methylmethacrylate monomer are compared in Table 5 with the triplet absorption maxima and half-lives obtained on laser flash photolysis. As expected 9-fluorenone does not operate as a photoinitiator under the excitation conditions used here However, all three substituted compounds are effective photoinitiators with the nitro, mono-t-butylperester (compound III).

TABLE 5 Rates of Photopolymerisation Rp (L^{-1} Moles s^{-1}) for Compounds I-IV (0.25% w/w) in Bulk Methyl Methacrylate, Triplet Absorbances and Half-Lives from Laser Flash Photolysis in Acetonitrile

Compound	Rp	Transient Spectra Absorbance	(Wavelength max)	Half-life
I	0.00	0.125	425 nm	1.2 us
II	5.33 x 10^{-4}	0.640	470 nm	220 ns
III	5.23 x 10^{-4}	0.140	410 nm	220 ns
IV	5.16 x 10-4	0.130	440 nm	110 ns

(II) being the most efficient followed by the mono-t-butylperester (III) and then the bis-t-butylperester (IV). As indicated above the presence of oxygen was found to completely inhibit photopolymerisation indicating that the lowest excited triplet state is the precursor in initiating free radical formation. However, this may simply be due to an inefficient radical initiation process. On comparison with the triplet data it is seen that photopolymerisation efficiency increases with increasing triplet absorption and half-life.

4 CONCLUSIONS

The t-butylperester derivatives of 9-fluorenone are highly effective initiators for inducing the photopolymerisation of vinyl monomers with near ultraviolet/visible radiation. Their efficiency depends not only on the absorption spectrum and extinction coefficients of the molecule but on their ability to produce active benzoyloxy free radicals via the lowest excited triplet state. The presence of a tertiary amine co-synergist competes with the latter radicals for the monomer and/or monomer radicals and depending upon the structure have a marked influence on the rate of photopolymerisation. Second-order derivative uv/visible absorption spectroscopy and fluorescence analysis are sensitive methods for detecting chromophoric end-groups where appropriate.

On laser flash photolysis of the t-butylperester

derivatives the triplet absorption, which is dependent on the structure of the molecule, decays by a first order process to give homolytic scission of the t-butylperester groups. In the presence of oxygen both the triplet state and photopolymerisation are effectively quenched with the concurrent formation of a new transient absorption. The latter is tentatively assigned to the formation of either an excited oxyplex, a benzoyloxy radical cation or more likely an aromatic peroxy radical.

ACKNOWLEDGEMENTS

The authors would like to thank the research and development staff of the Loctite Corporation, Connecticut, U.S.A for helpful discussions and advice and also a grant in support of one of them (SJH). The authors also thank the British Council under Acciones Integrades for a travel grant to support them.

5 REFERENCES

1) K. Yoshihara and D.R. Kearns, J. Chem. Phys., (1966), 45, 1991
2) L.A. Singer, Tetrahedron Lett., (1969), 923.
3) J.B. Guttenplan and S.G. Cohen, Tetrahedron Lett., (1969), 2125.
4) G.A. Davis, P.A. Carapellucci, K. Szoc and J.D. Gresser, J. Am. Chem. Soc., (1969), 91, 2264.
5) S.G. Cohen and G. Parsons, J. Am. Chem. Soc., (1970), 97, 7603.
6) A. Ledwith, J.A. Bosley and M.D. Purbrick, J. Oil. Col. Chem. Assoc., (1978), 61, 95.
7) D. Neckers, US Pat., 4,752,649, June 21st (1988).
8) L. Thijs, S. Gupta and D. Neckers, J. Org. Chem., (1979), 44, 4123.
9) L. Thijs, S. Gupta and D. Neckers, J. Polym. Sci., Polym. Chem. Ed., (1981), 19, 103; idem-ibid, (1981), 855.
10) R.W.R. Humphreys, US Pat., 4,604,295, August 5th (1988).
11) J.N. Demas and G.A. Crosby, J.Phys. Chem., (1971), 75, 91.
12) H.G. Heller and J.R. Lanagan, J. Chem. Soc., Perkin Trans., I., (1981), 341.
13) H.G. Heller, Brit. Pat., 7/1464603.
14) L.J. Andrews, A. Deroulede and H. Linschitz, J. Phys. Chem., (1978), 82, 2304.

Photoredox Induced Cationic Polymerisation of Divinyl Ethers

Per-Erik Sundell[1], Sonny Jönsson[1], and Anders Hult[2]

AB WILH BECKER, R&D, BOX 2041, S-195 02 MÄRSTA, SWEDEN
THE ROYAL INSTITUTE OF TECHNOLOGY, DEPARTMENT OF POLYMER
TECHNOLOGY, S-100 44 STOCKHOLM, SWEDEN

1 INTRODUCTION

Radiation curing is a technology that has grown rapidly during the last decades. Both technical and economical forces have driven the expansion to the present state. The advantages of radiation curing over traditional thermosetting curing include: reduced energy costs, increased productivity, reduced solvent emission and the coating of heat sensitive materials. A vast majority of radiation-curable systems in use today are based on crosslinking by a free radical mechanism, and acid-catalysed curing is less commonly used. Some drawbacks associated with free radical systems are absent in cationic polymerisation. Thus, is cationic propagation not sensitive towards oxygen and the longlived cationic species allows for extensive post-cure leading to almost complete conversion of reactive groups. The most important breakthrough in the field of radiation-induced acid-catalysed reactions was made in the mid 70´s by the discovery of various onium salts as latent sources of strong Brönsted acids that could be photochemically activated[1, 2]. Further advances in the field have shown that the latent acidity of these salts can be released not only by UV radiation but also thermally[3, 4] or by high-energy radiation[5, 6].

Cationic photopolymerisation has, so far, not been employed to any greater extent for coating applications. The poor absorption of onium salt photoinitiators above 300 nm limits the utilization of 313 and 366 nm light from medium-pressure and high-pressure mercury lamps. Several attempts have been made to overcome this limitation. Substituents on the aromatic ring that interact inductively or by resonance to delocalize the positive charge of the onium salt enhance the absorption at longer wavelengths[7]. Photosensitizers[8, 9] and free radical photoinitiators[3] have successfully been used to broaden the spectral response of the onium salts. Electron transfer from the excited state photosensitizer[10] or photogenerated organic free radicals to the onium salt results in the generation of the initiating cationic species, Scheme 1.

Another important factor that has limited the use of cationic photo-curing of coatings has been the slow polymerisation rate of epoxies which is the type of monomer that has so far been available. This problem has partly been solved by development of the more reactive vinyl ether[11, 12] and propenyl ether[13] oligomers as interesting alternatives to the epoxy monomers.

The primary purpose of this work has been to obtain detailed information concerning the cationic polymerisation of vinyl ethers induced by the reduction of iodonium and sulfonium salts. The interplay between modes of activation, onium salt structure and the vinyl ether structure was studied by photo-calorimetry.

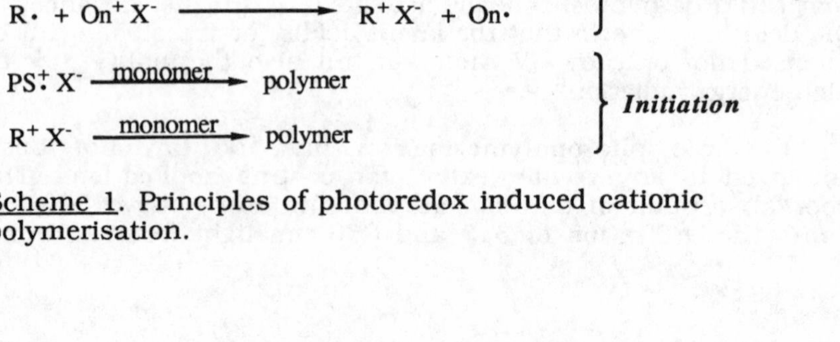

$$PS \xrightarrow{\;h\nu\;} PS^*$$

$$PI \xrightarrow{\;h\nu\;} R\cdot$$

$$\left.\begin{array}{c} \end{array}\right\} \; \textit{Excitation / Photolysis}$$

$$PS^* + On^+X^- \longrightarrow PS^{+\cdot}X^- + On\cdot$$

$$R\cdot + On^+X^- \longrightarrow R^+X^- + On\cdot$$

$$\left.\begin{array}{c} \end{array}\right\} \; \textit{Electron transfer}$$

$$PS^{+\cdot}X^- \xrightarrow{\;monomer\;} polymer$$

$$R^+X^- \xrightarrow{\;monomer\;} polymer$$

$$\left.\begin{array}{c} \end{array}\right\} \; \textit{Initiation}$$

Scheme 1. Principles of photoredox induced cationic polymerisation.

2 EXPERIMENTAL

Materials. Butanedioldivinyl ether, BDDVE, diethylene-glycoldivinyl ether, DEGDVE (obtained from GAF) were stored over KOH and distilled from calcium hydride prior to use. Cyracure UVI-6974 (Union Carbide) was used as recieved. Benzoinmethyl ether, BME (Aldrich), α-methylbenzoinmethyl ether, MBME (kindly supplied by H.J. Hageman, AKZO Chemie, Netherlands), 2,2-dimethoxy-2-phenylacetophenone, DMPA (Irgacure 651, Ciba-Geigy) and 1-benzoylcyclohexane-1-ol, BCHO, (Irgacure 184, Ciba-Geigy) were recrystallized twice from n-hexane. 2,4,6-Trimethylbenzoyldiphenylphosphine oxide, TMPDO (BASF), was recrystallized twice from ethanol/diethyl ether. 2,4,6-Trimethylbenzoylethoxyphenylphosphine oxide, TMPEO (BASF), benzophenone, BP (Aldrich, Gold Label) and phenothiazine (98% Aldrich) were used as received. Isopropylthioxanthone, ITX (Ward Blenkinsop and Co. Ltd.) was recrystallized three times from ethanol. The synthesis and characterization of the onium salts, Table 1, have been reported elsewhere[14].

Apparatus and procedures. The photo-DSC measurements were performed using a Perkin Elmer DSC-7 equipped with a differential photo accessory, 450 W XBO xenon-mercury lamp and a 366 nm narrow bandpass filter. Monomer containing $PTSSbF_6$ or Ph_2IPF_6 and a free radical photoinitiator was deoxygenated by bubbling argon for 15 minutes and 3 µl was placed in aluminium DSC pan using a calibrated GC syringe. The sample thickness (a drop) was calculated[15] to 570 µm. The concentration of free radical photoinitiators was adjusted to give an optical density at 366 nm of 0.3. From the standard heat of polymerisation of vinyl ethers[16], -60 kJ/mol, the heat of polymerisation was calculated to be -758 J/g which was used to calculate the conversions.

3 RESULTS AND DISCUSSION

The acid-generating mechanisms induced by irradiation of onium salts involves the reduction of the salts. This is realized by considering the oxidation state of the sulfur atom, which changes from +4 to +2. Diaryliodonium[17], triphenylsulfonium[18] and other sulfonium[19] salts undergo electrochemical reduction according to:

$$Ar_2I^+X^- + e^- \longrightarrow [Ar_2I\cdot]X^- \longrightarrow Ar\cdot + ArI + X^- \quad (1)$$

$$R_3S^+X^- + e^- \longrightarrow [R_3S\cdot]X^- \longrightarrow R\cdot + R_2S + X^- \quad (2)$$

These processes are usually irreversible due to the rapid decomposition of the reduced salts, i.e. the diaryliodine radical or the sulfur radical. The polarographic measurements of iodonium and sulfonium salts by a dropping mercury electrode and a calomel electrode as reference in aqueous solution were carried out[14] and a list of half-wave potentials, $E_{1/2}$ is given in Table 1. The $E_{1/2}$-values found for Ph_2IPF_6 and Ph_3SSBF_6 correlate well with the literature values for these salts. The structural influence on E^{red} was manifested for the triarylsulfonium salt in which the para-thiophenoxy substituent in PTPDS lowered the E^{red} value by 23 kJ/mol compared with that of the unsubstituted Ph_3SSbF_6. The benzylsulfonium salt with PF_6^- as counterion, $BTSPF_6$, gave an identical polarogram with $BTSSbF_6$ which proved that the counterion did not participate in the reduction process.

Other electron sources than electrochemical may be applied for the redox process. Electron-donating free radicals may transfer an electron to iodonium and sulfonium salts[3]. Whether a radical is able to transfer an electron or not depends on its oxidation potential, E_r^{ox}, and on the reduction potential, E_o^{red}, of the onium salt. Only if the radical has a lower oxidation potential than the reduction potential of the onium salt is the process thermodynamically feasible. The free energy of the reaction, ΔG, must be negative (exothermic):

$$\Delta G = E_r^{ox} - E_o^{red} < 0 \qquad\qquad \text{Eq. 1}$$

Electron transfer from excited state photosensitizers is also possible[8, 9]. The Rehm-Weller equation[20] in this case

$$\Delta G = E_{ps}^{ox} - E_o^{red} - E_{ps}^* < 0 \qquad\qquad \text{Eq. 2}$$

combines the oxidation potential, E_{ps}^{ox}, of the photosensitizer, the reduction potential, E_o^{red}, of the onium salt, and the excited state energy of the photosensitizer, E_{ps}^*.

<u>Table 1</u>. The half-wave potentials, $E_{1/2}$, in volts vs Standard Calomel Electrode (S.C.E.) also expressed as the reduction potential of the salts, E^{red}, in kJ/mol (obtained by multiplying the $E_{1/2}$-value by 97).

Onium Salt		$E_{1/2}$ (V vs S.C.E)	E^{red} (kJ/mol)
	Ph_2IPF_6	0.14	14
	$PTSSbF_6$	0.65	63
	$BTSSbF_6$	1.03	100
	$pCH_3OBTSSbF_6$	0.91	88
	Ph_3SSbF_6	1.15	112
	PTPDS	0.92	89
	$DBMSPF_6$	1.51	146

Eq. 2 states then that the energy of the excited state photosensitizer, E_{ps}^*, must be higher than the energy required for the oxidation of the I in the Rehm-Weller equation, an electrostatic correction term should also be included, but this term is often neglected in solvents with high dielectric constants. The dielectric constant in DEGDVE should be about four due to the ether structure and for simplicity the electrostatic correction term is neglected in the following discussion.

High-Energy Radiation-Induced Redox Initiation[21]

Electron beam induced cationic polymerisation of some divinyl ethers in the presence of onium salts requires very low dosages, typically less than 0.5 Mrads[5]. Any vinyl ether may be activated when exposed to high energy radiation but the product yields are strongly dependent on the molecular structure. For industrial radiation curing, electron beams are suitable for high speed production lines. The rate at which the total dose is delivered to the substrate is usually about 10^3-10^4 times higher for electron beams than for γ-radiation but, since the fundamental radiation effects are essentially the same, γ-radiation may be used to study the processes that are involved. The γ-radiolysis of BDDVE and DEGDVE yields only one detectable radical by ESR spectroscopy, i.e. an α-ether radical. A simplyfied description of the processes involved is shown below (3) which gives an α-ether radical, a solvated electron and a proton as products from the radiolysis. There are two alternative pathways of initiating polymerisation by these products: first, solvated electrons can reduce the onium salt as shown for Ph_2IPF_6 (4), liberating the counterion, PF_6^-, which forms a Brönsted acid with the protons derived according to reaction (3). Second, in the redox reaction between an α-ether radical and the onium salt, the radical can be oxidized to the corresponding cation, stabilized by PF_6^- (5). The resulting Brönsted acid and the ion pair are the true initiating entities in these systems. Any salt can be reduced by the first pathway (6) but, depending on the oxidizing ability of the onium salt, reduction may also proceed by the second pathway (7).

The oxidation potential for α-oxygen radicals derived from diethyl ether and tetrahydrofuran is -0.8 eV (-78 kJ/mol)[22]. Since the radicals derived from BDDVE and DEGDVE have similar structures, their oxidation potentials should be fairly equal, i.e. $E_T{}^{ox}$ = -78 kJ/mol. By calculating the free energy for the reaction (5) it is possible to predict whether radicals may transfer an electron to the onium salt or not.

$$\diagdown O \diagup R' \xrightarrow{\gamma} \longrightarrow \longrightarrow \diagdown O \diagup \overset{\bullet}{R'} + e^{-}_{s} + H^{+} \qquad (3)$$

$$Ph_2I^{+}PF_6^{-} + e_s^{-} \longrightarrow PhI + Ph\cdot + PF_6^{-} \xrightarrow{H^{+}} H^{+}PF_6^{-} \qquad (4)$$

$$\diagdown O \diagup \overset{\bullet}{R'} + Ph_2I^{+}PF_6^{-} \longrightarrow PhI + Ph\cdot + \diagdown O \diagup \overset{+}{R'} \; PF_6^{-} \qquad (5)$$

It is also possible experimentally to distinguish between the reduction pathways. If the samples are saturated with nitrous oxide, N_2O, solvated electrons are scavenged, while the concentration of reducing radicals remains constant. In this case, the onium salts that are reduced by solvated electrons (4) require a higher dose before inducing polymerisation. This effect is clearly illustrated in Figure 1, in which N_2O-saturation increases the doses for Ph_3SSbF_6 and $DBMSPF_6$. These salts cannot be reduced by α-ether radicals due to the unfavourable ΔG-value and the removal of solvated electrons by N_2O increases the dose. Ph_2IPF_6 and $PTSSbF_6$ are totally unaffected by the presence of N_2O revealing that these salts are essentially reduced by radicals (5). Oxygen-saturation of the samples results in scavenging of radicals, but also of some electrons. Therefore, O_2-saturation increases the required dosages for the samples that are initiated according to both mechanisms, Figure 1, but the effect is most pronounced for radical-induced decomposition (5) of Ph_2IPF_6 and $PTSSbF_6$. If a salt can be reduced in both ways, the radiation yield of oxidizable radicals and solvated electrons determines the reduction pathway. Since the stationary state concentration of radicals is much higher than that of solvated electrons, radical reduction of Ph_2IPF_6 and $PTSSbF_6$ is the dominating process. The exact value of the reduction potential at which electron transfer from radicals becomes less efficient and reduction by solvated electrons begins to dominate must lie in the vicinity of the reduction potential for α-ether radicals, i.e. -78 kJ/mol. The substituted benzylsulfonium salt, $pCH_3OBTSSbF_6$, was equally affected by N_2O and O_2, indicating that both free radicals and solvated electrons played important roles in the redox process, which was rather surprising since ΔG for this reaction is positive. This indicates that there was no abrupt change in the reduction mechanism, but rather that there seems to be a gradually changing process as the reduction potential changes.

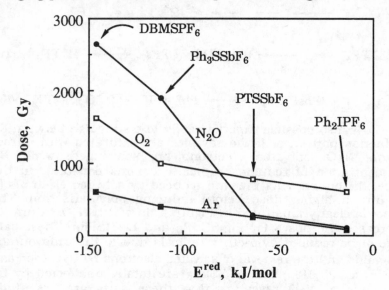

Figure 1. Cationic polymerisation of DEGDVE. Influence of oxygen-, O_2, and nitrous oxide-, N_2O, compared to argon-saturated solutions, Ar, of onium salts in DEGDVE.

Photochemically Induced Redox Initiation.[23]

Due to their more favourable reduction potential, diaryliodonium salts have been most frequently used as oxidizing agent. Dialkylphenacylsulfonium salts, possessing a somewhat weaker oxidizing capability, can also be used[24]. Previous polymerisation studies on the subject have generally been performed in solution and the efficiency has been expressed as gel doses or by determining the yields gravimetrically and the molecular weights of the polymer by GPC. Although valuable information of photopolymerisation of coatings has been obtained by these investigations it is desirable to study these processes as authentically as possible. Photocalorimetry is a convenient tool well suited for the study of photopolymerisation processes. Photoinitiator[25] and polymerisation[26] efficiency, gelation[27], oxygen[25, 28] and temperature[27, 29] effects as well as effects of additives[30] and light intensity[29] in free radical photopolymerisations have been evaluated by this method. Conditions similar to an industrial coating event were achieved since the atmosphere, temperature and sample thickness could easily be varied. Further, the number of double-bonds that are opened can be determined by the heat evolution during polymerisation, which is more informative than the gravimetric yield for the coating formulator.

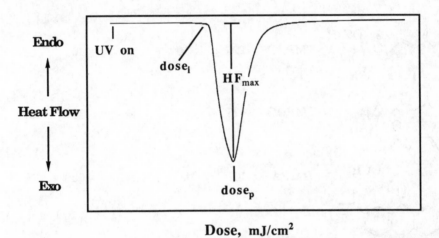

Dose, mJ/cm²

<u>Figure 2</u>. Polymerisation thermogram. The incident dose required for inducing polymerisation, dose$_i$, the incident dose at maximum rate of polymerisation, dose$_p$, and the maximum heat flow HF$_{max}$, are indicated.

Figure 2 shows a typical thermogram obtained from a photoinduced cationic polymerisation of DEGDVE. The dose at the peak, dose$_p$, was chosen as the parameter to display the reactivity of an initiator combination, i.e. free radical photoinitiator and onium salt. The maximum heat flow, HF$_{max}$, was reported since this reflected the maximum rate of polymerisation. For coating applications the dose$_p$-value is more informative than the dose$_i$-value since a lamp turn-off at dose$_p$ had no effect on the appearance of the exotherm.

<u>Table 2.</u> Results of the redox induced cationic polymerisation of DEGDVE at 25° C. [PTSSbF$_6$] = 5 mM.

λ = 366 nm. Light intensity = 0.81 mJ/cm^2. OD$_{366}$ = 0.3.

Free Radical Photoinitiator		ε_{366} M^{-1}cm^{-1}	Conc. mM	dose$_p$ mJ/cm^2	HF$_{max}$ J/g,s	conv. %
(structure)	BCHO	15	351	8	77	74
(structure)	TMPEO	220	34	13	110	59
(structure)	TMPDO	426	8	14	113	64
(structure)	DMPA	107	49	28	102	62
(structure)	ITX	4817	1	29	89	61
(structure)	BME	59	89	38	17	76
(structure)	MBME	77	68	41	7	74
(structure)	BP	44	120	246	7	72

The results of the polymerisation study of samples containing $PTSSbF_6$ are given in Table 2 together with the structure and abbreviations of the photoinitiators used. Slower polymerisation rates gave a higher conversion and produced clear, transparent cured samples whereas those giving larger exotherms were highly discoloured. The inhibition period before the onset of polymerisation is probably caused by nucleophilic impurities (e.g. H_2O) present in the system which compete with monomer for the initiating species. The difference in inhibition periods between combinations of free radical initiators and onium salts should depend on the quantum yields of radicals and the acid production efficiency by the redox pairs. The reactivity of samples containing Ph_2IPF_6 were almost as high as in direct photolysis of Ph_2IPF_6 with unfiltered UV radiation, giving no inhibition period at all. All samples containing Ph_2IPF_6 polymerized when exposed to 5 mJ/cm^2 or less, giving large exotherms (<100 J/g,s) and low conversions (<65%) except for the sample containing BP that required 13 mJ/cm^2 ($HF_{max}=44$ J/g,s and 73% conversion).

Generation of radicals. As in the case of high-energy radiation induced polymerisation of DEGDVE, α-ether radicals are responsible for the reduction of Ph_2IPF_6 and $PTSSbF_6$ resulting in the cationic polymerisation of the vinyl ether. The α-ether radicals were generated by exposing the free radical photoinitiators and photosensitizers to 366 nm UV-radiation, Scheme 2. The subsequent polymerisation was followed by DSC.

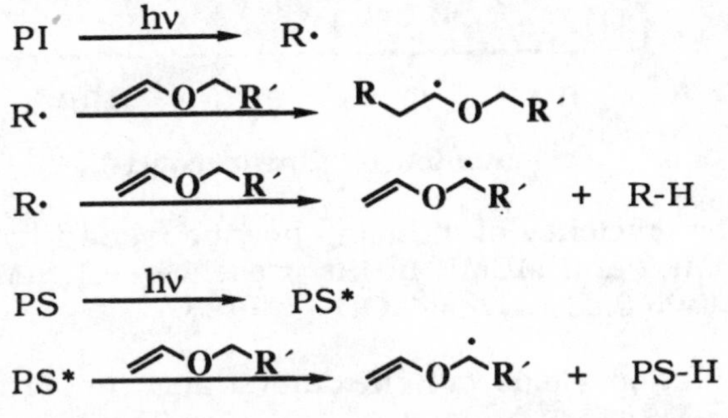

Scheme 2. Generation of α-ether radicals by the photolysis of free radical photoinitiators, PI, and photosensitisers, PS.

Photofragmenting free radical initiators yields α-ether radicals either by hydrogen abstraction or addition to the vinyl ether double bond. Intermolecular hydrogen abstraction by triplet excited BP and ITX from DEGDVE also yields α-ether radicals as well as the corresponding ketyl radical.

An additional effect of using free radical photoinitiators as radical sources is that the radicals derived from the photoinitiator can be oxidized by the onium salt. Thus α-hydroxyl radicals originating from BCHO, BP and ITX are even more easily oxidized than the α-ether radicals. The process with the α–hydroxylcyclohexyl radical and Ph_2IPF_6 is:

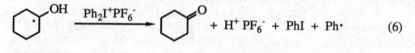

$$+ \; H^+ PF_6^- + PhI + Ph\cdot \qquad (6)$$

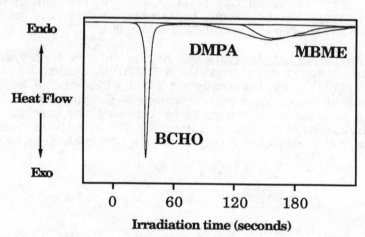

<u>Figure 3.</u> The efficiency of inducing polymerisation by photolysis of BCHO, DMPA and MBME in the presence of 1 mM $PTSSbF_6$. Light intensity = 0.81 mJ/cm^2, OD_{366} = 0.3.

The oxidation yields cyclohexanone and the Brönsted acid as the initiator. The ΔG-value for this process is -112 kJ/mol (E^{ox} = -126 kJ/mol) to be compared with -63 kJ/mol for the α-ether radical (5). The effect is illustrated in Figure 3, where BCHO, due to the formation of the α-hydroxylcyclohexyl radical, is much more efficient than MBME and DMAP, which gives only α-ether radicals. The benzoyl radical, which was the other initiator fragment derived from these photoinitiators, was not able to participate in the redox process with Ph_2IPF_6 [31] or $PTSSbF_6$, but is still useful as a source of α-ether radicals by H-abstraction and addition to the vinyl ether double bond. The reactions of the benzoyl, aromatic α-ether and the α-hydroxylcyclohexyl radical are summarised in Table 3.

Table 3. Radical Reactions.

	O ⊙ (benzoyl)	OR' R (α-ether)	OH (α-hydroxylcyclohexyl)
Abstraction	low	low	low
Addition	high	low	low
Oxidation	no	yes	yes

In view of that ethers are very good hydrogen donors in the photoreduction of BP which results in both α-ether and α-hydroxyl radicals, it was rather puzzling that BP required a much higher dose than the other initiators. The reason was found to be that the vinyl ether double bond effectively quenched the excited state triplet BP by energy transfer via an exciplex. Figure 4 illustrates the photoreduction of BP and and ITX in DEGDVE. The absorption of ITX is bleached due to the photoreduction process between excited triplet state ITX and DEGDVE, whilst the BP absorption remains unchanged indicating that no photoreduction of BP takes place in DEGDVE. For this reason BP seems to be unsuitable to used in combination with vinyl ethers. Also ITX was to some extent quenched by the vinyl ether double bond, but due to the lower π,π^* triplet energy of ITX compared to the n,π^* triplet BP energy transfer is less favourable allowing more hydrogen abstraction to take place, giving oxidizable α-ether and α-hydroxyl radicals. The very efficient initiation induced by ITX could then be attributed to effective generation of oxidizable free radicals, but also to an electron transfer process between excited state ITX and an onium salt giving a ITX radical cation as the initiating species, Scheme 1. Both processes seemed likely in this case, but no attempt was made to discriminate between them and to discern which one dictated the initiation efficiency.

Absorbance

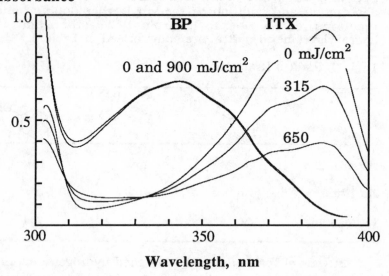

Wavelength, nm

<u>Figure 4</u>. UV Spectrum of BP and ITX in DEGDVE exposed to 366 nm UV radiation.

<u>Influence of onium salt structure.</u> Ph_3SSbF_6, $BTSSbF_6$, and $DBMSPF_6$ did not show any activity at all as an oxidizing agent together with BCHO and DMPA which reflects their unfavorable reduction potentials (Table 1). The $pCH_3OBTSSbF_6$ salt has a sufficiently low reduction potential for accepting an electron from an α-ether radical and initiated the polymerisation in combination with both BCHO and DMPA. The commercially available cationic photoinitiator, UVI 6974 (Ciba-Geigy), which is a mixture of 4-(phenylthio)phenyl-diphenylsulfonium bis-hexafluoroantimonate and 4-thiophenoxydiphenylsulfonium hexafluoroantimonate, PTPDS in propylene carbonate was also able initiate cationic polymerisation. Although initiation occurred even without a free radical photoinitiator, the presence of BCHO and DMPA strongly enhanced the efficiency, indicating the presence of a redox reaction between the salt

mixture and oxidizable radicals. The polarogram of UVI 6974 exhibited two waves at -0.92 V and -1.15 V vs S.C.E.. The second wave corresponds to the reduction of the difunctional salt and has the same value as that for pure triphenylsulfonium salt, Ph_3SSbF_6. The more favourable reduction potential of PTPDS then allows for efficient electron transfer reduction of the salt in combination with photolysis of BCHO and DMPA in DEGDVE. It has also previously been reported that PTPDS undergo electron transfer reduction by photolysis of DMPA[32] in THF which also would yield α-ether radicals as the reducing agent. A comparison between Ph_2IPF_6, $PTSSbF_6$ and UVI 6974 as redox initiators is illustrated in Figure 5. The extremely high reactivity of Ph_2IPF_6 in these systems is also a disadvantage since the storage stability in darkness at room temperature is less than one week. Even in the absence of a free radical photoinitiator the storage stability is less than one week due to the presence of adventitious radical sources such as hydroperoxides and peroxides[33]. Better storage stabilities (> one month) are obtained for $PTSSbF_6$ and UVI 6974 which reflects their lower reduction potentials.

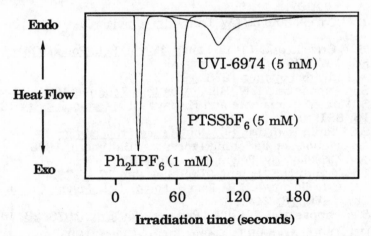

Figure 5. The efficiency of inducing polymerisation by photolysis of DMPA in combination with Ph_2IPF_6, $PTSSbF_6$ or UVI-6974. Light intensity = 0.81 mJ/cm^2, OD$_{366}$ = 0.3.

4 CONCLUSIONS

Free radical photoinitiators and photosensitisers can be used to broaden the spectral response of onium salts. A one-electron transfer reduction of onium salts by photogenerated free radicals and excited state photosensitisers is the source of cationic species. The reactivity of a redox system can be altered by the choice of onium salt, the free radical photoinitiator and the monomer as well as by changing the temperature or the light intensity. Electron donating free radicals can also be generated by exposure to high-energy radiation and UV radiation as well as thermally.

In view of the wide vareity of free radical sources that are available and the established corresponding techniques, redox initiated cationic polymerisation has a high potential as an acrylate free alternative to the free radical acrylate systems used today.

REFERENCES

1. J.V. Crivello and J.H.W. Lam Macromolecules 1977, 10, 1307.
2. J.V. Crivello and J.H.W. Lam J. Polym. Sci., Polym. Chem. Ed. 1979, 17, 977.
3. A. Ledwith Polymer 1978, 19, 1217.
4. S.P. Pappas and L.W. Hill J. Coatings Technol. 1981, 53, 43.
5. S. Mah, Y. Yamamoto and K. Hayashi Macromolecules 1983, 16, 681.
6. S.C. Lapin Radcure `86: Conference Proceedings, Association for Finishing Processes, Baltimore, 1986.
7. J.V. Crivello Adv. Polym. Sci., 1984, 62, 1.
8. G.H. Smith U.S. Patent 4 069 054, Jan 17, 1978.
9. J.V. Crivello and J.H.W. Lam J. Polym. Sci., Polym. Chem. Ed., 1978, 16, 2441.
10. S.P. Pappas and J.H Jilek Photogr. Sci. Eng. 1979, 23, 140.
11. R.R. Gallucci and R.C. Going J. Org. Chem. 1983, 48, 342.
12. S.C. Lapin Radtech `88: Conference Papers, New Orleans, 1988, 395.
13. J.V. Crivello and D.A. Conlon J. Poly. Sci., Chem. Ed. 1984, 22, 2105.

14. P-E. Sundell PhD Thesis, The Royal Inst. of Techn., Dep. of Polymer Techn., Stockholm 1990.
15. C. Dahlgren and Sunqvist J. Immun. Meth. 1981, 40, 171.
16. J. Brandrup and E.H. Immergut (eds.) "Polymer Handbook", 2nd edn. Wiley, New York, 1975, II-421.
17. H.E. Bachofner, F.M. Beringer and L.J. Meites J. Am. Chem. Soc. 1958, 80, 4269.
18. P.S. McKinney, S. Rosenthal J.Electroanal.Chem., 1968, 16, 261.
19. J. Grimshaw in "The Chemistry of the Sulfonium Group" C.J.M. Stirling and S. Patai (eds.), Wiley, New York, 1981, Ch. 7, 141.
20. D. Rehm and A. Weller Z. Phys. Chem. N.F. 1979, 69, 183.
21. P-E. Sundell, S. Jönsson and A. Hult in "Radiation Curing of Polymeric Materials" ACS Symp. Ser. 417. eds. C.E. Hoyle and J.F. Kinstle, Washington D.C. 1990, Ch. 32.
22. J. Henglein Electroanal. Chem. 1976, 9, 163.
23. P-E. Sundell, S. Jönsson and Hult, A. J. Polym. Sci. Chem. Ed. "Photoredox Induced Cationic Polymerisation of Divinyl Ethers" accepted for publication.
24. J.V. Crivello and J.L. Lee Polymer J. 1985, 17, 73.
25. C.E. Hoyle, R.D. Hensel and M.B. Grubb J. Radiat. Curing 1984, 11, 22.
26. G.R. Tryson and J. Schultz J. Polym. Sci. Polym. Phys. Ed. 1979, 17, 2059.
27. J.G. Kloosterboer Adv. Polym. Sci. 1988, 84, 3.
28. F.R. Wight J. Polym. Sci. Polym. Lett. Ed. 1978, 16, 121.
29. J.E. Moore in "UV Curing Science and Technology", ed. S.P Pappas Technology Marketing Corp., Norwalk, Conn., 1978, Ch. 5.
30. C.E. Hoyle, M. Keel and K-J Kim Polymer 1988, 29, 18.
31. Y. Yagci, J. Borbely and W. Schnabel Eur. Polym. J. 1989, 25, 129.
32. A. Ledwith, S. Al-Kass and A. Hulme-Lowe in "Cationic Polymerisation and Related Processes" ed. E.J. Goethals, Academic Press, London, 1984, 275.
33. P-E. Sundell, S. Jönsson and A. Hult J. Polym. Sci. Chem. Ed. "Thermally Induced Cationic Polymerisation of Divinyl Ethers using Iodonium and Sulfonium Salts" accepted for publication.

UV Photoinitiators in Pigmented Systems

Konrad Dorfner

E. MERCK, DARMSTADT, GERMANY

When UV-radiation is focussed on any medium or system, the most important effects which are encountered, are:

Reflection i.e., a change in direction and intensity of the incident radiation, either without penetrating the medium or after penetration at a component of the system

Refraction i.e., a change in direction of radiation while passing through the medium

Absorption i.e., a change of intensity of radiation while passing through the medium

Transmission i.e., also a change of intensity of radiation while passing through the medium.

Which of these effects occurs, and to what extent, depends on the kind of medium as well as on the wavelength of the irradiation.

It is often the practice in review articles and textbooks to demonstrate the light path of the incident radiation by diagrams which show that internal multiple reflections occur within the film, which effectively increase the intensity of radiation at any point inside the film, thus in all probability increase the radical yield.

Mainly with respect to quantitative modelling the optical conditions in a clear top coat are depicted in a way in order for Lambert-Beer's equation to be valid. It is postulated that the rate of generation of free radicals, and thus the rate of curing, is proportional to both the intensity of radiation at the surface and the photoinitiator concentration. Multiple internal reflections take place thus enhancing the rate of through cure. Obviously the thinner the film, the more multiple reflections occur.

Such schematic representations are, to a great extent, an over-simplification of the situation. In-depth investigations aimed at modelling the photoinitiator performance accurately in UV-curable clear coatings have shown that firstly, the interactions of photoinitiators with commercially available radiation sources (having numerous emission lines) are very complex, that secondly, the concentration of a photoinitiator affects the ratio of surface cure to body cure, thus meaning that a change of lamps can effect the choice of a photoinitiator. It is a misconception that the rate of photopolymerisation is proportional to the concentration of the initiator, and that optimum concentrations for some initiators do exist. A variety of other factors are also important, such as the average functionality of the formulation, the viscosity, the quantum yield of the initiator, the fraction of free radicals produced by the initiator which react with monomers. The presence of inhibiting agents such as oxygen, the presence of coinitiators such as tertiary amines are also significant.

We are dealing in this paper with UV-photoinitiators in pigmented systems, with the important question regarding their behaviour and their optimal application, i.e., their optimal concentration in the UV-curable formulation.

Comparing the passage of light in a clear film, it is common practice in the literature to demonstrate the light path of the incident radiation by diagrams as shown in Figure 1. This means that in a coating application with low pigmentation (e.g., rutile TiO_2) with high reflections, the effectiveness of the UV-irradiation is increased through an increase in the mean free path of the incident light, or in other words, that scattering increases the path length of the

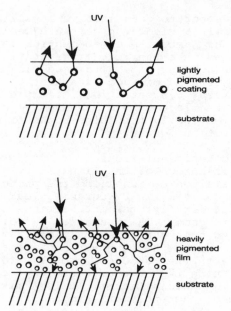

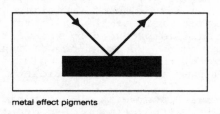

Figure 1

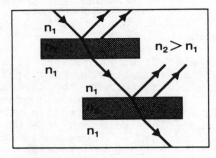

Figure 2

photons and hence increases the probability of photoinitiation. In a film of moderate thickness, with high
pigmentation, some scattered UV light will succeed
in penetrating through the film to the substrate. This
will depend on the particle size of the pigment. This
effect can be demonstrated by curing of a rutile pigmented film containing a photoinitiator absorbing even
in the region of maximum TiO_2 absorption, if a long
enough exposure to UV light is given.

The consequence of too long exposure times was certainly the reason why until recently the UV-curing of
white pigmented coatings was not a practical proposi-
tion.

In addition to this it is, in general, to be assumed
that the pictorial demonstration of pigmented systems
is also to a great extent an over-simplification of
what actually happens in practice.

The various pigments that would have to be dealt with
are

> Metallic pigments
>
> Nacreous and interference pigments
>
> Extender pigments
>
> Coloured inorganic pigments
>
> Coloured organic pigments
>
> White pigments
>
> Black pigments

of which the extender pigments are the easiest to
handle in UV-curing since they are substantially
transparent to UV light. What remains is the problem
of the optimally selected UV-photoinitiator with
respect to the system, often changing from the traditional but slow curing polyester-styrene to the faster
curing acrylic systems or combinations of the two
types.

It must otherwise be realized that an "ideal" photoinitiator has not been found yet so that one has to
deal with those which are available as commercial products.

In 1980 five classes of photoinitiators that had up to
then evolved with commercial success are shown below.

1980   1990

Benzoin ethers   (Benzoin ethers)
   Isopropyl benzoin ether   (Isopropyl benzoin
   Butyl/isobutyl benzoin   ether)
   ether etc.   (Butyl/isobutyl
                                         benzoin ether)

Benzophenones   Benzophenones
   Benzophenone   Benzophenone
   Michler's Ketone   (Michler's Ketone)
                                         4-Methyl benzophenone
                                         2,4,6-Trimethyl benzo-
                                         phenone

Thioxanthones   Thioxanthones
   Isopropyl thioxanthone   Isopropyl thioxanthone
   (Other thioxanthones)

Ketals   Ketals
   Benzil dimethyl ketal   Benzil dimethyl ketal

Acetophenones   Acetophenones
   Diethoxy acetophenone   (Diethoxy acetophenone)
   (2-Hydroxy-2-methyl-   2-Hydroxy-2-methyl-
   1-phenyl-propan-1-one),   1-phenyl-propan-1-one,
   (1-Hydroxy-cyclohexyl-   1-Hydroxy-cyclohexyl-
   phenyl-ketone)   phenyl-ketone

                                      Acylphosphinoxide

It can be seen that in 1990 one more very important photo
initiator , acylphosphinoxide, has to be added. Interestingly the benzoin ethers could not yet be completely
eliminated.

The curing of pigmented systems by UV-technology requires special care in order to ensure satisfactory
results. Initially, the formulator is able to run or
improve the performance of the selected resins by
using as a first choice one suitable photoinitiator
or if necessary a photoinitiator combination.

Some time ago we investigated the UV curing of films
pigmented with either metallic pigments or pearl lustre
pigments. Figure 2.

Metal effect pigments are platelet or flake pigments based on non-ferrous metals. The shiny pigment particles, which are opaque to visible light, consist usually of pure aluminum, pure copper or brass, i.e., of soft and ductile alloys or metals. The ratio of thickness to diameter in metal effect pigments is 1 - 50 or 1 - 250 µm/µm. Visible light falling on the pigment particles is reflected at the surface. If the pigment platelets in a coating are parallel to one another then directional illumination will produce a sensation in the observer which, depending on the particle size and the degree of parallel orientation, is known as metallic gloss or metallic effect.

Pearl lustre pigments are pigments of platelet shape with flat surfaces which, in contrast to metal effect pigments, are transparent and consist of highly refractive materials having a refractive index $n > 2$. In this situation only some of the incident light is initially reflected. The remainder is transmitted, and then similarly reflected at the boundaries. Under directional illumination, parallel oriented pearl lustre pigments in suitable binder systems create, as a result of the multiple reflection, an impression of pearl lustre and hence appear pearlescent. There are numerous synthetic pearl lustre pigments, including among others pigments based on bismuth oxychloride and basic lead carbonate, as well as metal oxide/mica. Transparent oxides, such as titanium oxide, iron oxide or chromium oxides, have been applied to mica platelets of suitable size, to achieve a similar result.

In systems containing pigments which are opaque to UV-radiation, the photochemically active UV-radiation can certainly no longer reach the deep-lying photoinitiator molecules by the direct route. If the UV-opaque pigment particles are of platelet shape as metallic pigments are, Figure 3, then the probability that the initiator molecules can absorb UV-radiation of suitable wavelength decreases with increasing layer thickness. This means that such systems are fully cured only in very thin layers.

Contrary to this are pearl lustre pigments transparent to UV-radiation, Figure 4. UV-radiation can pass through the transparent platelets having a parallel orientation, even down to the base in the case of a thick coating. It is not possible to show all results of the investigations here, so only essentials are summarized.

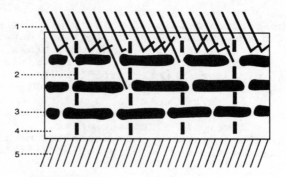

1 UV irradiation
2 photoinitiator molecules
3 UV opaque platelet pigment particles
4 UV-curable binder
5 substrate

Figure 3

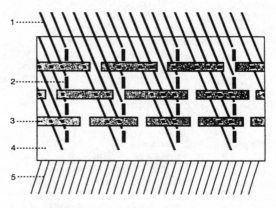

1 UV irradiation
2 photoinitiator molecules
3 UV transparent pearl lustre pigments (e.g. TiO$_2$/mica)
4 UV-curable binder
5 substrate

Figure 4

In an experimental binder system consisting of 75 parts by weight of an oligomeric epoxy acrylate and 25 % by weight of hexanediol diacrylate, 5 % of the UV photoinitiator HMPP (2-hydroxy-2-methyl-1-phenyl-propan-1-one) were incorporated. In a pass- or fail-experiment, a standard commercial silver pearl lustre pigment was taken. As Figure 5 shows up to a pigmentation of 20 % the pendulum hardness achieved does not effectively change, despite a higher belt speed, than   is required in practice. From these results it is obvious that pearl lustre pigments are transparent to UV-radiation, and hardening can be obtained almost to the same degree as with a clear varnish.

The conclusion of these experiments was that the photoinitiator behaves in a nacreously pigmented system as if it were a clear lacquer. The same cannot be expected from metallic effect pigments and therefore we continued with our investigations with photoinitiator combinations.

The manufacturers of metal effect pigments claim that UV-curing is possible with paints containing these pigments, but when details for evidence were requested no corresponding experimental or practical results were made available.

With the same experimental binder system which had been pigmented with

10 % of a titanium dioxide/mica pearl lustre pigment or
10 % of an aluminum-based metal effect pigment,

to the samples of the pigmented finishes were added various amounts of an initiator system comprising

5 parts by weight of a combination of aryl ketones, i.e., HMPP + ITX
1 part by weight of benzophenone and
1 part by weight of 4-(2-hydroxyethyl)morpholine.

Samples of the ready-to-use finishes were applied to glass plates in a thickness of 60 μm by means of a film applicator. The resulting coatings were then dried at a running speed of 5 m/min in a Beltron UV lab drier using 2 medium pressure mercury lamps each with an output of 80 W/cm. The hardness of the films was measured in accordance with DIN 53157.

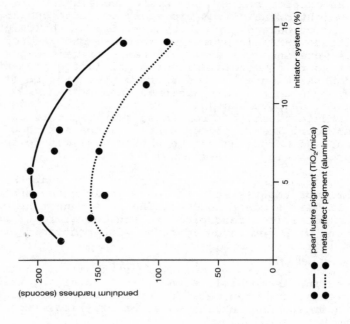

pendulum hardness (seconds)

initiator system (%)

● pearl lustre pigment (TiO₂/mica)
● metal effect pigment (aluminum)

Figure 6

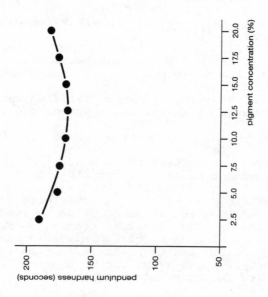

pendulum hardness (seconds)

pigment concentration (%)

Figure 5

The measured pendulum hardness values are seen,
Figure 6, against the level of initiator. Clearly,
films pigmented with the pearl lustre pigment are
always harder and more efficiently cured, i.e., have
higher pendulum hardness values, for every given level
of initiator than films pigmented with the metal
effect pigment.

In follow-up studies, it was possible to show that
coatings intended to have a certain hardness for a
given thickness can be cured at higher running speeds,
i.e., more economically, if they have been pigmented
with a pearl lustre pigment than if they have been
pigmented with a metal effect pigment.

The experimental work in these studies was done using
the same binder system and the same pigments in the
same concentration as in the studies before. The
photopolymerisation was initiated with HMPP + ITX
(4 : 1) + 1 % benzophenone + 1 % 4-(2-hydroxy-ethyl)
morpholine. The 60 μm thick layers were UV cured
under the same conditions at running speeds of 1 to
15 m/min.

The curing curves as shown in Figure 7 show the depend-
ence of pendulum hardness on the rate of curing.
Clearly, for a given rate, the layers containing the
pearl lustre pigment were harder than the coatings
pigmented with the metal effect pigment. It is clear,
moreover, that 60 μm thick layer pigmented with the
pearl lustre pigment gives a pendulum hardness of
180 seconds if cured at about 10 m/min. If the coating
contained the metal effect pigment, the running speed
hat to be lowered to somewhat slightly above 1 m/min
in order to obtain the same hardness.

Further studies have shown that UV-curable binder
systems can be pigmented to a much higher level with
pearl lustre pigments than with metal effect pigments
without markedly reducing the hardness of the resulting
coatings.

The experimental work in these first studies was done
using pearl lustre pigments and metal effect pigments
with pigment platelets of comparable particle size,
namely:

1.   a pearl lustre pigment made of titanium dioxide/
      mica which contains particle sizes of about 10 -
      50 µm,

2.   an aluminum-based metal effect pigment which,
      according to the manufacturer, contains particles
      having an average size of 25 µm,

3.   a pearl lustre pigment made of titanium dioxide/
      mica which contains particles having a size of
      about 5 - 20 µm,

4.   an aluminum-based metal effect pigment which,
      according to the manufacturer, contains particles
      having an average size of 50 µm.

These pigments were incorporated in proportions of
2.5 % to 20 % in the UV-curable binder system used so
far in these studies. Initially 5 % of the same photoinitiator combination had been added to the system.

The ready-to-use, pigmented finishes were again applied
to glass plates in a thickness of 60 µm by means of a
film applicator and were then cured in a Mini-Cure UV
laboratory drier at a running speed of 5 m/min using
two medium pressure mercury lamps, with an output of
40 W/cm and 80 W/cm, respectively.

Figure 8 shows the pendulum hardness values of the
films as a function of the pigment concentration. As
can be seen, the pendulum hardness for the films pigmented with the two pearl lustre pigments decreases
only slightly with increasing pigment content within
the given concentration range, namely from 195 sec
at a pigment concentration of 2.5 % to 180 or 185 sec
at the pigment concentration of 20 %. In contrast,
the pendulum hardness for films pigmented with the
metal effect pigments decreases drastically with
increasing pigment content. The pendulum hardness
which, for a pigment content of 2.5 % was as high as
170 and 180 sec, respectively, was a mere 60 sec for
a pigment content of 20 %.

UV-curable binder systems containing pearl lustre
pigments are also remarkable because they are curable
even in the form of relatively thick films. The studies
supporting this statement were carried out using the
same UV-curable binder system as in the preceding

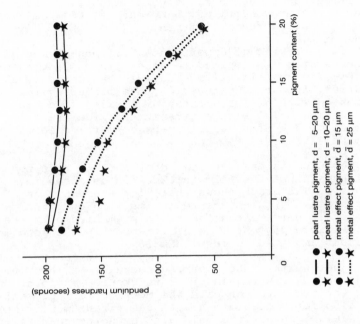

pearl lustre pigment, d =   5–20 μm
pearl lustre pigment, d = 10–20 μm
metal effect pigment, d̄ = 15 μm
metal effect pigment, d̄ = 25 μm

Figure 8

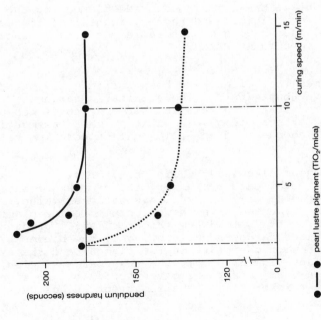

pearl lustre pigment (TiO₂/mica)
metal effect pigment (aluminum)

Figure 7

investigations. The coatings were likewise cured under the same conditions. Lamp output: 40 W/cm + 80 W/cm; curing speed: 5 m/min.

In a first series of experiments, the non-pigmented binder system was applied to glass plates in the form of 30 - 250 µm thick films and was then cured in a UV drier. In subsequent series, the binder system was pigmented with 5 % of a titanium dioxide/mica pearl lustre pigment, 5 % of a titanium dioxide (rutile) pigment or 5 % of an aluminum-based metal effect pigment.

The pendulum hardness values of the films have been plotted against the film thicknesses as shown in Figure 9. The graph shows that, for a given film thickness, the non-pigmented films were only insignificantly harder than the films pigmented with the pearl lustre pigment. The differences in the pendulum hardness amounted to less than 10 %.

It is noteworthy that a strong decrease in pendulum hardness with increasing film thickness was found not only for pigmentation with the metal effect pigment but also for pigmentation with the titanium dioxide (rutile) pigment. It should also be noted that films which contain the titanium dioxide pigment or the metal effect pigment and had a thickness of 150 µm or more, also having a measurable pendulum hardness, had not been fully cured.

In the case of UV-cured non-pigmented coatings, it is assumed that films have been sufficiently cured if they have about 75 % of the attainable hardness. It is assumed that this relationship also holds for pigmented coatings, this means that in this case curing has been adequate if the films have a pendulum hardness of about 150 sec.

This diagram likewise reveals that, of films pigmented with the metal effect pigment, those of a thickness of up to 30 µm will achieve this hardness value. Coatings pigmented with the titanium dioxide pigment could be up to 75 to 80 µm thick and still produce the required hardness. However, only the binder system pigmented with the pearl lustre pigment and the system without any pigment, i.e., the transparent finish, should be cured under the given conditions in significantly thicker films. The limit for an adequate cure

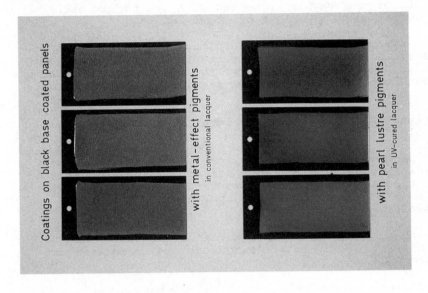

Figure 10

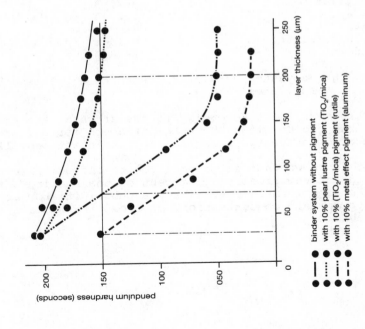

Figure 9

of the transparent finish was beyond 250 µm and in the
case of the system pigmented with the pearl lustre
pigment at a thickness of about 200 µm.

In conclusion for UV photoinitiators in pigmented
systems it may be shown in Figure 10 in combination
with pearl lustre pigments, the metal effects obtained
can be directed to a quite similar final appearance,
to metallic pigments.

With respect to extender pigments or matting agents
only one point can, because of the shortness in space,
be mentioned. If a matt surface is to be achieved, in
a photoinitiator system synergists such as tertiary
amines show a deleterious affect and should, therefore,
not be included in the formulation. In this case it
is to be recommended to use UV photoinitiators of the
α-cleavage type.

With respect to coloured pigments one could imagine
that, e.g., offset printing inks are formulated as
shown below.

Acrylate Oligomers*	30-35 %
Acrylate Monomers	0-10 %
Pigment	15-25 %
ITX	1-2 %
EPD	4-6 %
Other Initiators	6-10 %
Additives	5-10 %

* e.g. Polyester acrylates, Urethane acrylates, Epoxy
       acrylates

This means that according to the state of the art, the
first choice among UV photoinitiators for colour-pigmented formulations is isopropyl thioxanthone, ITX.

It is known that it is possible to characterize the
pigment numerically as percent of the running speed of
the same formulation, with one photoinitiator at an
appropriate concentration, as shown in the following
table.

Pigment	Running speed
Yellow	100 %
Magenta	86 %
Cyan	64 %
Black	28 %

This big difference especially from yellow to black cannot be compensated for by the currently available photoinitiators.

An increased proportion of the more reactive acrylates is required on moving from yellow to magenta to cyan to black. In doing that, the total concentration of the photoinitiator combination can be kept at approximately 12 % for yellow and magenta but must be increased up to 14 % for cyan and even higher for black.

A general behaviour of photoinitiators in colour-pigmented systems is that their concentration cannot be increased optionally in order to make the system more reactive, since beyond an optimal concentration of the photoinitiator the reactivity decreases again sharply. This is because the photoinitiators must absorb UV-radiation for being split into radicals, that is, that they are highly reactive UV absorbers. If they are present in a too high concentration they absorb already in the upper colour layers so much of the UV-radiation that the latter cannot reach the lower layers anymore with the consequence of no through cure.

In the general literature this is considered in relation to the transmission of the colour pigments, for instance yellow, magenta, orange, cyan, and carbon, in comparison to the absorption of the photoinitiator, e.g., benzophenone. Figure 11 may serve as a reminder.

There is in the field of pigmented UV-curable systems no satisfactory model or comprehensive systematic approach available to enable the enduser of this technology to make predictions in order to reduce the amount of experimental work.

We need a maximum photoinitiator efficiency in pigmented systems, to reduce the quantity of initiator present.

Two concepts are now discussed and applied. In the
first, it is concluded that the photoinitiator must
absorb light in the region where absorption and reflection of the pigment are minimal: the so-called
"optical window" concept.

If the photoinitiator has a low molar absorption
coefficient and is present in a lower concentration,
then a good penetration of the UV-radiation is possible. If the products of the photolysis are non-
absorbing species, the radiation can penetrate deeply
into a layer, thus providing a good through cure and
enabling one to cure as well satisfactorily also very
thick layers.

The second concept assumes that the photoinitiator
absorbs the irradiation in the same wavelength region
as the pigment does. Then there is competition between the initiator and pigment, and in order to absorb
a sufficient fraction of the radiation, the molar
absorption coefficient (or the concentration) must be
higher, with a rather low penetration of light. This
concept can, as has been demonstrated, be applied for
the curing of very thin films, as for instance, in
printing inks. It may be summarized that a higher
molar absorption coefficient will, per se, lead to
the application of a lower concentration.

In a real pigmented system a splitting of the incident light takes place. The incident light can be
reflected by the surface or be absorbed by the pigment.
Depending on other factors   the crucial point   is
that a substantial fraction of the radiation is still
transmitted and thus reaches the deeper layers of the
coating.

Reverting to ITX and its behaviour in colour-pigmented UV-curable systems, this molecule has the
possibility to shift absorption bands to longer wavelengths, beyond the absorption of the colorants. As
it has been shown in corresponding investigations by
Meyer and Zweifel in 1986 the bathochromic effect is
clearly related to the nature and position of the
substituents of thioxanthones. Electron donating
groups in 2 or 7 position and electron withdrawing
groups in 3 or 6 position cause shifts of the absorption bands to longer wavelengths. Unfortunately the
bathochromic effect of the sulphur atom is followed
by yellowing properties.

If ITX is considered as the photoinitiator, a photo-activator is needed for it to become photoactive. The question to be answered is whether the photoinitiator concentration can be kept at, or reduced to, a low level, or whether with a more suitable activator than those currently used the quantity of the total initiator combination can substantially be reduced and the problem of yellowing can be solved.

The following shows the photoinitiation mechanism of ITX with an aromatic tertiary amine.

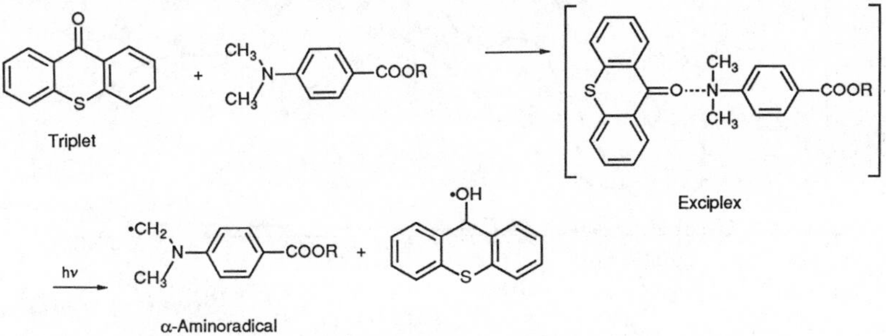

Triplet

Exciplex

α-Aminoradical

It is well known that aliphatic amines possess an unpleasant odour, and impart this to many cured films. Other potential problems of aliphatic amines such as bleeding of the surface and relatively high alkalinity are also known.

In view of colour-pigmented thicker coatings than those of an offset printing ink, as it might be expected that they are required by the furniture industry, it seemed appropriate to look into the influence of the various aromatic amines as well.

The following shows a number of aromatic amines

EPD

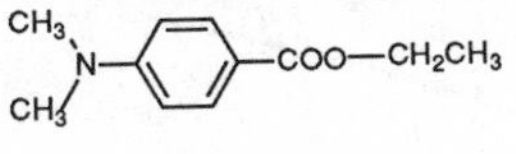

Ethyl *p*-dimethylamino benzoate

BEA

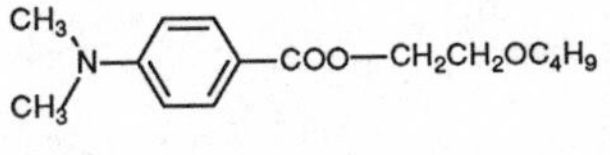

2-(n-Butoxyl) ethyl *p*-dimethylamino benzoate

DMB

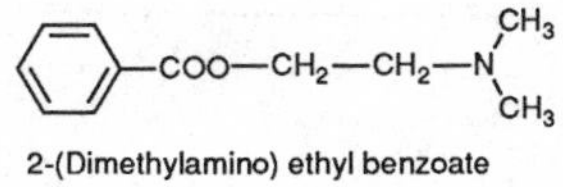

2-(Dimethylamino) ethyl benzoate

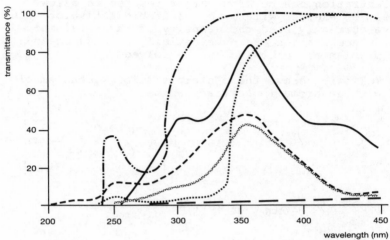

Figure 11

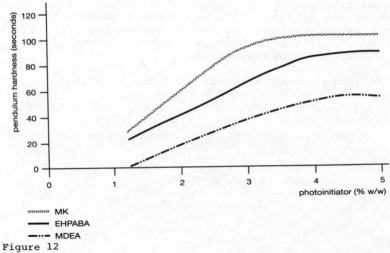

Figure 12

IPD — Isoamyl p-dimethylamino benzoate

EHPABA — 2-Ethylhexyl p-dimethylamino benzoate

MK — Michler's Ketone

of which EPD is nowadays mainly used, BEA has been recommended for a long time, DMB is covered by patents, IPD is said to have a very low activation energy, and EHPABA has a reactivity close to MK, Michler's Ketone, which would still be a good product but for the fact that it is carcinogenic, which has severely dampened enthusiasm for its use.

Not only the structural similarity of EHPABA to MK is evident, EHPABA offers a very appealing alternative to MK since it possesses no appreciable odour and it is, in fact, used topically as a sunscreen. Unlike EPD, it is liquid and miscible with benzophenone and many UV-curable resins. ITX can easily be added to such a formulation in order to obtain an initiator complex.

Figure 12 shows as only one example the application of EHPABA in a carbon-filled coating demonstrating that MK is superior in hardness at all tested levels. The cure with MDEA was considerably poorer. The results suggest the evaluation of EHPABA for application in coloured and even black inks and also coatings.

Up to only recently the UV-curing of white pigmented coatings had found only a limited industrial acceptance due to the facts:

1. The most widely used white pigment, TiO_2, which in its rutile modification shows a strong absorption between 300 and 400 nm, so UV radiation in this wavelength range is then no longer available to the initiator, was not applicable.

2. The addition of sensitizers helped in some cases to circumvent this deficiency but often caused yellowing so that an efficient sensitization without yellowing was never achieved.

3.   The well known thioxanthone-amine combinations
      showed substantial yellowing, especially after
      thermal post-treatment.

We could present results of investigations made in
cooperation with a manufacturer of white pigments for
the application of zinc sulfide white pigments for
base coats in can coating which proved quite promising but because of the space limit we have to post-
pone to do that.

Since the early 1980s Magnesiumtitanate found some
application in wood fillers, therefore, its transmission curve may in Figure 13 be compared with those
of TiO_2, anatase and rutile, respectively. From its
absorption curve Magnesiumtitanate is certainly suitable for UV curing.

Jacobi and Henne, and Jacobi, Henne and Böttcher,
respecitvely, published in 1983 and later their papers
on a novel class of UV photoinitiators, the so-called
acylphosphinoxides of which the 2,4,6-Trimethyl-
benzoyl diphenyl phosphinoxide, now named TPO, was
found as the most suitable one.

A look at the part of some UV photoinitiators in comparison with titanium dioxide, rutile, as white pigment,
as shown in Figure 14 reveals that only since TPO
became available was there a fair possibility to UV
cure coatings pigmented with this pigment.

From the transmission of titanium dioxide (rutile) it
can be seen, that below 380 nm the UV-radiation is
completely absorbed. The other initiators cannot
therefore, become reactive. contrary to this the
absorption spectrum of TPO shows a distinct absorption
also above 380 nm and overlaps with the transmission
curve of titanium dioxide, rutile, in the range of about
400 nm and beyond.

Reverting to TPO as a single UV-photoinitiator, it was
already found in the first studies, that an amine was
needed as a coinitiator because without an amine
present acylphosphinoxides were not reactive in pigmented acrylate resin systems, that rough surfaces,
because of the possibility to include air, which causes
an inhibition, are more difficult to be coated, and
that by these findings it became obvious that the
inhibition by oxygen in the case of TPO is higher than
with BDK or HMPP and that this is especially disad-
vantageous with flattened and pigmented coatings.

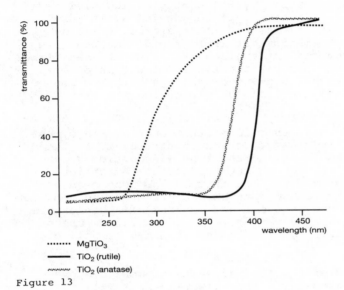

Figure 13

Figure 14

Regardless of the question what the reason is why
acylphosphinoxides, which are photoinitiators which
function by a Norrish Type I photocleavage giving two
photoinitiating radical species, needed a tertiary
amine, whether for instance the scission reaction
of the acylphosphinoxides is so rapid that there is
little chance for the excited states of the sensitizer
to be quenched. It was practically found in other
formulations that pigmented coatings with TPO showed
a too pronounced surface cure with all the known
consequences.

The general idea, therefore, that such coatings must
be hardened in a way from the substrate, i.e., the
bottom of the layer, to the top, was developed, and
followed in subsequent investigations. Attention may
here be drawn to the fact that it has long been an
empirical experience that HMPP shows a curing behaviour
more pronounced in the inner regions of a UV-curable
system than at the surface. The advantageous use of
a low pressure lamp in a first step may as well only
be cited.

We have, therefore, studied white pigmented epoxyacrylate systems with approximately 20 % TiO_2 rutile,
and TPO and HMPP as photoinitiators.

In Figure 15 the most important results with respect
to the chemistry are shown. Economic features were not
considered.

At an acceptable pendulum hardness and at an initiator
concentration of 3 % a higher cure speed of 40 fpm as
compared to 30 fpm can be applied.

The pendulum hardnesses that can be obtained with a
1:1 mixture of TPO and HMPP are shown in Figure 16 for
a pigmentation with 20 % TiO_2 rutile. At least for
wood coating these results are satisfactory.

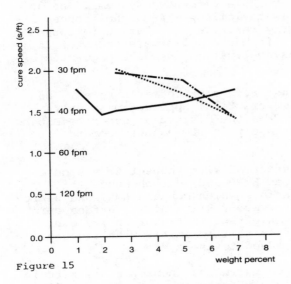

Figure 15

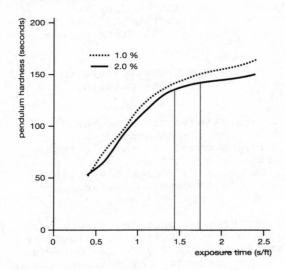

Figure 16

There have been very novel photoinitiators only re--
cently presented at scientific conferences which are
recommended for UV-curable pigmented systems. One of
them, 2-Benzyl-2-dimethylamino-1-(4-morpholinophenyl)-
butanone-2, is a very reactive one, but, unfortunately
as Figure 17 shows, it has a yellowness index of 25,
which prohibits its application in white pigmented
systems.

That an initiator has also a price may here only be
mentioned. But under practical application aspects
this is important. Then the mixture as described looks
very much different and pure TPO cannot compete with
it.

HMPP was also investigated with respect to a syner-
gistic effect between HMPP and benzophenone. As
Figure 18 shows coatings which contain HMPP and benzophenone in a ratio of about 9:1 exhibit surface cure
at speeds higher than either initiator on its own.
This, despite the fact that no amine, with a danger
for yellowing, was added to the system.

It can firmly be speculated that benzophenone acts in
the same way also in a three component mixture, TPO-
HMPP-Benzophenone. The most likely explanation for
this phenomenon as   has been shown already in the
1970s is that the peroxides which are formed in the
course of oxygen inhibition are attacked by the
excited benzophenone and form alkoxy or hydroxy radicals which are very efficient in initiating polymerisation.

For UV photoinitiators in black pigmented systems
enough has been said as far as offset printing inks
are concerned.

In the field of printing, manufacturers of ceramic
printing inks mainly for the black bands on wind-
screens of some automobiles requested early suitable
photoinitiators for such inks which utilize a spinel-
based black pigment, in order to change the printing
process to UV. We were able to develop suitable proprietary mixtures based on HMPP.

The same proprietary mixtures of photoinitiators with
photoactivators give in a binder system containing an
acrylated urethane and carbon black a curing curve as
shown in Figure 19. As the requests for suitable photoinitiators for black pigmented coatings come now from

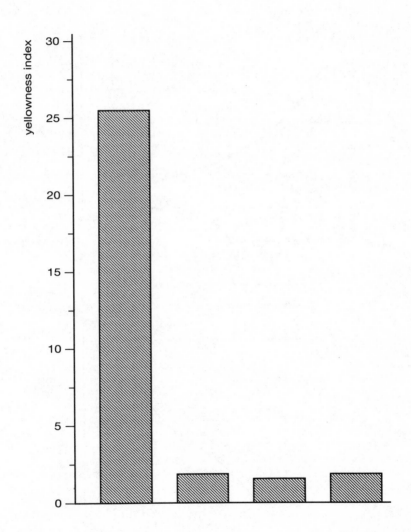

Figure 17

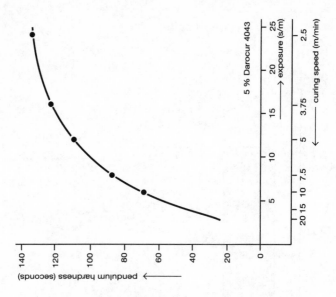

Figure 19

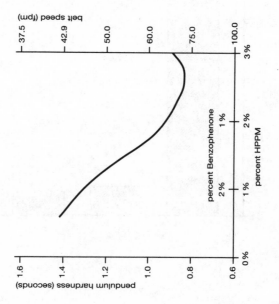

Figure 18

the furniture industry, a pendulum hardness of 110 sec at 5 m/min is an acceptable result. Further refinements with respect to several parameters of the corresponding UV curing process are certainly necessary and possible.

Conclusions. What can be concluded from looking into the state of the art of UV photoinitiators in pigmented systems?

The photoinitiator has always been and is the key substance of UV-curable systems either clear or pigmented.

One conclusion is that the full range of pigmented systems, that is from nacreous pigments through the colour range to white and black can be UV-cured, and that in both cases of printing inks and even thicker layers.

What is unsatisfactory is that such high concentrations of initiators are needed of which a high percentage remains unused. Better sensitization is most probably the key point for a satisfactory solution, this may have to start from in-depth studies of radical formation mechanisms.

The other way of synthesizing novel UV photoinitiators containing the properties of existing initiator combinations in one molecule bears inherently the risk of obtaining substances with too high odour forming and yellowing properties.

It is acknowledged with thanks that the preperation of this paper was supported by Dr. J. Ohngemach, E. Merck, Darmstadt, FRG, and Mr. M. Hanrahan, EM Industries, Hawthorne, NY, USA, with laboratory investigation results and fruitful discussions.

Substrate Chemistry

Performance of UV Curing Systems Through Monomer Selection

Robert C.W. Zwanenburg

CRAYNOR, VERNEUIL EN HALATTE, FRANCE

ABSTRACT

A range of 23 different monomers, ranging from monofunctional up to penta/
hexafunctional, were studied in model formulations based on a bisphenol A
epoxy acrylate oligomer.   Three different model formulations have been used:
one with constant composition, one with a constant acrylic double bond
concentration and one with constant viscosity.   From those formulations,
viscosity, tensile, elongation at break, konig hardness, pencil hardness, mandrel,
cure speed, and the glass transition temperatures were measured.   These data
have been used to establish the correlation between the chemical structure of the
used monomer, and the physical properties. of the formulation it is used in.

INTRODUCTION

Radiation curing is a well established industrial activity for more than 30 years
and regained a fast growth since the early eighties.   Acrylic monomers are an
essential part in virtually every radcure–formulation.   Nevertheless, there are only
a few studies published explaining the underlying logic in the use of acrylic
monomers (see bibliography).   Knowing that there is a wide range of monomers
offered to radiation curing industry, but hardly any formulator is in the position
to try them all, we felt the need to fill the gap and help the formulator in
selecting the right monomers for his purposes by performing a broad study,
revealing formulation guidelines.

EXPERIMENTAL

Materials

For this study a very wide range of monomers commercially available to the
radiation curing industry has been used.   This range comprised mono– and up

to penta/hexafunctional monomers (See Table 1).

In order to make it useful to as many formulators as possible, we selected first, second, and third generation monomers: both the widely used and relatively new developments in monomers, hardly known, tested, or used in commercial radcure applications.

Again, to make this study as usable as possible, we chose for our model-formulations an example of the real workhorse in radiation curing: oligomer Craynor CN 104A80, a bisphenol A type epoxy–acrylate, diluted with 20% TPGDA.

The photoinitiators used are Esacure TZT (Fratelli Lamberti), a liquid mixture of benzophenone and a substituted benzophenone, and Darocur 1173 (Merck).

Formulation

In this study, three different series of clear varnishes were made, all being as constant in composition as possible within one series.   In order to highlight as much as possible the differences, brought about by the variation of the used monomer, no additives of any sort were used.

In all three different series, we used 2 parts of Esacure TZT and 2 parts of Darocur 1173 added to 100 parts of formulation.

The three different approaches were:

*Series 1:   constant composition
*Series 2:   constant acrylic double–bond concentration
*Series 3:   constant viscosity

The idea behind these three series was in the first place, to see how the different monomers behave in three different types of formulations.

The series with constant composition was set up as the 'fingerprint' of the different monomers: the only thing changing here is the nature of the used monomer.

Formulations in this series looked like this:

62.5 p   epoxy–acrylate oligomer (80% in TPGDA)
12.5 p   TPGDA
25.0 p   monomer
2+2   p   photoinitiator

The series with constant acrylic double–bond concentration was set up in an attempt to avoid the influence of the crosslink–density of the system, caused by differences in both functionality and molecular weight of the various monomers.

TABLE 1 - USED MONOMERS

CHEMICAL NAME	ABBREVIATION	NAME	F	M W	TYPE
2-(2-ETHOXYETHOXY) ETHYL ACRYLATE	EOEOEA	SR 256	1	188	LINEAR ETHER
2-PHENOXYETHYL ACRYLATE	2 PEA	SR 339	1	192	CYCLIC HYDROCARBON /LINEAR ETHER
ISOBORNYL ACRYLATE	IBOA	SR 506	1	208	CYCLIC HYDROCARBON
ISODECYL ACRYLATE	IDA	SR 395	1	212	BRANCHED HYDROCARBON
LAURYL ACRYLATE	LA	SR 335	1	240	LINEAR HYDROCARBON
1,6-HEXANE DIOL DIACRYLATE	HDDA	SR 238	2	226	LINEAR HYDROCARBON
TRIETHYLENE GLYCOL DIACRYLATE	TIEGDA	SR 272	2	258	LINEAR ETHER
TRIPROPYLENE GLYCOL DIACRYLATE	TPGDA	SR 306	2	300	BRANCHED ETHER
TETRAETHYLENE GLYCOL DIACRYLATE	TTEGDA	SR 268	2	302	LINEAR ETHER
POLYETHYLENE GLYCOL 200 DIACRYLATE	PEG200DA	SR 259	2	302	LINEAR ETHER
PROPOXYLATED NEOPENTYL GLYCOL DIACRYLATE	NPGPODA	SR 9003	2	328	BRANCHED HYDROCARBON /BRANCHED ETHER
POLYETHYLENE GLYCOL 400 DIACRYLATE	PEG400DA	SR 344	2	508	LINEAR ETHER
ETHOXYLATED BISPHENOL A DIACRYLATE	BPADA	SR 349	2	556	BRANCHED HYDROCARBON /LINEAR ETHER
TRIMETHYLOL PROPANE TRIACRYLATE	TMPTA	SR 351	3	296	BRANCHED HYDROCARBON
PENTAERYTHRITOL TRIACRYLATE	PETIA	SR 444	3	298	BRANCHED HYDROCARBON
TRIS(2-HYDROXYETHYL)ISOCYANURATE TRIACRYLATE	THEICTA	SR 368	3	423	HETEROCYCLE/LINEAR ETHER
ETHOXYLATED TRIMETHYLOL PROPANE TRIACRYLATE	TMPEOTA	SR 454	3	428	BRANCHED HYDROCARBON /LINEAR ETHER
PROPOXYLATED GLYCERYL TRIACRYLATE	GPOTA	SR 9020	3	238	BRANCHED HYDROCARBON /BRANCHED ETHER
HIGHLY PROPOXYLATED GLYCERYL TRIACRYLATE	HPOGTA	SR 9021	3	573	BRANCHED HYDROCARBON /BRANCHED ETHER
HIGHLY ETHOXYLATED TMPTA	HEOTMPTA	SR 9035	3	1000	BRANCHED HYDROCARBON /LINEAR ETHER
PENTAERYTHRITOL TETRAACRYLATE	PETTA	SR 295	4	352	BRANCHED HYDROCARBON
DITRIMETHYLOLPROPANE TETRAACRYLATE	DTMPTTA	SR 355	4	438	BRANCHED ETHER
DIPENTAERYTHRITOL PENTAACRYLATE	DPETPA	SR 399	5	525	BRANCHED ETHER

Thus, ideally this series should highlight the influence of the chemical backbone of the monomer used. Of course we had to choose a workable model which is always disputable.

We chose to use the theoretical functionality and added the high molecular weight difunctional monomer, PEG 400 diacrylate to reduce the crosslinking and the tetrafunctional ditrimethylolpropanetetraacrylate to increase the crosslinking (both are very low in skin–irritancy). Thus, formulations of this second series looked like this:

46	p	epoxy–acrylate oligomer (80% in TPGDA)
20	p	monomer
x	p	polyethylene glycol 400 diacrylate
y	p	ditrimethylolpropane tetraacrylate
2+2	p	photoinitiator
$x+y$ = 34		

The ratio x/y can be found in Table 2.

The series with the constant viscosity has been set up bearing in mind that the effect of a monomer (and of an oligomer as well) is almost always judged as application–viscosity. For many applications the viscosity of a monomer (or better: its solvency), is the main driver to use it in the first place.

We chose a fixed ratio between the oligomer and the used monomer, and reduced the viscosity with TPGDA to 400 mPa s (at 20 ˙C), with a tolerance of *ca* 10%.

Formulations of the third series looked like this:

$2x$	p	epoxy–acrylate oligomers (80% in TPGDA)
x	p	monomer
y	p	TPGDA
2+2	p	photoinitiator/100 p formulation
2 $x+y$ = 100		

Test methods

One of the objectives of this study was to establish a correlation between widely used and accepted test methods in the coatings industry like mandrel, konig hardness, and pencil hardness on one hand, and tensile strength and elongation (both at break), on the other hand.

1.   Viscosity was measured with a Brookfield RVF–100 viscometer at 20 ˙C and reported in mPa s (cps).

2.   Cure speed was examined on a 10 μm draw–down on paper, by passing several times under an 80 W cm^{-1} Wallace–Knight labcure–unit.   The belt–speed during this test was 160 m min^{-1}.   Reported are the numbers of

TABLE 2 SERIES 2 : (CONSTANT ACRYLATE-DOUBLE BOND CONCENTRATION
RATIOS POLYETHYLENE GLYCOL 400
DIACRYLATE/DITRIMETHYLOLPROPANE TETRAACRYLATE

CHEMICAL NAME	RATIO	CHEMICAL NAME	RATIO
2-ethoxyethoxyethyl acrylate	10.9/23.1	trimethylolpropane triacrylate	29.5/4.5
2-phenoxyethyl acrylate	10.5/23.5	pentaerythritol triacrylate	29.2/4.8
isobornyl acrylate	9.0/25.0	tris(2-hydroxyethyl) isocyanurate triacrylate	17.8/16.2
isodecyl acrylate	8.6/25.4		
lauryl acrylate	6.5/27.5	ethoxylated tri-methylolpropane triacrylate	17.5/16.5
1,6-hexanediol diacrylate	24.5/9.5		
triethylene glycol diacrylate	20.3/13.7	propoxylated glyceryl triacrylate	17.5/16.5
tripropylene glycol diacrylate	16.1/17.9	highly propoxylated glyceryl triacrylate	10.6/23.4
tetraethylene glycol diacrylate	16.0/18.0	highly ethoxylated trimethylolpropane triacrylate	2.0/32.0
polyethylene glycol 200 diacrylate	16.0/18.0	pentaerythritol tetraacrylate	34.0/0.0
propoxylated neopentyl glycol diacrylate	13.9/20.1	ditrimethylolpropane tetraacrylate	25.6/8.4
polyethylene glycol 400 diacrylate	5.6/28.4	dipentaerythritol penta/haxaacrylate	34.0/0.0
ethoxylated bisphol A diacrylate	4.3/29.7		

passages, needed to get a fully–cured surface.   Fully–cured meaning: not marking after two subsequent finger–rubs.

3.   Pencil hardness was measured on 100 μm films on glass.   The tests were carried out with a Wolf–Wilburn pencil hardness testing set.   Recorded is the hardest pencil, not penetrating the surface.

4.   Konig hardness was also measured on 100 μm films on glass, using a Konig hardness tester, equipped with a counter.   Reproducible results were obtained on several films made from the same formulation.

5.   Mandrel tests were performed on 50 ∞m films on paper, using a set of cylindrical mandrels.   Reported are the lowest diameters in mm, not causing any cracks in the film.   Here also very reproducible results were obtained, using strips from several different draw–downs.

6.   Tensile strengths and elongation were measured using a JJ, type DVM 3 tensile machine.   Measurements were made on free films, prepared from 100 μm (nominal) draw–downs.

The samples were dumbbell–shaped having a 4 mm portion.   We found the best reproducibility using a jaw–separation speed of 2.3 mm min^{-1}.

In order to determine the exact tensile strength, the actual thickness of the middle section was measured.

We found it impossible to prepare good samples, giving acceptable reproducibility on using the 'harder', more brittle monomers.   Series 2 was designed in such a way that this one enabled measurement for all different monomers.

7   Glass transition temperatures were determined on 100 μm free films using either differential thermoanalysis (Mettler 20) using a temperature–programme running from –50 to 150 $^{\bullet}$C (10 $^{\bullet}$C min^{-1}) or differential scanning calorimetry (Perkin Elmer 2) using a temperature–programme running from 0 to 150 $^{\bullet}$C (10 $^{\bullet}$C min^{-1}).

For all these tests: formulations were made and equilibrated at least 24 hours at 20 $^{\bullet}$C.   For tests 3/7, several draw–downs were made on the various substrates.   These were fully cured by multiple passes under the 80 W cm^{-1} medium pressure mercury lamp.   Subsequently, the coatings were stored at exactly 20 $^{\bullet}$C and 20% humidity for one week, prior to further testing.   The glass transition temperatures were determined at some later point in time.

RESULTS AND DISCUSSIONS

The amount of data generated during this study is overwhelming.   Because of this, it is not possible to discuss these in any depth in this paper.   Therefore, I have tried to reveal general patterns highlighting some striking results.

Series 1: Constant Composition

The composition used for this series can be found in section 'Formulation'. Results can be found in the graphs following the text.

Viscosity

There are a few trends clearly present in the results.

Generally speaking, the lower functionality and the lower the molecular weight, the lower the viscosity. Steric hindrance substantially increases the viscosity. This is true for branching: compare, *e.g.* tetraethylene glycol diacrylate with TPGDA or, more hidden: propoxylated NPGDA with a molecular weight of 328 with PEG 400 diacrylate with a molecular weight of 508, whereas the viscosities are hardly different. Introduction of the sterically even more hindering cyclic groups increase the viscosity more dramatically: examples are 2-phenoxyethylacrylate, isobornyl acrylate, ethoxylated bisphenol A diacrylate and THEIC triacrylate.

The third feature is the lower viscosity using ether-type monomers, which have a higher freedom of movement, compared to the hydrocarbon-monomers.

Tensile Strength at Break

The general pattern followed here is: the lower the equivalent-weight, the higher the tensile strength. The monofunctional monomers have, generally speaking, the lowest tensile strengths because they lower the crosslink-density of the system. Notable is the exceptional high tensile strength found with isobornylacrylate. An explanation is the very bulky bicyclic side-group, directly linked to the polymeric network limiting the freedom of movement of this network as much as if it were higher crosslinked.

Another interesting feature is that the hydrocarbon-monomers can be found at the low end of tensile strengths, whereas the ether-type monomers can be found at the high end. The explanation is the polarity of the ether-linkage, giving some degree of dipolar interaction, leading to higher tensile strength.

Elongation at Break

In comparing elongation and tensile strength, it is most interesting to look at examples, where low tensile is coupled with low elongation and high tensile with high elongation, rather than discussing the general pattern of low tensile coupled with a high elongation and reverse.

Highly propoxylated glyceryl triacrylate is a monomer, having a very high elongation together with a higher than average tensile strength. This is also true for 2-ethoxyethoxyethylacrylate. More generally, the ether-monomers exhibit the

highest elongations, they combine a high freedom of movement around the
ether–linkage, enabling the stretch, with the polarity, giving the dipolar
interaction, which postpones breaking of the film to higher elongations.

Konig Hardness

The order found in testing the konig hardness parallels to a greater extent the
order found with the tensile, which is to be expected, because both methods
measure in a way the resistance to deformations.   Notice the results with the
monomers, which could not be measured by tensile testing: pentaerythritol
triacrylate and tris (2–hydroxyethyl) isocyanurate triacrylate have surprisingly low
konig hardness, whereas the others can all be found at the high end.

Pencil Hardness

Pencil hardness, the way we used it, gives an impression of the surface–
crosslinking.   It is not surprising though, to find the high functionalized
monomers at the high end.   Another important thing is nicely illustrated here:
the good surface hardness on using ether–monomers.   The explanation can be
found, bearing in mind that we omitted the use of an amine–synergist.
Therefore, the presence of abstractable hydrogens in the ether–monomers shows
its effectiveness very well.   Although more masked, this mechanism is also very
important in amine–synergist containing formulations.   This is one of the big
advantages of the second and third generation monomers.

Mandrel

The mandrel gives us a good impression of the flexibility of the system.
Although the amount of stretching is very small, the results found in
mandrel–testing nicely parallel those found in elongation–testing.   From
mandrel–testing we could predict which films would not allow us to make good
tensile–samples.   Hence, all these can be found at the high end.

Notice the very high flexibility of two third generation monomers – thus
having very low skin–irritancies – highly ethoxylated TMPTA and polyethylene
glycol 400 diacrylate.

Cure Speed

As to be expected, high functionality results in high reactivity.   Notice the
highest reactivity of all is found with dipentaerythritolpenta/hexaacrylate.

What was already mentioned discussing the pencil hardness shows here
again, the abstractable hydrogens found in ether–monomers increase the
reactivity.

Notice in this respect the remarkable reactivity found with
2–ethoxyethoxyethylacrylate!   The explanation for this, besides the presence of
altogether 10 abstractable hydrogens in such a small molecule, might also be the

extremely low viscosity found in this particular case. This enables crosslinking to an ultimate extent.

Glass Transition Temperature

The Tg very much reflects the crosslink density; therefore the lowest Tgs can be found amongst the monofunctional monomers or those having a high equivalent weight. Note in this respect the very low Tgs found for highly ethoxylated trimethylopropane and polyethyleneglycol 400 diacrylate.

Also the very high Tg of isobornylacrylate, a monofunctional monomer with a relatively high molecular weight, is remarkable.

The combination of the low polymerization–shrinkage with this high Tg makes isobornylacrylate so interesting for application on plastics.

Series 2: Constant Acrylate–Double Bond Concentration

As said before; this series has been set up in an attempt to avoid the influence of differences in crosslink density. Hence, the variation in results is much smaller than in series 1, and should be explained in terms of differences of the chemical backbones of the various monomers.

The compositions used in this series can be found by adding the ratios, found in Table 2 to the formulation, mentioned in section 'Formulations'. The results can be found in the graphs, following the text.

Viscosity

The viscosities, found in this series parallel those in series 1. This shows clearly that the nature of the used monomer is the decisive factor, rather than the ratio between polyethylene glycol 400 diacrylate and ditrimethylolpropane tetraacrylate.

Tensile at Break

A big advantage of this series is that it is complete. It shows very clearly the effect of polar ether–linkages, resulting in higher tensile. This is particularly true for ethoxylated (linear ether) monomers, where this effect is not obscured by steric hindrance, partially preventing this dipolar interaction. The importance of this dipolar interaction is shown in the best possible way by tris (2–hydroxyethyl) isocyanurate triacrylate, a highly polar monomer with a flat (polar) ring in the centre of the molecule. Low tensile can be found there, where these kind of interactions are impossible, *e.g.* with lauryl– and isodecylacrylate.

Elongation at Break

It is surprising to find here among the highly flexible ether–monomers the workhorses of radcure industry HDDA and TMPTA.

An explanation might be that these monomers have a higher tendency to copolymerize rather than forming blocks of homopolymer within the copolymer network.   Something recognized in other properties as well, making these products as important as they are.

Notice that again 2-ethoxyethoxyethylacrylate gives the most flexible systems.

Konig Hardness

It is very clear from the results that monomers containing ring-structures give rise to the highest hardness.   Branched monomers can be found in the middle range, whereas the linear monomers can be found on the low end.   Again, there is a fair resemblance between the tensile and konig hardness results.

Pencil Hardness

The big difference between the results here and in series 1 is that all systems here contain already high amounts of ether-monomers. Therefore the differences in surface-cure are small and we see here more the scratch-resistance as an intrinsic system property, rather than a surface-cure effect.

Notice in this respect the excellent results, obtained with ditrimethylol-propanetetraacrylate and dipentaerythritolpentaacrylate.

The intrinsic flexibility of ether-monomers places these at the low end in this series.

Mandrel

Mandrel results are very sensible for crosslink density.   That is the very first obvious conclusion that can be drawn from these results.

At the high end we find the real bulky monomers, *i.e.* monomers with multiple rings: isobornylacrylate and ethoxylated bisphenol A diacrylate.

Cure Speed

Even within this series dipentaerythritolpentaacrylate turns out to be the most reactive monomer.   This becomes even more valuable when you realize that we took 6 as the functionality in calculating its formulation.

Despite the fact that we did not notice its effect in the pencil hardness, in this test the ether-type of monomers show the highest reactivity, due to the presence of abstractable hydrogens.   This is the most obvious in those cases where these hydrogens are also sterically readily approachable.

Glass Transition Temperature

Note the extremely low Tg found for HDDA, which is very much according to other results found in this series for HDDA (*e.g.* tensile, elongation). This shows once more that one of the explanations of the special properties of HDDA is a result of its very low molecular weight. At the high end are isobornylacrylate and THEIC triacrylate, showing the influence of ring structures on the Tg. Note also the surprisingly low Tg found with dipentaerythritol-pentaacrylate, showing that this monomer combines its very high functionality, resulting in extreme reactivity with an intrinsically low Tg as opposed to, *e.g.* pentaerythritoltetraacrylate.

Series 3: Constant Viscosity

The composition used in this series can be derived from the corresponding graph. Unfortunately the viscosities obtained with 2-ethoxyethoxyethylacrylate and isodecylacrylate were too low (being much lower than 360 mPa s, the lower limit) to be compared with the others. The results can be found in the graphs following the text.

Solvency

From the results we obtained, it is obvious, that 2-ethoxyethoxyethylacrylate and to a lesser extent isodecylacrylate are unsurpassed within this monomer range as far as solvency is concerned. Laurylacrylate, showing compatibility problems at higher addition levels, is also an excellent reactive diluent, being completely compatible at this level.

The solvency of HDDA – well known for a long time of course – is also excellent. Furthermore, we see that generally speaking the ether-monomers, more specifically the linear ethers, are the best in solvency.

At the high end we find the monomers which are not used as reactive diluent, but as crosslinkers or modifiers.

Tensile at Break

Unfortunately, there are a lot of formulations here which were too hard to prepare proper samples from. Looking at the tensile we find here, as in the first series, a wide variation. The order parallels the one found in the first series quite closely with a few exceptions. The tensile found for TPGDA is much more in accordance with what one might expect than what was found in the first series. The same applies to ethoxylated bisphenol A diacrylate.

It is obvious that a combination of good solvency and an intrinsically high tensile (or any other property as well), shows clearly at application viscosity. However, it is also obvious from the results obtained, that if solvency is poor, but the intrinsic properties of the monomer are very distinct, this monomer is very effective as a modifier. Look, *e.g.* at isobornylacrylate, ethoxylated bisphenol A diacrylate and tris (2-hydroxyethyl)isocyanurate triacrylate.

Elongation at Break

These results follow the same pattern as we saw earlier in series 1. Two monomers show a remarkable high elongation without being monofunctional: polyethyleneglycol 400 diacrylate and highly ethoxylated TMPTA.

Konig Hardness

It is shown clearly by those results that laurylacrylate gives rise to by far the most flexible formulations. Furthermore, it is clear that ether–types of monomers, more specifically those having the highest molecular weights, yield the most flexible systems. Note that once more good hardness is found in formulations containing isobornylacrylate.

Pencil Hardness

Generally speaking, we find here at the low end high viscosity monomers containing no (or sterically hindered) abstractable hydrogens. The result obtained with pentaerythritol triacrylate seems a bit odd.

Mandrel

Even at constant viscosity, one encounters mandrel–values ranging from below 3 mm for the very flexible, up to above 25 mm for the (very) hard systems. Remarkably flexible are the highly ethoxylated trimethylolpropane triacrylate, laurylacrylate, and also polyethylene glycol 400 diacrylate.

Cure Speed

The fact that all monomers containing at least one ethylene glycol–unit can be found at the high end in terms of reactivity is illustrative. It is a bit surprising though, to find laurylacrylate being the highest in reactivity. However, this can easily be explained by the fact that in this particular formulation, the lowest amount of TPDGA and the highest amount of highly reactive epoxy–acrylate oligomer can be found. Keeping in mind that in the other two series, 2–ethoxyethoxyethylacrylate was found at the high end of reactivity, and given that it shows excellent solvency, it seems safe to assume that a formulation based on this monomer, would have turned out to be the most reactive in this series.

Glass Transition Temperature

The lowest Tg of all at constant viscosity gives PEG400 diacrylate, followed by highly ethoxylated TMPTA. This shows that ethyleneglycol units are very efficient in lowering the Tg, because it is exactly those two monomers, containing the highest number of those units. Also the very low Tg of dipenta-erythritolpentaacrylate is remarkable. Note at the high end THEIC triacrylate:

its flat and highly polar ring structure is responsible for this highest glass transition temperature, despite the fact that in order to overcome the high viscosity of THEIC triacrylate, the most TPGDA of all 23 formulations had to be used.

CONCLUSIONS

Again, the amount of data generated during this study is too big to enable a thorough discussion within the scope of this paper. It has been set up to show the formulator the general correlations between chemical structure, and performance of a given monomer. Thinking along the lines shown will help you solve your formulating problems. On the other hand, data given in the graphs for each individual monomer gives you a good feel for what this particular monomer would do in your formulation.

In terms of testing, it is obvious from this study that tensile and elongation at break, parallel, to a considerable extent, the konig hardness and mandrel. respectively.

In terms of flexibility the most flexible systems can be found among products having linear structures. Monomers containing ether–linkages are even much more flexible because of improved freedom of movement and also because of chain transfer reaction as a result of the presence of abstractable α–hydrogens. The harder systems on the other hand, can be formulated using monomers having limitation of freedom of movement built in the molecule, either by branching or, more severely, ring structures,

In terms of cure the importance of the presence of sterically approachable abstractable hydrogens in the monomer–molecule, were shown very clearly. Keeping in mind that those monomers are to be found among the second and third generation of monomers having low skin–irritancies, you find here another real performance advantage of the use of those monomers.

In terms of solvency/viscosity we found clearly the influence of both molecular weight and molecular structure: low molecular weight and the lack of mobility–limiting factors like branching or ring–structures increase the solvency of a monomer.

The Tg study revealed a strong correlation between crosslink density and Tg. The Tg-lowering effect of ethyleneglycol–units and the Tg–increasing effect of ring–structures was clearly shown .

ACKNOWLEDGEMENTS

The author would like to express his appreciation to Mr Max Holderbusch, who really did a magnificent job in preparing most of the formulations and collecting most data in this study, and Mrs Eva Gregorik and Mrs Christine Durant who did the DSC and DTA-study.

BIBLIOGRAPHY

1   S.R. Kerr, *I. Radiat. Curing*, 1984, **11** (3).
2   B.K. Christmas and E.G. Zey, *Speciality Chemicals*, February, 1988.
3   B.K. Christmas and E.G. Zey, *Speciality Chemicals*, August, 1988.
4   H.C. Miller, Radtech '88-North America Conference Papers.

Explanation of the Tables

f   =   Functionality
MW   =   Molecular Weight
Ratio   =   Polyethylene glycol 400 diacrylate/ditrimethylolpropane
               tetraacrylate

VISCOSITY

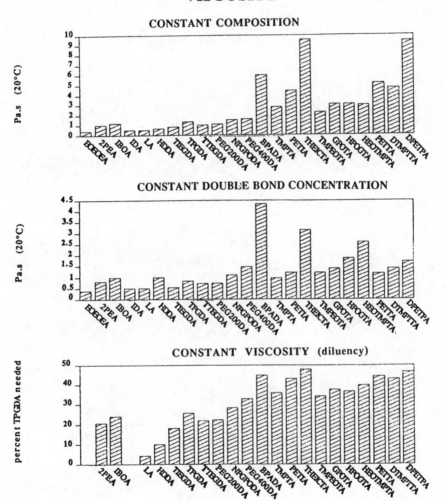

CONSTANT COMPOSITION

CONSTANT DOUBLE BOND CONCENTRATION

CONSTANT VISCOSITY (diluency)

TENSILE STRENGTH

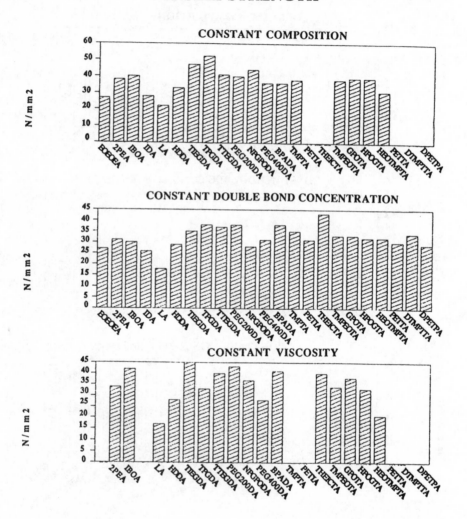

ELONGATION AT BREAK

CONSTANT COMPOSITION

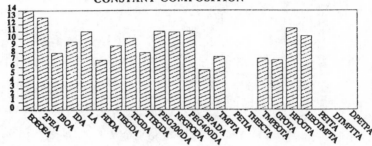

CONSTANT DOUBLE BOND CONCENTRATION

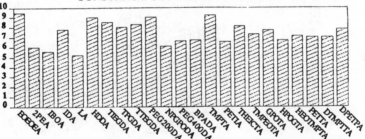

CONSTANT VISCOSITY

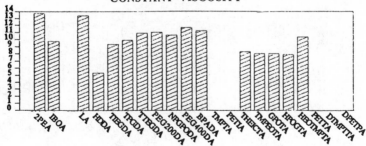

KONIG HARDNESS

CONSTANT COMPOSITION

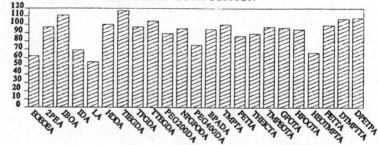

CONSTANT DOUBLE BOND CONCENTRATION

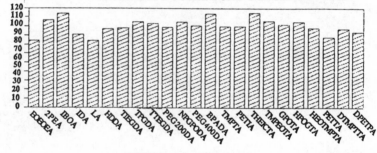

CONSTANT VISCOSITY

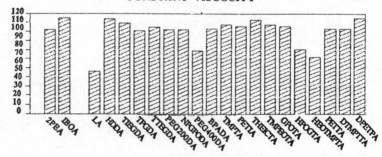

PENCIL HARDNESS

CONSTANT COMPOSITION

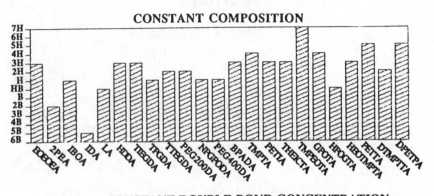

CONSTANT DOUBLE BOND CONCENTRATION

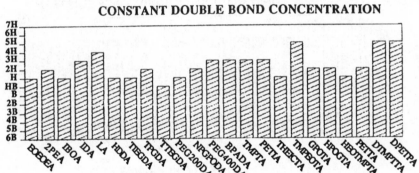

CONSTANT VISCOSITY

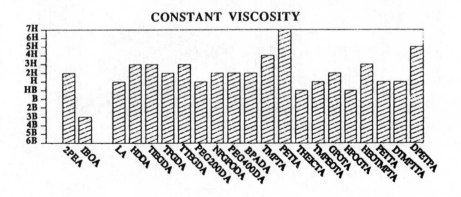

MANDREL

CONSTANT COMPOSITION

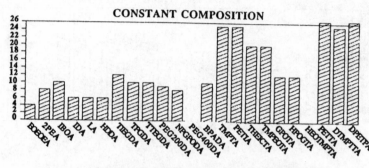

CONSTANT DOUBLE BOND CONCENTRATION

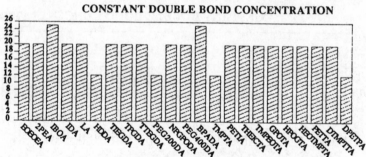

CONSTANT VISCOSITY

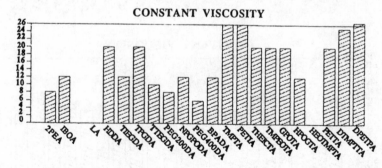

CURE SPEED

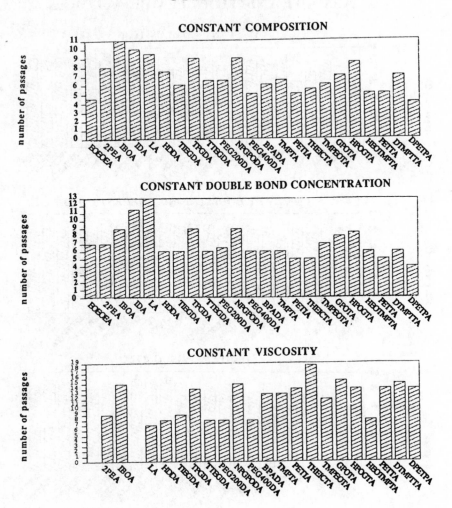

GLASS - TRANSITION TEMPERATURES

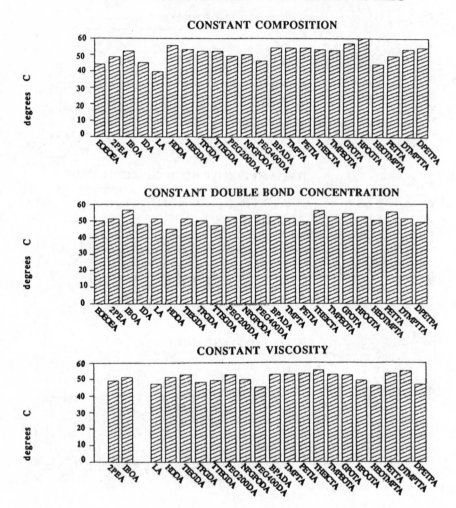

Trends in Low Viscosity Acrylate Resins for Radiation Cured Formulations

A. Howell

ANCHOR CHEMICAL (UK) LTD., CLAYTON, MANCHESTER M11 4SR, UK

Radiation Curing techniques have developed considerably since the process became commercially viable in the early 1970's. Raw material development has also advanced during this time, partly as a result of new technology, but more importantly due to restrictions put on many materials relating to Health and Safety aspects.

Present day formulations still rely heavily on the principle that a system comprises oligomers, monomers, photoinitiators and additives. In many cases the oligomer components are comparable to those available 10-15 years ago, these include examples such as epoxy acrylates and urethane acrylates. Traditionally, these types of material are very viscous and require dilution with reactive monomers to achieve the desired viscosity and reactivity.

A number of monomers previously utilised have been eliminated due to factors such as toxicity, odour and skin irritation. Although monomers used at present perform very well, in certain areas high levels of safer monomers are required to achieve workable viscosities and this can detract from the film properties imparted by the base oligomer.

New opportunities are now available to examine novel systems specifically designed to use either a minimum amount of monomer or eliminate the use of monomers completely. These include low viscosity monomer free oligomers, water thinnable oligomers and in special cases oligomers containing volatile, non-reactive solvents.

This paper is designed to give an insight into these novel materials and stimulate ideas as to where these products can be used.

In the USA and Japan, the markets and the requirements for raw materials differ considerably, as does the balance between UV and EB

Curing techniques. These markets however, along with Europe, share the same growth potential over the next few years.

EUROPEAN MARKET - 1989

		TONNES
EPOXY ACRYLATE CONSUMPTION	-	3,500
URETHANE ACRYLATE CONSUMPTION	-	1,500
POLYESTER ACRYLATE CONSUMPTION	-	2,500
MONOMER CONSUMPTION	-	9,500
ESTIMATED GROWTH POTENTIAL OVER NEXT 3 - 5 YEARS	-	8 - 12% BY VOLUME

For 1989, the European market analysis indicated that 3,500 tonnes of epoxy acrylate was consumed, the bulk of this material was used on either wood or paper substrates.

Urethane acrylate usage was less than half that for the epoxy acrylate whereas polyester acrylate consumption was mid-way between the two.

Monomer consumption was greater than all three of the oligomer types combined at 9,500 tonnes. The bulk of this usage was made up of materials such as TPGDA, TMPTA, HDDA and alkoxylated trifunctional monomers.

WHY LOW VISCOSITY SYSTEMS?

LIMITATIONS ON MONOMER DEVELOPMENTS

- RAW MATERIALS

- EUROPEAN LEGISLATION ON MONOMERS AND POLYMERS

- COSTS INVOLVED

Although it would be presumptuous to say that raw material sources have been exhausted, the scope for innovative monomer development is limited, even though an infinite number of ethoxylated or propoxylated polyols can still be promoted as novel materials.

Concerns in the UK relating to skin irritancy and the SBPIM guidelines mean that many monomers perfectly acceptable a few years ago, cannot now be used in radiation curable inks. This has had a knock-on effect in the varnish industry. The trend to toxicologically safe products will continue to restrict the use of existing monomers.

European legislation which sets guidelines for what constitutes a monomer or polymer, makes it more attractive to have a material classified as the latter rather than the former. As an estimate, it is likely to cost in excess of $60,000 to test a new development monomer satisfactorily. This is mandatory on products that are classified as new monomers by the European Commission.

In the case of polymers, this is not mandatory though it is still essential that tests be carried out to ensure that these materials are safe.

OPTIONS OPEN TO RESIN MANUFACTURERS
AND FORMULATORS

- MONOMER DILUENTS
- VOLATILE SOLVENTS
- LOW VISCOSITY MONOMER
- FREE OLIGOMERS
- WATER THINNABLE OLIGOMERS

Manufacturers of resins and monomers have certain options and avenues of approach to ensure their products are received enthusiastically. To attain low viscosities, simple measures can be taken such as diluting with monomers or even organic non-reactive solvents. A novel approach is to look at specifically designed monomer free oligomers or even more revolutionary, the manufacture of water thinnable oligomers.

HISTORICAL DEVELOPMENTS

MONOFUNCTIONAL	DIFUNCTIONAL	POLYFUNCTIONAL
HEA	BDDA	TMPTA
EOEOEA	NPGDA	PETA
PEA	HDDA	EOTMPTA
EHA	TPGDA	GPTA
IDA	ALKOXYLATED BPADA	EO PETA
IBA	ALKOXYLATED NPGDA	

Traditionally, monofunctional monomers were classed as having unacceptable odours and high skin irritation, this is misleading as many of the newer types of monofunctional diluents are considerably safer than their predecessors.

Monomers such as phenoxy ethyl acrylate (PEA), ethyl hexyl acrylate (EHA) and ethoxy ethoxy ethyl acrylate (EOEOEA), though very efficient diluents, all have problems with odour, skin irritation and can be potentially toxic.

Monomers such as isodecyl acrylate (IDA) and isobornyl acrylate (IBA) are more acceptable monomers for use in todays applications. IDA is a good example of a relatively inexpensive monomer with good dilution characteristics yet incapable of improving the cross linked density of cured formulations. Though not as fast to cure as polyfunctional monomers it can give better flexibility and adhesion on many difficult substrates.

There has been a number of diacrylate monomers introduced onto the market over the years. Many of the earlier ones also had problems with odour and skin irritation or were classed as carcinogens. Butane diol diacrylate (BDDA) and neopentyl glycol diacrylate (NPGDA) were both ruled out on a combination of these factors.

Hexane diol diacrylate (HDDA) on the other hand is an excellent monomer especially on difficult substrates such as plasticised PVC, with good viscosity cutting characteristics and fast cure speed. However, its high Draize rating severely restricts its use. HDDA is still used extensively in some areas of application in Europe but in the UK it cannot be used in inks and it has an uncertain future.

Tripropylene glycol diacrylate (TPGDA) has been available for many years, it has a relatively low Draize rating and performs well. It is one of the few monomers which continues to meet current Health and Safety requirements.

Alkoxylation has been an option open to produce safer monomers and is used with both neopentyl glycol and Bisphenol A.

However, the poor viscosity cutting characteristics of the latter coupled with its aromatic nature, restricts the use of this product in many new applications.

Polyfunctional materials are generally used for their speed of cure and ability to enhance the cross link density. Many problems were apparent with trimethylol propane triacrylate (TMPTA) and the tri and tetra acrylate esters of penta erythritol (PETA). However, both are still used in some areas with large tonnages of TMPTA used especially outside the UK.

Alkoxylated aliphatic monomers are now far more common and have more acceptable Draize ratings. Many can also be used without problems in the ink industry especially in litho inks where they are ideal for achieving the desired water balance.

Varying degreees of ethoxylation and propoxylation on trimethylol propane, glycerol and pentaerythritol (EOTMPTA, GPTA, EOPETA) ensure that many options are open to the monomer manufacturer. Conversely the action of the monomer as an efficient diluent can be severely diminished by higher degrees of alkoxylation and again this can detract significantly from the excellent properties imparted by the base oligomers.

PROBLEMS WITH MONOMER DILUENTS

- ODOUR
- SKIN IRRITANCY
- CARCINOGENIC?

- HIGH FILM SHRINKAGE
- LOSS OF ADHESION
- VARIABLE CURE SPEEDS
- CHANGE IN OVERALL FILM PROPERTIES

To summarize, skin irritation and toxicology must be acceptable before monomers can even be considered for use in this technology. Low odour and low taint are becoming increasingly important demanding that high odour products be avoided.

On cure monomers generally have quite high shrinkage values resulting in poorer adhesion, they can also have an adverse effect on their cure speeds depending on their functionality and molecular weight.

In future, no doubt many new monomers will become available. The evidence suggests the majority of these novel or specialised monomers may be expensive and this will prohibit their use in many areas of industry where competition is fierce.

New ways have to be found to overcome the present restrictions.

USE OF VOLATILE ORGANIC SOLVENTS

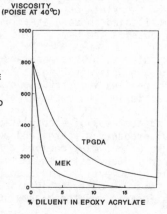

- RAPID VISCOSITY REDUCTION

- MAY NOT REQUIRE REMOVAL STAGE

- OLIGOMER PROPERTIES EMPHASISED

- HIGH MWT EPOXIES/NOVOLACS
  CAN BE USED

- OLIGOMERS CAN BE SUPPLIED
  IN SOLVENT

The use of non-reactive volatile organic solvents as diluents is found in special application areas. The graph shows the rapid viscosity reduction found when MEK is compared to TPGDA in combination with an epoxy acrylate resin. Although the effects on viscosity reduction are excellent, the negative aspects are more important. These include fire hazards, an extra solvent removal stage, problems with solvent emissions and odour.

In cases where coatings are applied by spray, where low viscosities are required of less than one poise, small amounts of solvent can be used in preference to the use of large proportions of reactive monomers.

A major advantage in the use of non-reactive solvents is that an oligomer, tailor made for a specific application, can be applied to a substrate and after solvent removal, impart exactly the properties that the resin has been designed to offer. This is especially applicable to high molecular weight solid acrylate esters of epoxy resins and epoxy novalacs. With these types of material it is common to supply the oligomers in a solvent, though glycol ethers would normally be used rather than MEK. Examples of this are in developable etch resists and solder resists for the electronics industry.

DEVELOPABLE ETCH AND SOLDER RESISTS

- USUALLY SUPPLIED AT 50-70% SOLIDS

- GLYCOL ETHER SOLVENT

- SCREEN PRINTED COATING

- TOUCH DRY UNCURED FILM AFTER SOLVENT REMOVAL

- GOOD DEFINITION EXPOSURE FROM PHOTOMASK

- UNCURED RESIN DEVELOPABLE BY SOLVENT OR BASE

- MAY REMAIN INTACT (SOLDER) OR SOLVENT

  REMOVED (ETCH) RESISTS

These are good examples where monomer free coatings are applied at normally 50-70% solids. The glycol ether solvents can be changed to give the required evaporation rate. The systems are normally screen printed and after solvent removal, touch dry films are produced when completely uncured.

Following exposure through a photomask, the unexposed areas can be removed by either solvent or base and the cured resin forms the basis of a solder resist. It can be removed at a later stage by an aggressive solvent in the case of an etch resist.

In less specialised areas where large quantities of materials are utilised, this type of approach cannot be used. Solvents are not viable and other ways have to be found to produce monomer free oligomers. In general terms the bulk oligomers are usually of the three following types.

CONVENTIONAL OLIGOMERS

	EPOXY ACRYLATE	URETHANE ACRYLATES	POLYESTER ACRYLATES
VISCOSITY	HIGH	HIGH	VARIABLE
DILUTION WITH MONOMERS	EASY	EASY	EASY
VISCOSITY CUTBACK	GOOD	FAIR	GOOD
CURE SPEED	VERY FAST	VARIABLE	VARIABLE
FLEXIBILITY	POOR	GOOD	VARIABLE
CHEMICAL RESISTANCE	EXCELLENT	GOOD	GOOD

The standard epoxy acrylate is a well known and established raw material. In its undiluted form it is extremely viscous although it is soluble in most monomers and the rate of viscosity reduction is very rapid. The resin itself shows fast cure and has excellent chemical resistance though the cured oligomer is very brittle.

With urethane acrylates many different types are available, they are all high viscosity but can be diluted with most monomers, however the viscosity reduction rate is usually inferior to that of epoxy acrylates due to the chemical nature of the oligomer. Cure speeds vary considerably but chemical resistance and flexibility are generally good.

Even more variables are possible with polyester acrylates and they can be tailor made to suit an application. Many materials have been examined in the past to see whether they can be compared to epoxy acrylates on formulating but combinations of cure speed, chemical resistance and overall performance are difficult to match.

In designing a suitable yet novel alternative, certain criteria had to be kept in mind:

MONOMER FREE LOW VISCOSITY
EPOXY ACRYLATES

- TOXICOLOGICALLY ACCEPTABLE
- EASY HANDLING
- FAST CURE SPEEDS
- EXCELLENT GLOSS
- LOWER SHRINKAGE ON CURE
- BETTER ADHESION TO SUBSTRATES
- REDUCED SOLVENT LEVELS IN SPRAY COATINGS

The materials have to be acceptable toxicologically, this also covers areas such as odour before and after cure in an attempt to produce low odour, taint free films. Ease of handling is also a major consideration, an easy flowing monomer free oligomer has excellent potential if it also conforms to the fast cure speeds and good gloss coatings associated with epoxy acrylates. A further advantage is the promise of lower shrinkage on cure and hence most likely better adhesion. If spray coatings are required, it is still likely that volatile solvents may be needed but only in very low concentrations.

LOW VISCOSITY OLIGOMERS - DEVELOPMENT

	RESIN 1	RESIN 2	RESIN 3	RESIN 4
MOLECULAR WEIGHT	530	520	450	470
VISCOSITY (POISE AT 25° C)	1000+	12	7	25
CURE SPEED (fpm)	100+	40	60	90
CHEMICAL RESISTANCE (MEK RUBS)	200+	120	200+	200+
TENSILE STRENGTH (Nmm^{-2})	-	1	2	9.3
TENSILE MODULUS (Nmm^{-2})	-	32	57	223
ELONGATION (%)	-	4	3	7.5

A series of resins has been examined to produce a good low viscosity material capable of comparing favourably with commercially available epoxy acrylates.

Resins 1-4 inc. are all approximately the same molecular weights. The variation in viscosity of the undiluted resins is, however, significant. Resin 1 is a standard aromatic epoxy acrylate, it has a viscosity in excess of 1,000 poise at ambient temperature and shows excellent speed of cure and resistance to MEK. Unfortunately, due to the brittle nature of the cured film the tensile strength, modulus and elongation cannot be measured easily.

Resin 2 was manufactured to try to overcome many of the restrictions found with the Resin 1 type. This product was, however, much slower to cure and the chemical resistance was inferior. Although the modulus showed this oligomer to have good flexibility, it had poor properties highlighted by a negligible tensile strength and elongation. The resin was of poor colour with an acrid odour. Although many of the restrictions from the standard epoxy had been overcome, they were replaced by others.

Resin 3 was aimed at improving the product colour and residual odour, this was successful yet the tensile strength and elongation were still below par. The modulus figure showed a further improvement on flexibility with the chemical resistance improving.

Finally Resin 4 has all the best facets of the resins under study. Cure speed is around 90 fpm and the cured film has excellent chemical resistance. It has a useful tensile strength of 9.3 Nmm^{-2}, good flexibility and a good elongation at 7.5%, not normally associated with an epoxy acrylate. The product has a colour value of 4-6 Gardner and a low viscosity at 25 poise on the monomer free form. The product has a much better odour than the lower viscosity resins and promises to be very useful in a number of areas especially low odour and low taint coatings.

COMMERCIALLY AVAILABLE RESINS

	A	B	C
EPOXY ACRYLATE A	50	-	-
LOW VISCOSITY EPOXY ACRYLATE B	-	80	-
POLYESTER ACRYLATE C	-	-	90
TPGDA	50	20	10
PHOTOINITIATORS	5	5	5
RESIN VISCOSITY (POISE AT 60° C)	50	-	-
RESIN VISCOSITY (POISE AT 25° C)	-	25	11.8
FORMULATION VISCOSITY (POISE AT 25° C)	6.3	7.2	5.4
CURE SPEED (fpm)	80	100	90
FLEXIBILITY (180° CREASE)	POOR	GOOD	GOOD
CHEMICAL RESISTANCE (MEK RUBS)	200+	185	168
GLOSS LEVEL (60° GLOSSMETER)	96.4	95.4	100.6

In the table above, a comparison is made of the performance of a standard epoxy acrylate and a polyester acrylate specifically designed for use in overprint varnishes, with the low viscosity monomer free oligomer.

Formulation A is comparable to many simple overprint varnishes available without the added flow aids, slip aids, etc. To achieve a viscosity of 5-8 poise, the monomer level has to be at 50%.

The polyester acrylate based formulation C had a low initial viscosity. The level of monomer required was also low and its performance was good. Cure speed at 90 fpm was better than the epoxy acrylate at 50% dilution. The gloss level was the best of the three examples but the resistance to MEK was the poorest.

The use of this polyester acrylate to emulate the performance of an epoxy acrylate, has the major drawback that it is available commercially but is considerably more expensive than the standard epoxy acrylate or low viscosity epoxy acrylate.

Formulation B using the new low viscosity resin has a maximum of 20% monomer added, cure speed and performance are good and the added flexibility is a bonus, adhesion is also marginally better. The base resin is easy to use initially and does not require warming prior to formulation, also wastage is minimal. The gloss level reached was marginally worse than the standard epoxy acrylate. This product however excelled on flexibility, adhesion and cure speed.

WATER THINNABLE RESINS

- AN INEXPENSIVE YET EFFICIENT DILUENT
- PERFORMANCE COMPARABLE TO UNDILUTED OLIGOMERS
- EXCELLENT GLOSS
- GOOD CHEMICAL RESISTANCE
- VERSATILITY IN ROLLER COATING/SCREEN PRINTING
- BETTER CLEAN UP PROCEDURES
- REDUCED ECOLOGICAL PROBLEMS

Another way of eliminating monomers from formulations, is to use a cheap, efficient, colourless and odourless diluent - water.

The concept of water as a diluent is not new and although litho processes would not be responsive to water as a diluent, many other processes would be.

The main advantage of water thinnable systems is that the oligomers can be tailor made to suit special applications. This class of product offers high reactivity and the materials can be cured either undiluted or with variable amounts of water added as a diluent. They perform well as overprint varnishes exhibiting high gloss levels plus good water and chemical resistance. Materials are available which can be used in other low viscosity or higher viscosity applications.

An added advantage is that these resins can be used in water free formulations, where the affinity of these products for water means they can be used to ensure easy clean up of machinery and eliminate the use of organic solvents in this procedure.

SOLUBILISING UV CURABLE RESINS GIVES
SCOPE FOR DIFFERENT APPLICATIONS

OVERPRINT VARNISH	SCREEN PRINT VARNISH
LOW WATER UPTAKE	MODERATE-HIGH WATER UPTAKE
FAST CURE SPEED	INTERMEDIATE CURE SPEEDS
HIGH GLOSS	VARIABLE
GOOD CHEMICAL RESISTANCE	- GLOSS
GOOD WATER RESISTANCE	- CHEMICAL RESISTANCE
	- WATER RESISTANCE

Obviously, requirements vary considerably. Taking the two extremes; a normal type of overprint varnish uses a low viscosity in the region of 3-7 poise, low water uptake in the oligomer is advisable and the prime requisites are fast cure, high gloss and good resistance to water and MEK. These can be achieved easily.

Conversely, for screen printing the requirements are very different. A moderate or high water uptake enables a lower film weight to be coated this enables screen to move into process printing. Fast cure speeds are not as important and gloss, chemical resistance and water resistance although important are not as critical as in overprint varnishes.

COMMERCIALLY AVAILABLE
WATER THINNABLE UV CURING RESINS

	RESIN X	RESIN Y
INITIAL VISCOSITY (POISE AT 25° C)	37	550
MAXIMUM WATER UPTAKE (%)	25	60
MINIMUM VISCOSITY (POISE AT 25° C)	1.5	0.5
CURE SPEED (fpm)	100	80
WATER RESISTANCE (DOUBLE RUBS)	200+	90
MEK RESISTANCE (DOUBLE RUBS)	200+	200+

In the above table, the first product on this slide, Resin X, is on examples of a water thinnable oligomer. At ambient temperature it flows easily and contains no water or added monomers. It can be diluted to approximately 75% solids at which stage the homogeneous solution will have a viscosity of 1.5 poise. A typical cure speed on a low viscosity varnish will be 100 fpm and it exhibits excellent resistance to both MEK and water.

Resin Y is designed for use in areas such as screen printing. It has a high initial viscosity at 550 poise yet will readily dilute to a minimum of 40% solids at which stage the viscosity will be less than one poise. Cure speed of a typical varnish at say 20-30 poise is 80 fpm, resistance to MEK is excellent and to water it is still good.

RESIN X IN A ROLLER COATING VARNISH FORMULATION

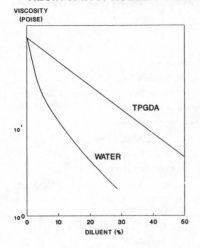

VISCOSITY
(POISE)

TPGDA

WATER

DILUENT (%)

RESIN X	80
WATER	20
IRGACURE 184	2
BENZOPHENONE	2
COAT THICKNESS (microns)	6
CURE SPEED (fpm)	100
ADHESION ON BOARD (%)	100
180° CREASE	GOOD
WATER RESISTANCE (DOUBLE RUBS)	200+

The above graph shows that Resin X can be diluted with conventional monomers such as TPGDA as well as water. In the case where the formulation is used for better clean up facilities then Resin X is compatible with a host of conventionally used additives.

Water, when used as a diluent, shows a dramatic viscosity cutting effect. Water is easily incorporated and the formulation shown with 20% added has a viscosity in the range of 3-5 poise. Irgacure 184 and Benzophenone have been found to perform as well as any photoinitiators in this type of system and a level of 2.5% of each on the total solids is recommended as a good starting formulation.

When coated onto litho printed board and cured with one medium pressure 80 Wcm^{-1} Mercury Lamp, a cure speed of 100 fpm is typical.

The cured film produces excellent gloss and adhesion. The flexibility as characterised by the 180° Crease test is very good - this would be impossible to achieve on a standard epoxy acrylate without the use of modifying monomers or flexibilisers. Chemical resistance to both water and MEK is excellent.

Resin X is designed for use as a low viscosity water thinnable oligomer primarily as a varnish on paper and board substrates. Alongside this, there are possibilities for its use in the wood coating industry. The potential for this material in inks has not yet been fully studied.

RESIN Y IN A SCREEN PRINT VARNISH FORMULATION

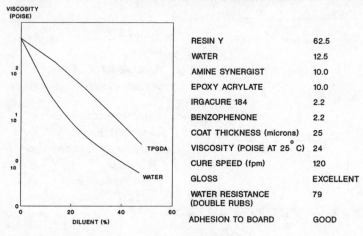

VISCOSITY
(POISE)

RESIN Y	62.5
WATER	12.5
AMINE SYNERGIST	10.0
EPOXY ACRYLATE	10.0
IRGACURE 184	2.2
BENZOPHENONE	2.2
COAT THICKNESS (microns)	25
VISCOSITY (POISE AT 25° C)	24
CURE SPEED (fpm)	120
GLOSS	EXCELLENT
WATER RESISTANCE (DOUBLE RUBS)	79
ADHESION TO BOARD	GOOD

Resin Y shows the same viscosity cutting trends as Resin X. With water this is very rapid and Resin Y can take up much more water, bringing the solids content down to 40%, and thereby producing a very low viscosity. Again this material is compatible with reactive diluents such as TPGDA but the viscosity cutting is less efficient.

The formulation shown was designed as a screen print varnish for paper and board. A product viscosity of 24 poise was chosen and the use of a small amount of a standard epoxy acrylate plus a synergist gave a faster cure speed at 120 fpm. Adhesion to board was good and gloss was excellent. Water resistance was also good though not as high as with the lower viscosity type of materials such as in the cure of Resin X.

Resin Y has been designed for different application areas than those intended for Resin X. In screen print applications the advantages are obvious and variable viscosity can be manipulated effectively to suit requirements. It is anticipated that the use of this material will again be best rewarded on paper, board and possibly wood.

In the ink area it has potential as a water based screen ink vehicle though this has yet to be optimised. At the other end of the scale it could be useful as a flexo ink vehicle in an effort to eliminate the use of volatile organic solvents and their environmental problems.

CONCLUSIONS

THERE IS A HIGH LEVEL OF INTEREST IN LOW TOXICITY,
LOW VISCOSITY RADIATION CURABLE MATERIALS.

- LOW VISCOSITY OLIGOMERS ARE NOW COMMERCIALLY
  AVAILABLE OFFERING FAST CURE, EXCELLENT FILM PROPERTIES
  AND LOW SKIN IRRITATION.

- WATERBASED UV CURABLE OLIGOMERS ARE AVAILABLE
  ELIMINATING THE NEED FOR MONOMERS AND SOLVENTS
  IN FORMULATIONS

- WATER THINNABLE SYSTEMS MAY BE USED TO ELIMINATE THE
  USE OF ORGANIC SOLVENTS FOR WASHDOWN PROCEDURES

There is a high level of interest in low viscosity, low toxicity, radiation curable resins. They offer many advantages over conventional systems currently available.

With low viscosity systems, whether water thinnable, or, specially designed low viscosity epoxy acrylates, they offer excellent performance and versatility.

Elimination, or the reduction in the use of monomers offers a new concept. Areas of interest are likely to be mainly in the paper and board industry with wood coatings also a possibility.

Cationic UV Curable Coatings: Contribution of Acrylate Oligomers to Final Film Properties

P.J-M. Manus

UNION CARBIDE CHEMICALS AND PLASTICS EUROPE SA, VERSOIX, SWITZERLAND

1.   ABSTRACT

Commercial acceptance of UV curing cationic technology is now well established in various areas of the radiation curing industry.

Hybrid systems based on cycloaliphatic diepoxide / polycaprolactone polyols and acrylated oligomers, either epoxy acrylate or urethane acrylate constitute a novel approach to the formulation of high performance coatings.

Optimal cure and final film characteristics in terms of solvent resistance and low film shrinkage can be achieved using the hybrid approach. These film properties will be compared to both cationic and free radical systems.

2.   INTRODUCTION

Cationic UV - initiated polymerization of cycloaliphatic diepoxide has enjoyed a significant growth over recent years in the radiation curing industry, in particular in the graphic arts and in metal can end coating. Based on its specific characteristics such as:

- no oxygen inhibition
- low volume shrinkage
- excellent adhesion to most plastics and metals
- high cure speed[1]
- absence of odor of the cured film
- low level of extractables[2]
- low irritancy / skin sensitization potential

cationic UV-initiated systems have proved to constitute a complementary technology to conventional acrylate based systems cured by free radicals. Even more significantly, however, cationic UV initiated technology has led to a breakthrough in converting some of the solvent based gravure inks and overprint lacquers to radiation curing technology for packaging applications.

Despite the continuous growth of the overall radiation curing industry[3] UV and EB coatings are still restricted to the following main market areas:

- wood coatings
- graphic arts
- electronics
- plastic coatings.

However, there is some indication that radiation curing is currently in an introductory phase in some large volume industrial sectors such as coil coatings.[4]   Other new applications have been presented at recent conferences and look promising, for example stereolithography.

Nevertheless, novel formulation approaches combined with the development of new functional oligomers[5] are required to expand further the scope of radiation curing technology and meet the technical requirements of new areas of application.   In this context, this paper outlines the coating performance obtained with hybrid systems based on cycloaliphatic diepoxide / polycaprolactone polyols and acrylated oligomers such as epoxy acrylates and urethane acrylates.

In recent years technical papers have been presented giving details of hybrid systems combining cationic and free radicals polymerization mechanisms, however, the majority of the presented data were related to the incorporation of acrylated monomers[6] and acrylated oligomers[7] to vinyl ethers or cycloaliphatic diepoxide / acrylated monomers systems.[8] [9]   Sauerbrunn[10] reported on the rapid cure response of a cycloaliphatic diepoxide / TMPTA hybrid system confirmed by differential photocalorimetry.

The purpose of this paper is to develop and expand the scope of formulations for hybrid systems through the use of acrylated oligomers incorporated into formulations based on cycloaliphatic diepoxide and polycaprolactone polyols.   Coatings produced by curing cycloaliphatic diepoxide alone tend to be brittle; however, polyols, in particular polycaprolactone polyols, have been demonstrated to improve the flexibility properties as well as the cure speed of such[11] systems through a chain transfer mechanism polymerization (fig.1).   The evaluation of acrylated oligomers will be undertaken in the context of this bench mark formulation so as to maximize the benefits of the dual ring opening of the epoxy functionality and the acrylic polymerization.   The epoxy acrylate modification of the cationic system is mainly focussed on metallic substrates whereas the urethane acrylate / cycloaliphatic diepoxide evaluation is targeted to fulfill the requirements of plastic coatings.

3.   EXPERIMENTAL

a)   Cationic Bench Mark Formulation

In this study, the reference UV curing cationic system is based on 3, 4 - Epoxycyclohexyl methyl - 3, 4 epoxycyclohexane carboxylate (ECC) in conjunction with a 300 molecular weight trifunctional polycaprolactone polyol - at an epoxy / hydroxy ratio of 5:1.   The curing performance and physical properties of this system are well documented.[12]

The cationic photoinitiator selected throughout the study is a triarylsulfonium hexafluoro antimonate salt in solution in propylene carbonate (50% active ingredient). Earlier work has also demonstrated the efficiency of this photoinitiator to polymerize cycloaliphatic epoxide / acrylate monomers without the presence of either an intermolecular H - abstraction or a Norrish type 1 photocleavage free radicals photoinitiator[13].

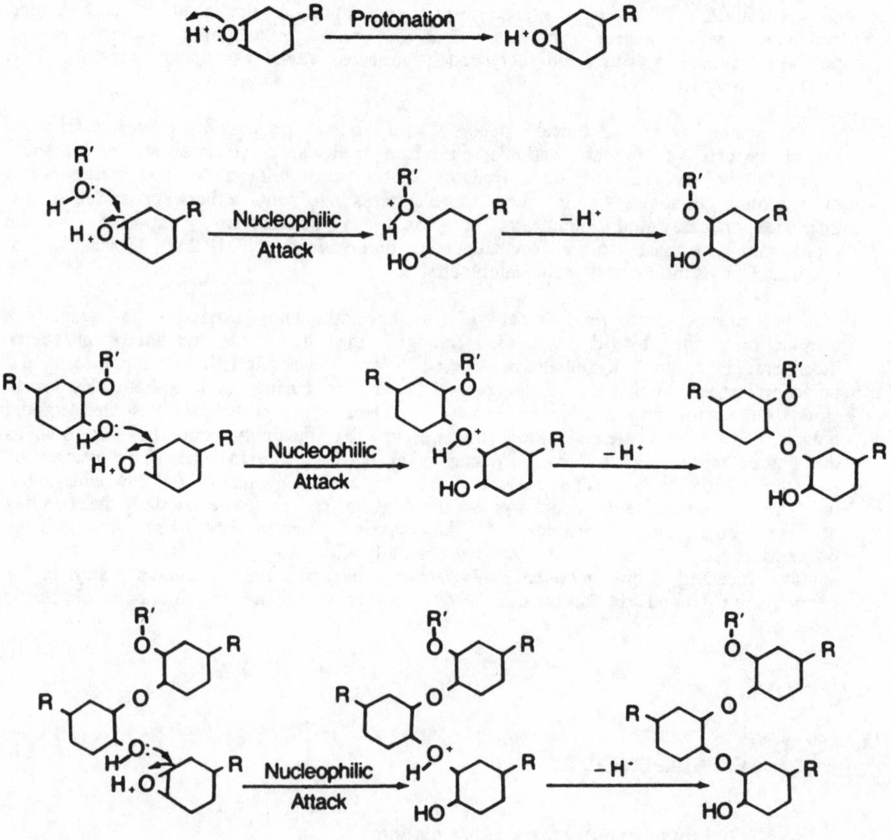

Figure 1

ECC [1]   85.00
Polycaprolactone triol 300 MW [2]   12.60
Triarylsulfonium salt [3]   2.00
Surfactant [4]   0.40

                                                   100.00

Viscosity: 440 cps at 25°C
             Brookfield LV2, spindle 2

(1)   UVR 6110:   Union Carbide   Cycloaliphatic diepoxide resin

(2)   TONE^R 0301: Union Carbide   Polycaprolactone triol

(3)   CYRACURE^R UVI 6974 :   Union Carbide   Photoinitiator

(4)   Silwet L720:   Union Carbide polyalkeneoxide - modified dimethyl
          polysiloxane surfactant.

Control Formulation UV Curing Cationic System

b)   Acrylated Oligomers

Epoxy Acrylate Oligomers.   These commercially available epoxy acrylate
oligomers were evaluated in combination with the control cationic system.
These products are listed in Table 1 along with their respective suppliers.
Epoxy A has the characteristic of having a low viscosity for this kind of
oligomer, and does not contain functional acrylate monomer as a diluent.

Table 1   Epoxy acrylate oligomers

Epoxy A   Low viscosity aromatic epoxy diacrylate
                        Viscosity: 30 000 cps at 25°C
                        Derakane XZ-86799 from Rahn AG.

Epoxy B   Aromatic epoxy diacrylate diluted with 20% aliphatic acrylate
                        monomer.
                        Viscosity: 500 - 1000 cps at 60°C
                        Photomer 3104 from Harcros Chemicals UK Ltd.

Epoxy C   Aromatic epoxy diacrylate diluted with 20% propylated glycerol
                        triacrylate monomer.
                        Viscosity: 80 000 - 120 000 cps at 20°C.
                        Setacure AP 570 from Akzo Synthese BV.

These acrylate epoxy oligomers were incorporated at three concentrations (i.e. 10, 20, and 40%) in the control cationic formulation containing 1% per weight of benzildimethyl ketal, Norrish Type I photocleavage photoinitiator (Irgacure 651 from Ciba Geigy) as indicated in Table 2. At the different levels of epoxy acrylate modification the cycloaliphatic diepoxide / polycaprolactone triol ratio was adjusted to maintain the epoxy / hydroxy ratio at a volume of 5:1. An alteration in the balance of the cationic ingredients would affect the coating performance, mainly with regard to cure speed.

Table 2

Test formulation for epoxy acrylate / cationic systems

	Min.	Max.
ECC	75.4	49.3
Polycaprolactone triol	11.2	7.3
Epoxy acrylate oligomer	10 -	40
Triarylsulfonium salt (50% solids)	2	
Benzildimethyl ketal	1	
Surfactant	0.4	
	————	
	100.00	

Epoxy / hydroxy ratio = 5/1*

*   Epoxy/hydroxy   ratio   was   calculated   as   follows:

A: Weight of Epoxy (ECC) in formulation/135
B: Weight of Polyol in formulation/100

Urethane Acrylate Oligomers.   Urethane acrylate oligomers were evaluated in similar conditions as a modification of the UV initiated cationic control system. S. Lapin[14] demonstrated   the   inhibition   effect   of   polar   functional   groups,   in particular with urethane linkages on the cationic polymerization of cycloaliphatic diepoxide   resin.   Other   published   data[15]   indicates   that   certain   urethane   acrylate oligomers would crosslink when combined with cycloaliphatic diepoxide resin, providing   that   the   cationic   photoinitiator   level   is   increased   to   overcome   the inhibition   effect.   An   excess   of   proton   production   with   a   higher   level   of cationic photoinitiator during the UV exposure partially compensates the "trapping" effect resulting from the urethane linkage.

As a result the photoinitiator system selected throughout the evaluation of urethane acrylate oligomers in hybrid formulations contains 4% per weight of triarylsulfonium salt and 1% per weight of benzildimethyl ketal.

Table 3

Test formulation for urethane acrylate / cationic systems

	Min.	Max.
ECC	73.6	47.6
Polycaprolactone triol	11.0	7.0
Urethane acrylate oligomer	10	- 40
Triarylsulfonium salt (50% solids)	4	
Benzildimethyl ketal	1	
Surfactant	0.4	

	100.00

Epoxy / Hydroxy ratio = 5/1.

Urethane acrylate oligomers are listed in Table 4. They were selected from the aliphatic class of urethane oligomers to retain the weatherability characteristics of the cycloaliphatic diepoxide resin system.

Table 4

Urethane acrylate oligomers

Urethane A

Pentafunctional aliphatic urethane acrylate diluted with 40% Monigomer[R] PPTTA.
Viscosity: 10 000 cps at 25°C
Genomer RCX-88 990 LV from Rahn AG.

Urethane B

Aliphatic polyester urethane triacrylate diluted with 25% TPGDA.
Viscosity: 10 0000 cps at 25°C
Photomer 6250 from Harcros Chemicals UK Ltd.

Urethane C

Aliphatic urethane hexafunctional acrylate, 100%.
Viscosity: 22 000 cps at 25°C
Photomer 6316 from Harcros Chemicals UK Ltd.

c)   Application and Curing Conditions

Application

Films were drawn down to a thickness of 12 microns on Bonder 132 panels for the general assessment of the physical properties of the coatings. Film properties were tested after 1 hour for MEK double rubs resistance and 24 hours for other coatings characteristics including a comparative MEK double rubs resistance to monitor the progression of the crosslinking density.

However, since the epoxy acrylate modification is mainly designed for metallic substrates, these formulations were also tested on tin plate reference Ancrolyt euronorm from Rasselstein GmbH and galvanized steel reference: Granodine 108.

A post bake schedule in an oven at 180° for 3 minutes was applied after the UV exposure for the latter two substrates.

The galvanized steel was degassed at 180°C for 3 minutes to remove any trace of humidity prior to the coating application. As a general rule, to minimize the testing variables, the metallic substrates were soaked in isopropanol and rinsed with acetone before being air dried.

Curing Conditions

Coatings were cured by exposure to a single 120 watt per cm ultra violet light source. This source is an electrodeless, air cooled, microwave activated mercury lamp (H bulb) produced by Fusion Systems. Rate of cure was measured for two coating thicknesses, 12 and 6 microns respectively, as the maximum conveyor belt speed in meters/minute to give a tack free surface (cotton ball test).

4. RESULTS AND DISCUSSION

a) Epoxy Acrylate Oligomers in Cycloaliphatic Diepoxide / Polycaprolactone Triol

Table 5 gives the physical properties on Bonder 132 panels of these hybrid systems compared to the reference UV-curing cationic formulation. The incorporation of epoxy acrylate oligomers increases the viscosity of the overall formulation, although Epoxy A which is a low viscosity oligomer without the presence of an acrylate monomer as diluent, has less influence on the viscosity of formulations than the other epoxy acrylate tested.

A feature of cationic UV curing technology is the extended storage stability of formulated systems which obviates the need for stabilizers. The presence of 40% epoxy acrylate oligomers together with the dual photoinitiator system leads to the formation of gels or significant increase of viscosity during storage over a period of one month at 50°C and room temperature.

Cure Speed. In all cases, curing occurs in the presence of oxygen; however, some differences in cure rate exist between the three epoxy acrylate oligomers evaluated. A marked cure speed enhancement is noticeable when Epoxy A is used at a level of up to 20%, compared to the cationic bench mark system (see Figure 2). However, addition of more than 20% of all the epoxy acrylate oligomers reduces the cure speed, despite the presence of the free radical photoinitiator in the system. It is worth noting the limited cure speed difference obtained between the two coating thicknesses evaluated, i.e. 6 and 12 microns. This may indicate that further optimization of the photoinitiator system (triarylsulfonium salt and benzildimethyl ketal) is necessary to boost the cure speed, especially for thin coatings.

Solvent Resistance. In common with all cationic systems, the

OLIGOMER	%	Viscosity @ 25° (cps)	CURE SPEED (m/mn)		MEK DOUBLE RUBS		PERSOZ HARDNESS	CROSSHATCH ADHESION		PENCIL HARDNESS	IMPACT[*] RESISTANCE		SHELF[**] LIFE STABILITY	
			6µ	12µ	+1H	+24H		+1H	+24H		Direct	Reverse	RT	50°C
Standard Formulation		410	30	25	5	60	420	100%	100%	2H	10	5	St.	St.
Epoxy A	10%	780	40m	30m	30	>100	423	100%	100%	3H	20	10	St.	St.
	20%	1110	35m	30m	40	>100	420	100%	100%	3H	30	10	St.	St.
	40%	2390	20m	20m	>100	>100	425	100%	100%	4H	35	10	G	G
Epoxy B	10%	840	35m	30m	20	70	422	100%	100%	3H	20	10	St.	St.
	20%	1335	30m	30m	35	>100	422	100%	100%	4H	15	5	St.	St.
	40%	3600	20m	20m	100	>100	425	20%	40%	4H	20	5	G	G
Epoxy C	10%	850	35m	25m	15	65	415	100%	100%	3H	25	10	St.	St.
	20%	1310	25m	20m	30	70	423	100%	100%	4H	45	10	St.	St.
	40%	3320	20m	20m	65	>100	425	70%	100%	4H	45	15	St.	St.

<u>Table 5</u>   PROPERTIES OF HYBRID SYSTEMS (EPOXY ACRYLATE / CYCLOALIPHATIC EPOXIDE) COMPARED TO STANDARD U.V. CURING CATIONIC FORMULATION

*   ISO TR 6272 Impact expressed in cm. kg.

**   St = Stable. No significant viscosity increase after one month storage.

G: Gelled:- 100% viscosity increase or gel formation

polymerization of hybrid formulations continues over time after the initial UV exposure. The evolution of the MEK double rubs resistance measured one hour and 24 hours after the UV exposure is illustrated in Figure 3.

The MEK double rubs resistance of hybrid systems is optimal after 24 hours lead time at room temperature. However, the major advantage of such hybrid systems is the improvement of solvent resistance one hour after the UV exposure. Although the "dark" cure continues for 24 hours with the cycloaliphatic diepoxide / epoxy acrylate combination, solvent resistance is significantly increased immediately after the UV exposure as compared to the bench mark cationic system.

This improvement in solvent resistance may be useful for certain applications, particularly if further processes are to be carried out on the cured films or if improved mar and scuff resistance are required.

Fig. 2   EPOXY ACRYLATE / CYCLOALIPHATIC EPOXIDE
                Compared to Standard UV Curing Cationic Formulation
                            MEK DOUBLE RUBS

MEK DOUBLE RUBS

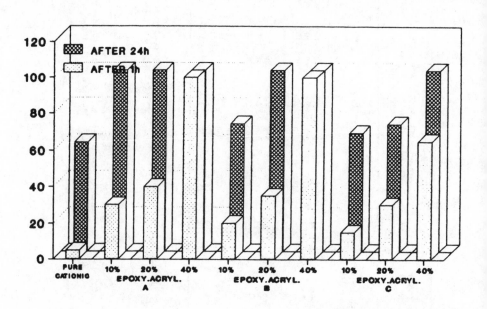

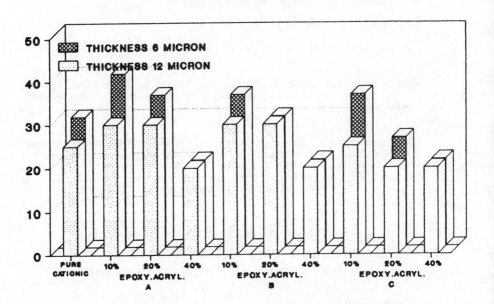

Fig. 3 EPOXY ACRYLATE / CYCLOALIPHATIC EPOXIDE
Compared to Standard UV Curing Cationic Formulation
CURE SPEED
UV FUSION 1 LAMP 120 W

Comparison With a Free Radical Epoxy Acrylate System. For the purpose of illustration, the bench mark cationic system and the hybrid formulation containing 20% of Epoxy A were evaluated against a 100% free radical epoxy acrylate coating. The acrylate formulation was selected from reference literature[16] on UV curing formulations for printing inks and coatings.

Fig. 4

EPOXY ACRYLATE FREE RADICAL FORMULATION

Epoxy A	50
TPGDA	30
TMPTA	10.6
Benzophenone	3
Benzildimethylketal[1]	3
N-Methyl diethanolamine	3
Surfactant	0.4
	100.0

(1) Irgacure 651 from Ciba Geigy

Viscosity Brookfield RVT, spindle 5   450 cps
at 25°C

Cure Speed m/min 1 lamp 120 w/cm   45 m/min
(film thickness: 12 microns on Bonder 132)

MEK double rubs resistance

      1 hour   20
      24 hours   50

Although the cure speed of the acrylate system is higher than for both
hybrid and cationic bench mark formulations, the MEK double rubs resistance
is inferior especially in comparison to the hybrid system as illustrated in
Fig. 5.

When assessed for adhesion properties on difficult metallic substrates such
as galvanized steel and tin plate, only the cationic formulation gave excellent
crosshatch adhesion on both substrates.   The combination Epoxy A /
cycloaliphatic diepoxide had good adhesion on galvanized steel whereas the
100% acrylate system did not adhere to either substrate (Table 6).

Fig. 5

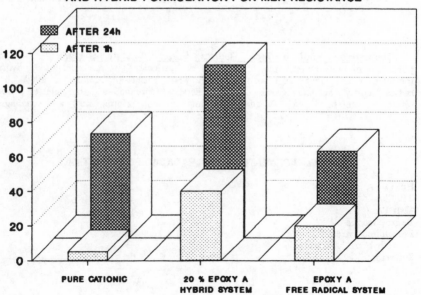

EPOXY ACRYLATE FREE RADICAL SYSTEM
COMPARED TO STANDARD UV CATIONIC FORMULATION
AND HYBRID FORMULATION FOR MEK RESISTANCE

Table 6   **Adhesion Characteristics on Metallic Substrates**

| | Galvanized Steel Granodine 108 | Tin Plate | |
	Crosshatch adhesion	Crosshatch adhesion	T-bend* in mm
UV Cationic bench mark formulation	100%	90%	10/100
Hybrid system 20% Epoxy A	100%	40%	30/100
Free radical system based on Epoxy A	0%	0%	------

Coatings were cured under UV at 20 m/min and post baked at 180°C for 3 minutes.

* ASTM D3281 - 73

To conclude, the incorporation of epoxy acrylate oligomers and particularly Epoxy A to cationic technology based on cycloaliphatic diepoxide / polycaprolactone triol leads to:

- superior solvent resistance characteristics compared to both pure cationic and pure free radical systems.

- superior adhesion characteristics on metallic substrates compared to the free radical formulation, although the predominant part of the hybrid formulation is of a cationic nature.

- a maximized crosslink density through continuing or "living" polymerization.

Differential Scanning Calorimetry measurements. These findings were confirmed by Differential Scanning Calorimetry (D.S.C) measurements carried out on cured films. The cationic bench mark formulation (ECC / polycaprolactone triol), the hybrid system containing 20% Epoxy A and the free radicals formulations based on Epoxy A were applied at a thickness of 12 micron on a release paper. These samples were cured in the same conditions under a single UV exposure at a cure speed of 20 m/min. After four hours ageing at room temperature, typically at 25°C, the cured samples were cut into plane discs and placed in a hermetic aluminium cell. The cell was heated from 25°C to 300°C at a heating rate of 10°C per minute. Each sample was tested twice. Fig. 6 represents the DSC plot of the three samples, four hours after the UV polymerization.

The cationic bench mark formulation shows an exotherm peak at 62°C, an enthalpy change of 31 joule/g. This peak represents the residual polymerization of unreacted species after the UV exposure which is typical of the living polymerization of UV-curing cationic systems. The lower residual enthalpy change obtained with the hybrid systems correlates well with the superior solvent resistance of the hybrid cured coating, achieved one hour after the polymerization.

In contrast, the free-radical cured film does not contain species or functionalities which undergo further polymerization with heat. The final crosslink density of an acrylate system is obtained almost immediately after the UV exposure.

D.S.C. plots on UV cured coatings of the cationic bench mark formulation and the hybrid systems did not show any enthalpy change for the samples which were kept at room temperature for three days. This confirmed that the optimal crosslink density developed during the post-cure period.

b) Urethane Acrylates

Table 7 presents the physical properties of urethane acrylate modified cycloaliphatic diepoxide / polycaprolactone triol systems. Formulation based on both Urethane B and C were unstable during storage without the presence of stabilizers. Improvement in impact resistance is noted with both urethane A and C without a detrimental effect on the surface hardness of the coatings.

Cure Speed. Coatings based on Urethane B did not cure in the presence of oxygen when a level of up to 20% urethane oligomer was incorporated to the cationic system. In terms of reactivity Urethane A and C gave superior cure speed compared to the bench mark cationic system, although this benefit is mainly achieved with a slight urethane acrylate oligomer modification. These results are illustrated in Fig. 7.

Solvent Resistance. The "living" polymerization of cationic chemistry still prevails as an increase in MEK double rubs resistance is achieved over a period of 24 hours for urethane A and urethane C. (see fig. 8)

As expected with urethane acrylates, the inhibition effect of polar groups, such as the urethane linkages, on the cationic mechanism limit the polymerization rate; however this could be overcome with a higher concentration of triarylsulfonium salt. Moreover, a careful selection of the urethane oligomer is necessary, to develop hybrid systems with adequate cure characteristics and film properties. The latter would also imply a knowledge of the additives and catalysts used during the urethane acrylate oligomer synthesis. Amine catalysts and inhibitors of a basic nature are often used in urethane chemistry.

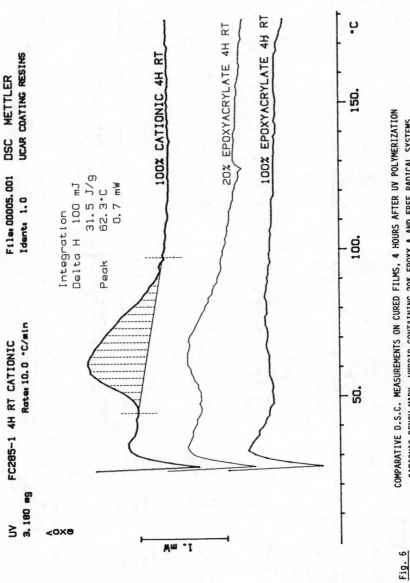

Fig. 6

COMPARATIVE D.S.C. MEASUREMENTS ON CURED FILMS, 4 HOURS AFTER UV POLYMERIZATION
CATIONIC BENCH MARK, HYBRID CONTAINING 20% EPOXY A AND FREE RADICAL SYSTEMS

OLIGOMER	%	VISCOSITY @ 25°C (cps)	CURE SPEED (m/min)		MEK DOUBLE RUBS		PERSOZ HARDNESS	CROSSHATCH ADHESION		PENCIL HARDNESS	IMPACT[‡] RESISTANCE		SHELF[**] LIFE STABILITY	
			6U	12U	+1H	+24H		+1H	+24H		Direct	Reverse	RT	50°C
Standard Formulation	-	410	30m	25m	5	60	420			2H	10	5	St.	St.
Urethane Acrylate A	10%	620	40m	25m	20	40	410	100%	100%	3H	20	10	St.	St.
	20%	810	20m	15m	30	>100	410	100%	100%	3H	20	10	St.	St.
	40%	1312	10m	5m	>100	>100	405	50%	100%	4H	20	10	St.	St.
Urethane Acrylate B	10%	750	15m	10m	7	20	410	100%	100%	H	30	10	G	G
	20%	1320	5m	N.C.	3	20	408	100%	100%	3H	30	10	G	G
	40%	3500	N.C.	N.C.	-	-	-	-	-	-	-	-	G	G
Urethane Acrylate C	10%	590	45m	35m	5	8	410	100%	100%	3H	20	10	G	G
	20%	840	20m	15m	10	70	415	100%	100%	3H	20	10	G	G
	40%	1440	10m	5m	>100	>100	415	100%	100%	4H	30	10	G	G

Table 7   PROPERTIES OF HYBRID SYSTEMS (URETHANE ACRYLATES / CYCLOALIPHATIC EPOXIDE) COMPARED TO STANDARD U.V. CURING CATIONIC FORMULATION

*   ISO TR 6272 Impact expressed in cm. kg.

**   St = Stable. No significant viscosity increase after one month storage.
     G = Gelled - 100% Viscosity increase or gel formation.
     N.C. - does not cure under UV exposure.

5.   CONCLUSION

Hybrid systems based on cycloaliphatic diepoxide / polycaprolactone polyols and modified with acrylated oligomers offer new formulation opportunities.   In this context, the incorporation of epoxy acrylates into a basic cationic formulation contributes to the improved solvent resistance of the cured coatings.

This study has demonstrated that a level of epoxy acrylate of up to 20% increases the cure response compared to pure cationic formulation.   Unlike free radical coatings, these hybrid coatings retain excellent adhesion characteristics on difficult substrates.

Selection of urethane acrylates needs fine tuning since this hybrid approach is more prone to cure inhibition and gel formation during storage; however, specific film properties can be achieved with this type of formulation.

Fig. 7   URETHANE ACRYLATES / CYCLOALIPHATIC EPOXIDE
COMPARED TO STANDARD UV CURING CATIONIC FORMULATION
CURE SPEED
UV FUSION 1 LAMP 120 W

CURE SPEED M/MIN

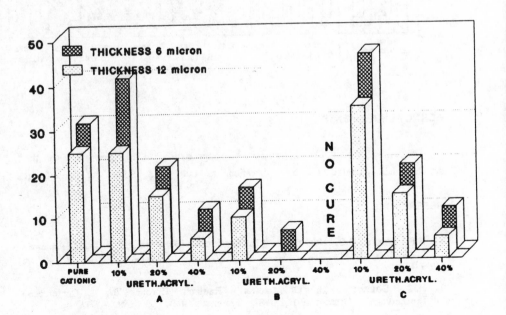

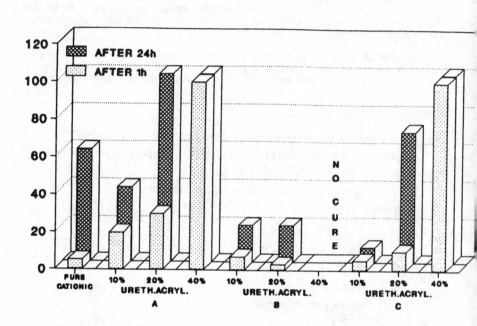

Fig. 8 URETHANE ACRYLATES / CYCLOALIPHATIC EPOXIDE
COMPARED TO STANDARD UV CURING CATIONIC FORMULATION
MEK DOUBLE RUBS
MEK DOUBLE RUBS

ACKNOWLEDGEMENTS

The author would like to express his appreciation to F. Chomienne for the detailed data gathered for this publication, as well as S. Bohler for the editing.

REFERENCES

1. **R.F. Eaton; B.D. Hanrahan; J.K. Braddock,** "Radtech '90 - North America, Conference Proceedings," March 25 - 29, 1990, Chicago, U.S.A., pp. 384 - 401.

2. **H.G. Gaube; J. Ohlemacher,** "Radtech Europe '89, Conference Proceedings," October 9-11, 1989, Florence, Italy, pp. 699 - 711.

3. **P. Dufour;** "Radtech Europe '89, Conference Proceedings," October 9 - 11, 1989, Florence, Italy, pp. 57 - 71.

4. **C. Lowe,** "Radtech Europe '89, Conference Proceedings," October 9 - 11, 1989, Florence, Italy, pp. 143 - 162.

5. **J-P. Ravijst,** "Radtech '90 North America, Conference Proceedings," March 25 - 29, 1990, Chicago, USA, pp. 278 - 291

6. **F.J. Vara; J.A. Dougherty,** "Radtech Europe, '89, Conference Proceedings," October 9- 11, 1989, Florence, Italy, pp 523 - 533

7. **J.A. Dougherty; F.J. Vara,** "Radtech '90, North America, Conference Proceedings," March 25 - 29, 1990, Chicago, USA, pp 402 - 409

8. **W.C. Perkins,** Journal of Radiation Curing, 1981, 8 (1), 16

9. **P. J-M. Manus,** "Radtech Europe '89, Conference Proceedings," October 9 - 11, 1989, Florence, Italy, pp 535 - 553

10. **S.R. Sauerbrunn; D.C. Armbruster,** "Radtech '90 North America, Conference Proceedings," March 25 - 29, 1990, Chicago, USA, pp 303 - 312

11. **J.V. Koleske,**" Radtech '88, North America, Conference Proceedings," April 24 -28, New Orleans, USA, pp 353 - 371

12. **R.D. Hanrahan; P. J-M. Manus; R.F. Eaton,** "Radtech '88, North America, Conference Proceedings," April 24 - 28, New Orleans, USA, pp. 14 - 27

13. **W.R. Watt,** Journal of Radiation Curing, 1986, 13 (4) (7), 7

14. **S.C. Lapin,** Polym. Mat. Sci. Eng., 1989, 61, 302

15. **C.E. Warburton,** et. al., European Patent Application 89 302 993.4, 3/89.

16. **R. Holman,** "UV & EB Curing Formulation for Printing Inks and Coatings,", SITA-Technology, London, 1984, 149.

Hybrid Cure of 'Half Acrylates' by Radiation

Jean-Pierre Ravijst

RADCURE SPECIALISTS SA, DROGENBOS, BELGIUM

ABSTRACT

The photochemically initiated curing of coatings and inks by a free radical polymerization has been well known for more than 20 years, and applications continue to grow in printing, metal-, wood-, plastic-, and paper finishes. The complementary cationic photoinitiation technology has application in a much smaller segment of the total radiation cure industry.

Each photoinitiation technology has some drawbacks. Pure free radical systems often show poor adhesion on non-porous substrates due to the high shrinkage associated with acrylic polymerization. Cationic systems generally suffer from slow cure rates. This study was set up to look at the radiation curing performance of a resin with both acrylic and epoxy functionality in a hybrid system combining both free radical and cationic photoinitiators.

INTRODUCTION

As with acrylate based radiation cure systems, the commercial use of pure cationic curing coatings developed very slowly, with the first significant industrial applications developing only in the past few years. In fact, cationic curing represents only a small part of the total radiation curing market, chiefly limited to metal can end coating applications.

In recent years technical papers have been presented explaining cationic cure, dual cure and hybrid cure systems in great detail.

Irgacure 184 (Ciba-Geigy):
1-hydroxycyclohexylphenylmethanone
 Irgacure 651 (Ciba-Geigy):
2,2-dimethoxy-2-phenylacetophenone
Irgacure 500 (Ciba-Geigy):50% 1-hydroxycyclohexylphenyl
ketone 50% benzophenone
FX 512 (3M): triphenylsulfonium hexafluorophosphate
UVE 1014 (GE): triaryl sulfonium hexafluoroantimonate
CG 24-61 (Ciba-Geigy): (cyclopentadien-1-yl)
(1,2,3,4,5,6-n)-(1-methylethyl)-benzene-iron(1+)-hexa-
fluorophosphate
Uvecryl P104 (Radcure Specialties): reactive amine
synergist
Ebecryl 169 (Radcure Specialties): acrylated phosphate
adhesion promoter
TTEGDA (Radcure Specialties): tetraethylene glycol
diacrylate
NVP (GAF): N-vinylpyrrolidone
DCPA (Radcure Specialties): dicyclopentenyl acrylate
IBOA (Radcure Specialties): isobornyl acrylate
Tone M100 (Union Carbide):hydroxyl functional
caprolactone monoacrylate
EDA: epoxidized dicyclopentadiene monoacrylate(Viking
Chemicals)
BZO: benzophenone

1. INFLUENCE OF PHOTOINITIATORS ON THE CURE RESPONSE OF DGEBA

Three different cationic cure photoinitiators were
evaluated with DGEBA to determine their cure efficiency
on aromatic epoxy resins, CG 26-61, UVE 1014 and FX 512.
Epoxy resin ER 510 was blended with 3% photoinitiator and
applied on Bonderite 1000 steel panels at a thickness of
0.5 mil (approximately 12 microns). Curing was done
using UV light at various doses (expressed in mJ/cm^2),
and the cure response is shown in Figure 1.

The ratings between 0 and 3 represent a subjective
evaluation of surface cure and the hardness of the
coatings. No solvent resistance was measured for these
systems. As expected, the aromatic epoxy resins have a
slow cationic cure response compared to the well-known
cycloaliphatic epoxy resins, and almost no cross-linking
occurred on UV exposure under these conditions. However,
it was evident that FX 512 was the best choice to
continue our experiments.

Hybrid curing systems are defined as those which simultaneously undergo more than one curing mechanism on exposure to ultraviolet (UV) or electron beam (EB) radiation. Dual cure systems are those which can polymerize or cross-link in two separate stages, via either the same or different reaction mechanisms.

Most of these early papers described dual and hybrid radiation curing systems based on blends of epoxy acrylates and/or urethane acrylates with acrylic monomers, cycloaliphatic epoxy resins, glycidyl ethers, divinyl ethers, etc. This paper describes a different way in which both acrylic and epoxy functionality can be used in a hybrid or dual cure technique. The base resin used in this study is a bisphenol A mono acrylate/mono epoxide called Ebecryl 3605. It was designed to enable the free epoxy group to react with cationic photoinitiators or a thermal catalyst, while the residual acrylate group remains available to undergo free radical polymerization.

In this study, the following parameters will be discussed:

1. Influence of different photoinitiators on the cure response of the diglycidyl ether of bisphenol A (DGEBA).
2. Reactivity of DGEBA.
3. Cure response of a fully acrylated DGEBA.
4. Cure response of half-acrylated DGEBA.
5. Influence of combined photoinitiator systems.
6. Influence of inert atmosphere during curing.
7. Surface hardness as a function of UV dose.
8. Influence of diluting monomers on the cure response.
9. Influence of thermal treatment after UV curing.
10. Metal adhesion.

Materials Used:

Epi-Rez 510 (Hi-Tek Polymers): diglycidyl ether of bisphenol A (DGEBA)
Ebecryl 3700 (Radcure Specialties): diacrylate ester of DGEBA
Ebecryl 3605 (Radcure Specialties): half-acrylate ester of DGEBA
Rapidcure DVE-3 (GAF): triethylene glycol divinyl ether ($CH_2=CH-O-CH_2CH_2OCH_2CH_2OCH_2CH_2-O-CH=CH_2$)

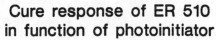

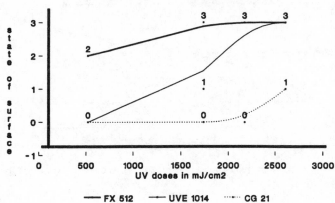

Figure 1 (GRAPH HB 1)

2. CURE RESPONSE OF DGEBA

Epoxy resin ER 510 was blended with 3% FX 512, coated on Bonderite 1000 steel panels and cured at 2 different doses (1300 mJ/cm^2 and 5200 mJ/cm^2). All coatings were tack free after curing, and the pencil hardness was "F" for the lower UV dose and "H" for the higher one. Thermal treatment of the coating for 30 minutes at 65°C (150°F) one hour after curing produced only a slightly higher coating hardness but increased the MEK resistance to more than 200 single rubs. Figure 2 depicts the MEK resistance as a function of the UV dose and the post cure time after the thermal treatment.

As with all cationic curing systems, polymerization continues over time after photoinitiation, and final properties are obtained 24 to 96 hours following UV curing.

MEK-resistance of ER 510
with FX 512 as photoinitiator

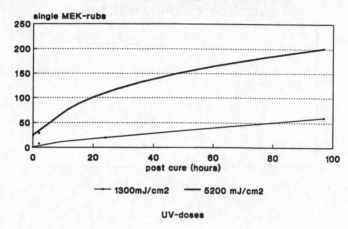

Figure 2 (GRAPH HB 3)

Reactivity of EB 3700
with 2 photoinitiators

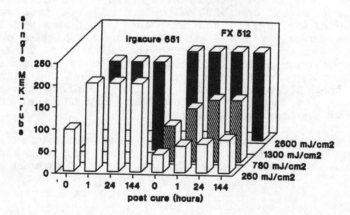

Figure 3 (GRAPH HB 8)

3. CURE RESPONSE OF A FULLY ACRYLATED DGEBA

Although FX 512 is considered to be a cationic photoinitiator, it is marginally efficient in curing acrylate systems. Ebecryl 3700 was combined with 3% FX 512 and cured at two different UV doses (780 and 2600 mJ/cm^2) and showed relatively good coating hardness and solvent resistance even with this "purely cationic" photoinitiator. When Ebecryl 3700 is mixed with 5% Irgacure 651 to promote free radical polymerization, cure speed is significantly higher.

Figure 3 illustrates the difference in solvent resistance at different UV doses as a function of post cure time for blends of Ebecryl 3700 with FX 512 and Irgacure 651.

4. CURE RESPONSE OF A HALF-ACRYLATED DGEBA EPOXY RESIN

Ebecryl 3605 contains both acrylate and epoxy functionality in the same resin molecule. The epoxy content is 2.2 meq/g, and the acrylic content is 2.2 meq of vinyl unsaturation (C=C)/gr. The remainder of this paper covers study of this resin in hybrid cure systems.

Respectively, 3% FX 512 and 5% Irgacure 651 were added to Ebecryl 3605 to find an optimum curing speed. Only marginal MEK resistance was measured when Ebecryl 3605 was cured in the presence of Irgacure 651, but much better solvent resistance was obtained with photoinitiator FX 512. Figure 4 shows the difference in solvent resistance with the two photoinitiators as a function of time.

A comparison between Ebecryl 3700 and Ebecryl 3605 when cured in presence of each type of photoinitiator shows that Ebecryl 3605 is much more sensitive to cationic photoinitiators than it is to free radical photoinitiators where very little cross-linking is observed. The fully acrylated epoxy resin shows excellent cure with Irgacure 651 and only marginal cure with FX 512. This comparison is shown clearly in Figure 5.

Table 1 shows the pencil hardness of coatings based on Ebecryl 3700 and Ebecryl 3605 as a function of the photoinitiator.

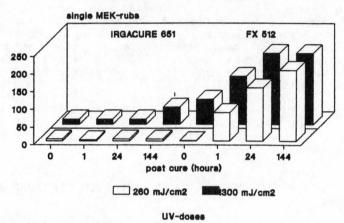

Figure 4 **(GRAPH HB 9)**

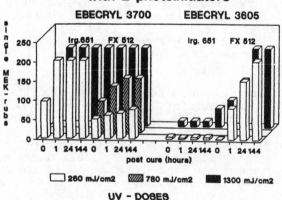

Figure 5 **(GRAPH HB 10)**

Table 1.
Coating Hardness As A Function Of Photoinitiator Type

UV doses mJ/cm²	Ebecryl 3700 Irg. 651	FX 512	Ebecryl 3605 Irg. 651	FX 512
260	2H	-	HB	F
780	2H	F	HB	F
1300	2H	H	HB	F
2600	2H	-	HB	-
5200			H	

It is clear that both resins cured with the same photoinitiator systems show differences in hardness as a function of the UV dose. Ebecryl 3605 behaves more as a monofunctional oligomer, and it is evident that these coatings cure softer than coatings based on Eb 3700 even after four days of post curing.

5. INFLUENCE OF DIFFERENT PHOTOINITIATORS ON CURE BEHAVIOR OF EBECRYL 3605

Nine different photoinitiators or photoinitiator combinations were tested at different ratios with Ebecryl 3605 :

1. Irgacure 651	5%
2. FX 512/BZO/Uvecryl P104	3% / 11% / 3%
3. FX 512/Irgacure 651	2% / 2%
4. FX 512	3%
5. FX 512/Irgacure 651	3% / 2%
6. FX 512/Irgacure 651	1% / 4%
7. FX 512/Irgacure 651	3% / 1%
8. FX 512/Irgacure 651	4% / 1%
9. FX 512/Irgacure 500	3% / 1%

Figure 6 gives an overview of the cure behavior of these systems as a function of the post cure time.

The combination FX 512/Benzophenone/Uvecryl P104 exhibits a pot life of less than 24 hours at room temperature. With this photoinitiator system, the curing was likely inhibited due to the presence of organic nitrogen. Increasing the amount of Irgacure together with FX 512 results in a much faster cure but, as we will see later, it negatively affects other properties (i.e adhesion on metals). The best results were obtained with the combination FX 512/ Irgacure 500, providing the best solvent resistance with the least amount of photoinitiator.

MEK - Resistance of EB 3605
cured with different photoinitiators

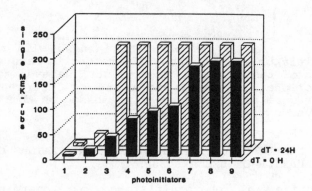

Figure 6 (GRAPH HB 2)

6. INFLUENCE OF ATMOSPHERE DURING CURING

Ebecryl 3605, blended with 3% FX 512 and 1% Irgacure 500 as the photoinitiator system was cured under an inert atmosphere and under ambient conditions. Less than 800 ppm oxygen was measured during the inert atmosphere trials.

After UV curing at different doses and post-curing, the coatings were rated between 1 and 6, based on a combination of solvent resistance, scratch resistance and coating dryness after curing. A rating of 6 represents a dry and hard coating after curing. A rating of 1 indicates a tacky surface after curing and coating softness after post curing. Figure 7 illustrates the difference in performance of the same coating when cured in air versus under nitrogen.

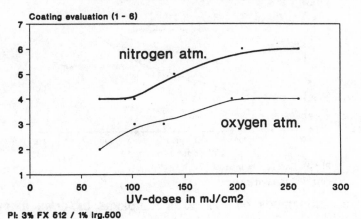

Figure 7   **(GRAPH HB 4)**

As with most free radical polymerizations, the influence of the inert atmosphere on surface cure is as important as the UV dose.

7. SURFACE HARDNESS OF CURED EBECRYL 3605

The surface hardness of coatings based on Ebecryl 3605 containing 3% FX 512 and 1% Irgacure 500 as photoinitiator system and cured at three different UV doses was measured with a Sward Hardness meter Model C. All hardness measurements were taken one hour after UV curing, since the irradiated films still were tacky immediately after UV exposure, and a hard surface developed only with time. This behavior is typical of cationic curing systems. The results are shown in Figure 8, which indicates little differences in hardness values between curing under nitrogen and oxygen atmospheres at higher UV doses. The difference appears to become larger at lower UV doses.

Surface hardness of EB 3605
f(UV-doses)

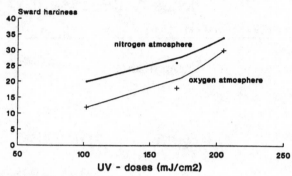

PI = 3% FX 512 / 1% Irg.500

Figure 8 (GRAPH HB 5)

8. INFLUENCE OF REACTIVE MONOMERS ON CURE BEHAVIOR

A range of monomers were blended with EB 3605, and the blends were cured at different UV doses. Solvent resistance was measured as a function of the UV dose and the time after curing. In this study we distinguished the effect of the different types of curing mechanism, functionality and the presence of functional groups on curing and post-cure behavior. Specifically, the following experiments were done:

A. Comparison between TTEGDA and Rapidcure DVE-3

B. Influence of three different monofunctional monomers NVP, DCPA and EDA

C. Difference between two monoacrylates containing functional groups EDA and Tone M100

D. Influence of the concentration of monofunctional monomers on cure speed

A. TTEGDA versus Rapidcure DVE-3

Formulations tested:

```
# 8.1.1          EB 3605              70 p
                 Rapidcure DVE 3      30 p
                 FX 512                3 p
                 Irgacure 651          3 p


# 8.1.2          EB 3605              70 p
                 TTEGDA               30 p
                 FX 512                3 p
                 Irgacure 651          3 p
```

EB 3605 was mixed with 30% of each monomer and cured under UV light at different doses in the presence of 3% FX 512 and 3% Irgacure 651. The ambient relative humidity was approximately 50 percent. Figure 9 shows the differences in cure response.

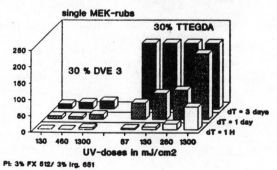

Figure 9 **(GRAPH HB 11)**

EB 3605 blended with 30% DVE-3 and cured at different doses showed poor performances compared with TTEGDA, a monomer having the same backbone but acrylate functionality. The DVE-3 formulation was stable for less than 24 hours at 60°C. The DVE-3 formulation cured at up to 1300 mJ/cm^2 (Fusion lamp 300W/inch at 10 ft/min) remained wet or tacky after curing and became tack-free only after 24 hours post-cure. Coatings cured at higher doses (2600 mJ/cm^2) were tack-free after curing. Generally, cross-hatch adhesion on Bonderite steel reached 100% only after 24 hours, but the solvent resistance never exceeded 100 single MEK rubs.

A blend of EB 3605 with 30% TTEGDA containing the same photoinitiator system showed a completely different cure response. With this formulation and the same UV doses, MEK resistance increased with the dose, post curing was much more evident than with formulation 8.1.1 and solvent resistance was much higher. The coatings were tack free seconds after curing and showed much better scratch resistance than coatings based on formulation 8.1.1.

Typically, the presence of TTEGDA gave less adhesion on Bonderite steel, because the cross-link density of this formulation (3.3 meq C=C/g) and shrinkage during curing (6%) is much higher for 8.1.2 than for 8.1.1. Surprisingly, the cure speed of the vinyl ether formulation is much slower than the acrylate version. We consistently have had difficulty realizing the rapid cure speeds claimed for vinyl ether systems and plan future work to develop a better understanding of this discrepancy with published literature.

B. Influence of mono-functional monomers on cure speed

Three different mono-functional monomers were evaluated for cure response in combination with EB 3605 in the presence of the photoinitiator combination FX 512/Irgacure 651:

# 8.2.1	EB 3605	70 p
	NVP	30 p
# 8.2.2	EB 3605	70 p
	DCPA	30 p
# 8.2.3	EB 3605	70 p
	EDA	30 p

These formulations were cured under UV light at different doses, and the MEK resistance after 24 hour post cure is shown in Figure 10 as a function of the UV dose.

Formulation 8.2.1 didn't cure well even at a high UV doses, almost certainly due to the presence of an alkaline nitrogen site in the NVP. At the same time, this formulation showed a short shelf life (48 hours at room temperature). All the coatings were tack free after irradiation, but no measurable solvent resistance could be obtained even after 24 hours post-curing. Furthermore, none of these coatings adhered on Bonderite steel.

Formulation 8.2.2 cured better under UV exposure compared with 8.2.1, but coatings cured at doses less than 260 mJ/cm^2 remained wet or tacky and showed an unacceptable odor after curing. Above 260 mJ/cm^2, the coating cured as a function of the UV dose and behaved overall as a hybrid cure system with only limited solvent resistance after curing.

The low viscosity half acrylate half epoxidized dicyclopentadiene monomer used in formulation 8.2.3 contains 4.5 meq C=C/gr and 4.5 meq epoxy/gr. As expected, it appears to have exceptional utility in systems which combine free radical together with cationic polymerization.

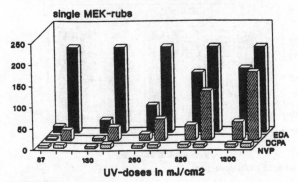

Influence of monofunctional monomers. on curing speed

70% EB 3605 / 30% monomer

Figure 10 (**GRAPH HB 13**)

Figure 10 shows the efficiency of this monomer when used in combination with EB 3605. Even at very low UV doses (87 mJ/cm^2) more than 200 single MEK rubs were measured 24 hours after curing. The coating is totally dry and tack-free only 72 hours after curing at 130 mJ/cm^2. All UV doses above 130 mJ/cm^2 had a tack-free surface after curing.

C. Presence of functional groups in monomers

In this study Tone M100 was compared against EDA in the following formulations:

# 8.3.1			# 8.3.2		
EB 3605	70 p		EB 3605	70 p	
EDA	30 p		Tone M100	30 p	
FX 512	3 p		FX 512	3 p	
Irg. 651	3 p		Irg. 651	3 p	

meq C=C/gr:	2.90	4.45
meq epoxy/gr:	2.90	1.55
meq prim OH/gr:	0	.87
meq sec OH/gr:	1.55	1.55

In different studies it has been shown that the presence of OH groups in cationic cure systems improves cure behavior.

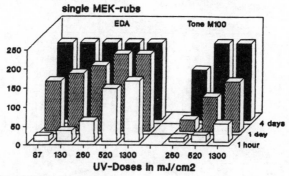

**Cure performance of
Ep-DCP-Acr / Tone M100**

70% EB 3605 / 30% monomer

Figure 11 **(GRAPH HB 12)**

With formulation 8.3.2 we expected a considerably faster curing system due to the presence of a primary OH at a ratio of 1 meq epoxy / 0.56 meq primary hydroxyl, and at the same time containing 53% more acrylic unsaturation.

Figure 11 shows the difference in cure response (MEK resistance) between formulation 8.3.1 and 8.3.2. Even at very low doses, 8.3.1 cures better than 8.3.2. The post cure (after 1 and 4 days) gives an even better picture of the total cure behavior. However, formulation 8.3.2 gave a harder cured film. While coatings based on 8.3.1 remained scratchable at all UV doses and after post curing, coatings based on 8.3.2 were only slightly scratchable at 520 mJ/cm^2 and completely scratch free at 1300 mJ/cm^2.

D. Influence of the concentration of monofunctional monomers on cure speed

Isobornyl acrylate was blended with EB 3605 at four different concentrations (10% - 20% - 30% - 50%). All the formulations were cured in the presence of 3% FX 512 and 1% Irg. 500 with 205 mJ/cm^2 (100 ft/min/300W/in) under a nitrogen atmosphere. After curing, the coating surfaces were dry and tack-free, but the coating containing 50% IBOA remained very scratchable. Figure 12 shows the single MEK rubs after curing as a function of the IBOA concentration and after 24 hours post curing.

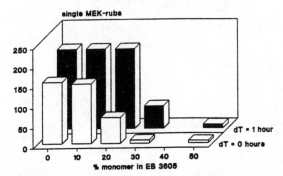

Figure 12 **(GRAPH HB 14)**

It is clear that the solvent resistance and the surface hardness of the coating decreases as a function of an increased concentration of monoacrylate monomer.

9. THERMAL TREATMENT OF COATINGS

A formulation comprising: EB 3605   80 p
                           IBOA   20 p
                           FX 512   3 p
                           Irg. 500   1 p

was coated on Bonderite steel panels and cured with 205 mJ/cm^2. The films were subjected for different durations to thermal treatments at either 65°C or 180°C, either 1 or 24 hours after the initial UV curing. (In the following figures, the term "dT" indicates the post-curing time at room temperature before the film was subjected to the thermal treatment.) Figure 13 shows MEK resistance as a function of the time and temperature of the thermal treatment.

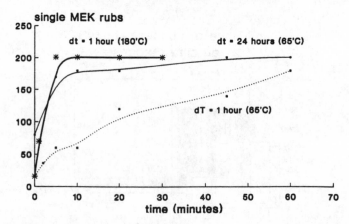

EB 3605/IBOA 80/20
MEK-resistance f(time & temperature)

Figure 13 **(GRAPH HB 7)**

The lower 65°C temperature clearly was below a critical level needed to achieve rapid and complete cure. The 24 hour ambient temperature post-curing period provided a substantial contribution to building polymer molecular weight and cross-link density. Rapid and complete cure could be achieved by raising the temperature of the thermal treatment to 180°C; the higher temperature treatment for 5 minutes immediately after UV curing produced MEK resistance of more than 200 single rubs. Coatings treated for 5 minutes or more at 180°C or 10 minutes or more at 65°C became totally scratchfree following treatment.

10. METAL ADHESION.

Bonderite 1000 steel panels were chosen as the metal substrate for this work. Coatings of 0.5 mil thickness were applied and cured with UV light, evaluating the following parameters:

A) Influence of the UV-dose
B) Influence of different photoinitiator systems
C) Influence of different diluting monomers
D) Influence of additives: Epi-Rez 510 and Ebecryl 169
E) Thermal treatment

A) UV DOSE

The formulation: Ebecryl 3605   100 p
                      FX 512   3 p
                      Irg. 500   1 p

was cured at six different UV doses. The solvent resistance as a function of the UV dose and the time after curing is given in Figure 14. This figure also shows cross-hatch adhesion, expressed in percent. This cross-hatch adhesion remained the same at dT=0 hours as at dT=24 hours.

Adhesion & Solvent resistance of EB 3605 in function of UV-doses

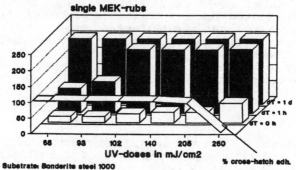

Substrate: Bonderite steel 1000

Figure 14   (GRAPH HB 15)

Table 2
Optimum Conditions to Obtain 100% Cross-hatch Adhesion on Bonderite 1000 Steel

Composition	Photoinitiator System	min. UV-doses (mJ/cm2)	time postcure (hours)	single MEK rubs
A				
EB 3605	3 FX 512/ 1 IRG.500	68	24	201
EB 3605	1 FX 512/ 4 IRG.651	130	24	201
EB 3605	2 FX 512/ 2 IRG.651	130	24	201
EB 3605	3 FX 512/ 1 IRG.500	130	24	201
EB 3605	4 FX 512/ 1 IRG.651	130	24	201
EB 3605	3 FX 512	195	24	201
EB 3605	3 FX 512/ 1 BZO/ 3 UvP104	260	96	30
EB 3605	5 IRG.651	1300	144	45
B				
70 EB 3604/ 30 EDA	2 FX 512/ 2 IRG.651	87	72	201
80 EB 3605/ 20 IBOA/ 5 EB 169	3 FX 512/ 1 IRG.500	1025	96	201
70 EB 3605/ 30 DVE 3	2 FX 512/ 2 IRG.651	2600	96	100
70 EB 3605/ 30 DVE 3	3 FX 512/ 3 IRG.651	260	72	20
50 EB 3605/ 50 IBOA	3 FX 512/ 1 IRG.651	205	1	10

Table 3
Influence of Functionality on Metal Adhesion

Formulation	Composition			Functionality			Treatment	Adhesion
	EB 3605	IBOA	ER 510	C-C/g	epoxy/g	shrinkage		%
10.D.1	100	0	0	2.2	2.2	4		100
10.D.1	100	0	0	2.2	2.2	4	15 min @ 65°C	100
10.D.2	80	20	0	2.7	1.8	4.6		0
10.D.2	80	20	0	2.7	1.8	4.6	60 min @ 65°C	0
10.D.3	80	10	10	2.2	2.3	4.1		max 50
10.D.3	80	10	10	2.2	2.3	4.1	45 min @ 60°C	100
10.D.3	80	10	10	2.2	2.3	4.1	5 min @ 100°C	100

Almost every composition has a critical UV dose above which overexposure can lead to a loss of adhesion on the substrate. As shown in Figure 14, 100 percent adhesion was obtained up to the critical UV dose for this formulation on Bonderite 1000, about 210 mJ/cm^2 (or 100 ft/min @ 300 W/inch).

B) PHOTOINITIATORS

Table 2-A shows the conditions for a series of photoinitiators under which Ebecryl 3605 gives 100% cross-hatch adhesion for a maximum number of single MEK rubs.

C) DILUTING MONOMERS

Because the presence of fully acrylated monomers increases the total shrinkage of the coating during curing and decreases the amount of epoxy groups necessary for the cationic curing, it was difficult to obtain 100% adhesion on Bonderite steel panels with formulations containing any diacrylate. Only EDA monomer proved to be acceptable (see Table 2-B) and still required a post-cure time of 72 hours at room temperature to obtain more than 200 single MEK rubs. With 87 mJ/cm^2 (150 ft/min @ 300 W/inch), the coating remained slightly tacky for 24 hours (130 single MEK-rubs); better results were obtained at 260 mJ/cm^2, where the coating was tack-free after curing, but the adhesion was borderline.

D) ADDITIVES

i) Epoxy resins

Table 3 shows the effect of acrylate monomer additives on adhesion and compensation for shrinkage using epoxy resins. The three formulations each contain 3% FX 512 and 1% Irgacure 500 as the photoinitiator package and were cured at the same dose (205 mJ/cm^2). Formulation 10.D.1 had excellent adhesion either with or without a thermal treatment. Formulation 10.D.2 contained added mono-functional acrylic monomer and had no adhesion, even with a thermal treatment. However, loss of metal adhesion for a formulation containing an acrylated diluting monomer can be corrected if some of the acrylate monomer is replaced by an epoxy resin and a thermal post treatment of the coating is allowed. In formulation 10.D.3, half of the IBOA was replaced by Epi-Rez 510 to maintain the same ratio of epoxy and acrylic groups as formulation 10.D.1.

Table 4
Influence of the Adhesion after a Thermal Treatment

Composition	Photoinitiator	dT (h)	UV-dose (mJ/cm2)	MEK rubs	% Adh	Treatment	MEK rubs	% Adh
ER 510	3% FX512	1	1300	8	90	30m @ 75°C	201	100
ER 510	3% FX512	1	5200	25	90	30m @ 75°C	201	100
EB 3605	3% FX512	1	260	16	100	30m @ 75°C	201	90
EB 3605	3% FX 512	1	260	16	100	10m @ 180°C	201	100
EB 3605	3% FX 512	1	1300	100	0	30m @ 75°C	201	100
EB 3605	5% Irg.651	1	260	4	100	30m @ 75°C	4	100
EB 3605	5% Irg.651	1	1300	15	100	30m @ 75°C	15	100
EB 3605	5% Irg.651	1	1300	15	100	10m @ 180°C	55	100
EB 3605	3% FX512/ 1% Ir.500	1	205	25		15m @ 65°C	201	100
EB 3700	3% FX512	1	780	120	100	30m @ 75°C	140	10
EB 3700	3% FX512	1	2600	201	100	30m @ 75°C	201	0
EB 3700	5% Irg.651	1	260	201	0	30m @ 75°C	201	0
EB 3700	5% Irg.651	1	1300	201	0	30m @ 75°C	201	0
EB 3605/ 20 IBA	3% FX512/ 1% Ir.500	1	205	16	0	5m @ 65°C	60	0
EB 3605/ 20 IBA	3% FX512/ 1% Ir.500	24	205	80	0	5m @ 65°C	170	0
EB 3605/ 20 IBA	3% FX512/ 1% Ir.500	1	205	16	0	60m @ 65°C	185	0
EB 3605/ 20 IBA	3% FX512/ 1% Ir.500	24	205	80	0	60m @ 65°C	201	0
EN 3605/ 20 IBA	3% FX 512/1% Ir.500	0	205	5	0	1m @ 180°C	70	10
EB 3605/ 20 IBA	3% FX512/ 1% Ir.500	0	205	5	0	5m @ 180°C	201	100
EB 3605/ 20 IBA/ 5 EB 169	3% FX512/ 1% Ir.500	0	205	40	100	1m @ 180°C	150	80
EB 3605/ 20 IBA/ 5 EB 169	3% FX512/ 1% Ir.500	24	205	201	100	1m @ 180°C	201	100
EB 3605/ 20 IBA/ 5 EB 169	3% FX512/ 1% Ir.500	0	205	40	100	5m @ 180°C	201	100
EB 3605/ 10 ER510/ 10 IBA	3% FX 512/1% IR.500	0	205	8	20	5m @ 100°C	201	100
EB 3605/ 10 ER510/ 10 IBA	3% FX 512/1% IR.500	0	205	8	20	45m @ 65°C	201	100

Table 5
100% Water Resistance
(14 days immersion)

Composition	Photoinitiator	dT (days)	UV-dose (mJ/cm2)	Treatment	MEK rubs	% Adh.
EB 3605	4% FX 512/ 1% Irg.651	4D	130		201	100
EB 3605	1% FX 512/ 4% Irg.651	4D	130		201	100
EB 3605	2% FX 512/ 2% Irg.651	4D	130		201	100
EB 3605	3% FX 512/ 1% Irg.500	0	155	nitrogen cure	42	100
EB 3605	3% FX 512/ 1% Irg.500	1D	102	nitrogen cure	201	100
75 EB3605/ 20 IBA/ 5 EB 169	3% FX 512/ 1% Irg.500	4D	158		100	100
75 EB3605/ 20 IBA/ 5 EB 169	3% FX 512/ 1% Irg.500	4D	410		110	100
75 EB3605/ 20 IBA/ 5 EB 169	3% FX 512/ 1% Irg.500	4D	1025		200	100
75 EB3605/ 20 IBA/ 5 EB 169	3% FX 512/ 1% Irg.500	1D	205	1 min @ 180oC	201	100
80 EB3605/ 10 IBA/ 10 ER510	3% FX 512/ 1% Irg.500	0	205	5 min @ 100oC	201	100

ii) Acrylated phosphate

Ebecryl 169 acrylated phosphate was evaluated as an adhesion promoter in formulations containing Ebecryl 3605 diluted with IBOA.

Formulations tested:

	# 10.D.4	# 10.D.5
Ebecryl 3605	80 p	75 p
IBOA	20 p	20 p
Ebecryl 169	0 p	5 p
FX 512	3 p	3 p
Irg. 500	1 p	1 p

Figure 15 shows the comparison in curing speed between formulations 10.D.4 containing only IBOA and 10.D.5, which has the acrylated phosphate additive. It appears that the curing speed and post curing are influenced by the presence of the strong acid like Ebecryl 169. All of the coatings were dry seconds after curing, and the coatings containing only IBOA became scratch-free at doses higher than 400 mJ/cm^2. Formulation 10.D.4 is identical to 10.D.2 and thus had no adhesion even after thermal treatment. In contrast, formulation 10.D.5 showed 100% metal cross-hatch adhesion even with UV curing doses up to 1025 mJ/cm^2, far above the 210 mJ/cm^2 critical dose observed for Ebecryl 3605 alone.

E) THERMAL TREATMENTS

Table 4 shows how thermal treatment of coatings on metal affects the adhesion and solvent resistance. It shows the crosshatch adhesion and single MEK rubs resistance after curing and after thermal treatment. All of the basic principles observed in the previous discussion are evident in this data. A new observation is a loss of adhesion in a few cases after thermal treatment. This occurs chiefly where acrylate functionality is high.

Curing speed of EB 3605 with
20% IBOA / 20% IBOA + 5% Ebecryl 169

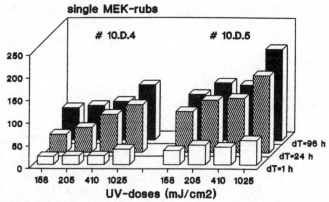

PI: 3% FX 512/ 1% Irg.500

Figure 15(GRAPH HB 16)

CONCLUSION

Ebecryl 3605 offers new and interesting opportunities for hybrid cure applications. It can be used in formulations where adhesion, intercoat adhesion, post-cure, and low shrinkage are important. Ebecryl 3605 doesn't behave like a classical aromatic epoxy acrylate. Formulated with monomers and adequate photoinitiators, it cures slower under UV light than the fully acrylated resin but has much better adhesion to rigid, non-porous substrates. With a thermal post-treatment, good metal adhesion and more than 200 single MEK rubs were measured with coatings cured at 100 ft/min. Good adhesion was obtained on metal even without thermal treatment, but formulations containing Ebecryl 3605 are sensitive to the presence of acrylated monomers, the nature of photoinitiators and the UV dose, so adjustment to specific application conditions is necessary to obtain full benefits. The slower cure speed measured with most formulations is probably because the resin responds chiefly as a mono-functional oligomer with each of the two different photoinitiator systems.

FUTURE WORK

The preliminary work to date was directed chiefly at measuring the cure speed and adhesion of hybrid curing systems containing Ebecryl 3605. In future work, we will examine other important performance properties, such as tensile strength and flexibility, and the cure response to electron beam radiation. The work will cover the cure response and adhesion of pigmented coatings. The resin technology also will be extended to include new partially acrylated/partially epoxidized oligomers, including urethanes, rubber-based resins and acrylic resins.

Effects of Light on Cyclic Vinyl Ethers

C. Lowe[1] and A.E. Wade[2]

[1] CROWN INDUSTRIAL PRODUCTS, DARWEN, LANCASHIRE, UK
[2] ICI ADVANCED MATERIALS, BLACKPOOL, LANCASHIRE, UK

Introduction

Since the discovery that divinyl ethers polymerise rapidly (1) when initiated by photosensitive catalysts which produce hydrogen ions as one of their photolysis products, a substantial amount of research into this particular area of radiation curing has been carried out (2)(3)(4). Much of the work has involved the synthesis of monomeric (2) and oligomeric (3) linear vinyl ethers as the initial vinyl ethers studied appeared to cure as fast as or faster than acrylates (1) and they do not have the same health and safety problems (3).

Unfortunately the preparation of linear divinyl ethers is dependent upon either acetylene chemistry, which requires very special precautions, or very expensive raw materials and catalysts (1). Cyclic vinyl ethers have also been reported (5) to be responsive to photogenerated acids and these compounds can be made quite simply by heating the necessary reactants to quite modest temperatures at atmospheric pressure (6). In an attempt to evaluate this chemistry two cyclic vinyl ethers were prepared and the cure response compared to a typical epoxide resin and linear vinyl ethers. FT-IR was the main technique used to follow cure, backed up by the resistance of the cured coating to MEK double rubs. Tentative mechanisms for polymerisation and a potentially detrimental side reaction will be proposed.

Experimental

The cyclic vinyl ethers used in this study were prepared using acrolein as the starting raw material. A tetramer of acrolein was the first compound to be synthesised the structure of which is shown and compared to that of a cycloaliphatic epoxy resin in figure 1. The procedure followed is detailed here: Redistilled acrolein (Aldrich) (420.0g) and hydroquinone (4.2g) [1% w/w] were placed in a stainless steel autoclave (1l) and heated rapidly to 170°C.

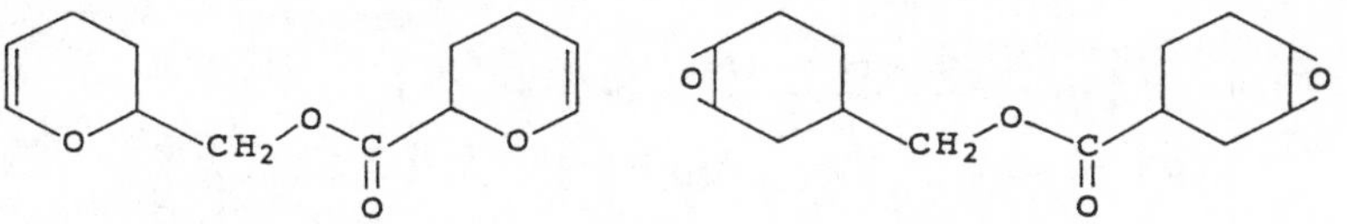

Tetramer (Vinyl Ether I)   Cycloaliphatic Epoxy Resin

Figure 1   Comparison of the structure of Vinyl Ether I with that of the Epoxy Resin.

The temperature was maintained between 170-175°C for 2 hours in order to encourage the thermally activated Diels - Alder 4 + 2 cyclo-addition reaction between two molecules of acrolein. Upon cooling the reaction mixture was recovered and distilled at atmospheric pressure to yield unreacted acrolein (b.p. 52°C, 108g) and further distilled under reduced pressure to give the acrolein dimer (b.p. 42-45°C at 10mm Hg, 141g (35%)]. A higher boiling residue was also obtained. This reaction was carried out several times under slightly different conditions but no improvements on yield could be achieved.

The distilled dimer was then converted into the tetramer using the Tischenko reaction (7). This involved placing the dimer (224.0g) in a 500ml round bottomed flask equipped with a stirrer and thermometer. Aluminium isopropoxide (5.6g) was added in small amounts with stirring at room temperature, an ice cold water bath was used to maintain the temperature of the reaction mixture below 35°C.

After the temperature had started to fall the mixture was allowed to stand overnight at ambient conditions. The product, 2 - (3, 4 - dihydro - 2H pyranyl) methyl 3, 4 - dihydro - 2H - pyran - 2 - carboxylate (I) was isolated by distillation [b.p. 125-127°C 2mm Hg 166g (74%)].

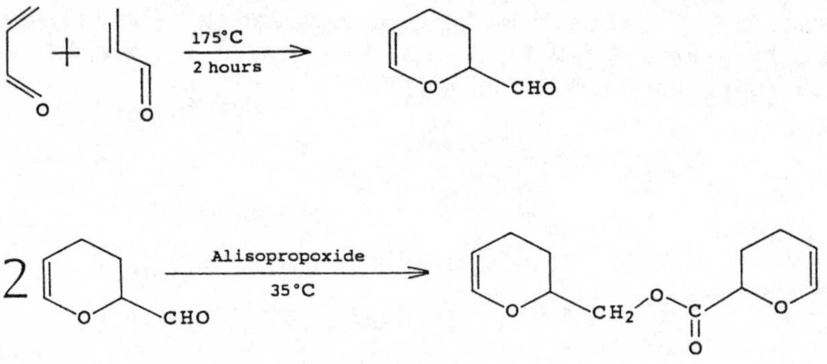

Scheme 1   Formulation of Vinyl Ether I.

1, 4 - Bis (3', 4' - dihydro - 2'H pyran - 2'H oxy) butane (II) was the second cyclic divinyl ether used in this study, its structure can be seen in figure 2. The preparation of this compound is described here: Acrolein (274.0g) was placed into a stainless steel autoclave (1l) together with butane diol divinyl ether (GAF) (347.0g) and hydroquinone (0.5g).

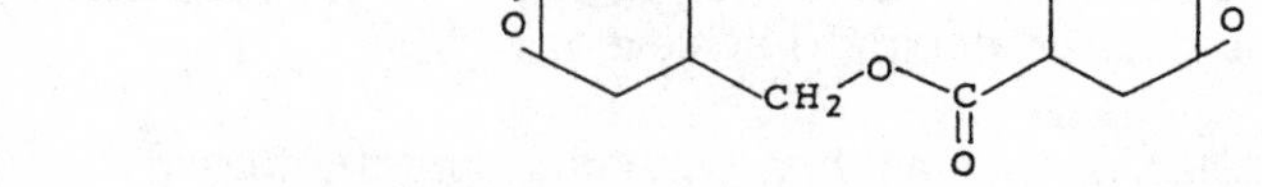

Vinyl Ether II   Cycloaliphatic Epoxy Resin

Figure 2   Comparison of the Structure of Vinyl Ether II with that of the Epoxy Resin

The autoclave was sealed and rapidly heated to 185°C and then held there for 2 hours, it was allowed to cool to room temperature before the reaction mixture was recovered and distilled. Initial fractions were unreacted acrolein (53.0g) and butane diol divinyl ether (72.0g), the main fraction contained the required product [b.p. 136 - 142°C at 1.0mm Hg 435.0g (70%)].

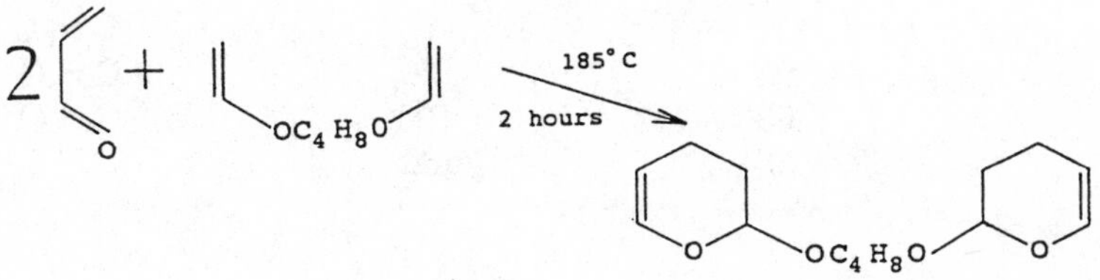

Vinyl Ether II 70% yield

Scheme 2 Formation of Vinyl Ether II.

The product was further characterised by IR and MS.

i)   IR (See figure 3)

Bond	Intensity	Assignment
3063 cm^{-1}	M	H-C=C Stretch
2929 cm^{-1}	S	H-C- Stretch
1757 cm^{-1}	S	C=O Stretch
1650 cm^{-1}	S	C=C Stretch
1242 cm^{-1}	S	C=C-C-O
1197 cm^{-1}	S	C-O of C=O
1110 cm^{-1}	M	
1070 cm^{-1}	S	C-O of O-CH$_2$
1046 cm^{-1}	M	
728 cm^{-1}	S	

MS

Empirical Formula $C_{12}H_{16}O_4$ Vinyl Ether I

Measured Mass	% Base Intensity	Assignment
224	18	M^+
110	17	$C_7H_{10}O^+$
96	26	$C_6H_8O^+$
83	100	$C_5H_7O^+$
55	60	$C_3H_3O^+$

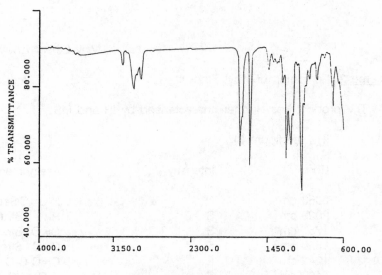

Figure 3   IR Spectrum of Vinyl Ether I

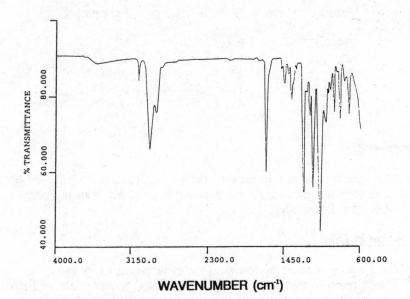

Figure 4　　　IR Spectrum of Vinyl Ether II

ii)　　　IR (See figure 4)

Bond	Intensity	Assignment
3064 cm⁻¹	M	H-C=C Stretch
2935 cm⁻¹	S	H-C-
1651 cm⁻¹	S	C=C Stretch
1223 cm⁻¹	S	C=C-C-O
1117 cm⁻¹	S	C-O-C
1057 cm⁻¹	M	
1037 cm⁻¹	S	
728 cm⁻¹	S	

MS

Empirical Formula $C_{14}H_{22}O_4$ Vinyl Ether (II)

Measured Mass	% Base Intensity	Assignment
254	0.9	M^+
125	1	$C_7H_9O_2^+$
99	30	$C_5H_7O_2^+$
83	100	$C_5H_7O^+$
55	81	$C_3H_3O^+$

Instrumentation

IR spectra were obtained using a Nicolet 5ZDX FT-IR spectrophotometer, mass spectral data were obtained from a Finnigan 1020 at Salford University.

Evaluation of Cure

Cure was evaluated by following the disappearance of the 1651 cm⁻¹ (C=C) band in the case of vinyl ether based formulations and the 910 cm⁻¹ band in the case of epoxide based formulations. Nicolet software was used to integrate areas under these particular bands. Cure was also determined by evaluating the number of MEK double rubs required to break the surface of a 6 micron thick coating cured on a mild steel Q-panel.

Acid Hydrolysis of Vinyl Ether

A sample of vinyl ether was placed in a test tube, shaken with HCl and then left over night. An IR spectrum was then taken of the top layer.

Formulation Used

The vinyl ethers prepared were compared to an epoxy resin of similar structure and to two linear vinyl ethers which are commercially available. In order to do this in a consistent way each was included in a typical formulation.

Table 1   Formulations Used

Formulation	A	B	C	D	E
Epoxy Resin	50	-	-	-	-
I	-	50	-	-	-
II	-	-	50	-	-
BDDVE (GAF)	-	-	-	50	-
TEGDVE (GAF)	-	-	-	-	50
Polyol	25	25	25	25	25
Diluent	20	20	20	20	20
Photoinitiator	4.5	4.5	4.5	4.5	4.5
Additives	0.5	0.5	0.5	0.5	0.5

Curing Conditions

Formulations drawn down on steel were cured by passing under two 80 W/cm medium pressure mercury lamps at a speed of 20 m/min. The procedure was repeated until a solid film was produced.

For spectroscopic analysis the formulations were applied to polythene film and cured at the same line speed. In the former case the results were given as the number of MEK double rubs required to break the surface, the latter case the degree of cure is expressed as a percentage of the original band area.

Results

Table 2   Solvent Resistance of Final Film

Formulation	No. of Passes	MEK Double Rubs
A	3 passes	20
B	5 passes	7
C	> 20 passes	Uncured

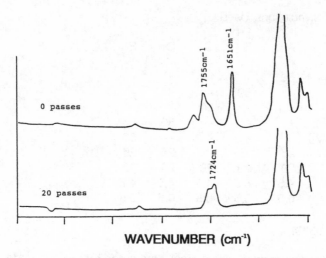

<u>Figure 5</u>   Spectra of Formulation B before and after cure

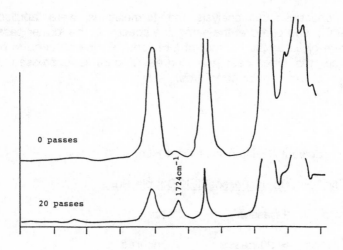

WAVENUMBER (cm⁻¹)

<u>Figure 6</u>   Spectra of Formulation C before and after cure

Table 3   FT-IR Analysis of UV Cured Formulations

No. of Passes	Percentage of area under band remaining for various formulations		
	A (910cm^{-1})	B (1651cm^{-1})	C (1651cm^{-1})
0	100	100	100
1	2	98	66
2	1	95	43
5	-	54	15
10	-	13	3
20	-	4	1

The above results describe what happens to the functional groups on irradiation however there is another interesting feature of the spectra taken of formulations B (figure 5) and C (figure 6). That is a band appears around 1720cm^{-1} - 1730cm^{-1} during irradiation and continues to grow with further passes. This is more visible in the spectra of C because there is no contribution from an ester carbonyl as in the case of B.

Discussion

The results clearly show that cyclic vinyl ethers cure much slower than the cycloaliphatic epoxide. From the infra red data it would appear that cyclic vinyl ether II cures faster and further than cyclic vinyl ether I, solvent resistance indicates the reverse.

This dichotomy could be explained if vinyl ether I reacts chiefly via an inter molecular mechanism which would give a cross linked coating where as vinyl ether II may react via predominately intra molecular mechanism which would give rise to a product of low molecular weight with few cross links. Alternatively vinyl ether II could be undergoing some sort of cleavage reaction instead of polymerisation. This second explanation would appear to be more likely as the film of the formulation based on viny ether II is still "wet" when the reaction is complete and even a small amount of polymerisation would lead to at least a tacky film at the very least.

A clue to what is occurring in the film comes from the IR spectra. The appearance of a band between 1720 - 1730cm^{-1} is consistent with a saturated aldehyde and it is being produced in both formulations. A mechanism which is consistent with the protonation of vinyl ethers and produces a polymer with aldehyde functionality is given in scheme 3.

Protonation occurs preferentially at the β - carbon because the resulting carbocation is resonance stabilised by the lone pairs of electrons on the oxygen atom. The ring opening step is driven by a relief of strain and thus this mechanism occurs faster than the more straight forward cationic polymerisation reaction normally invoked for linear vinyl ethers. However the final macroscopic end result is the same i.e. a solid film.

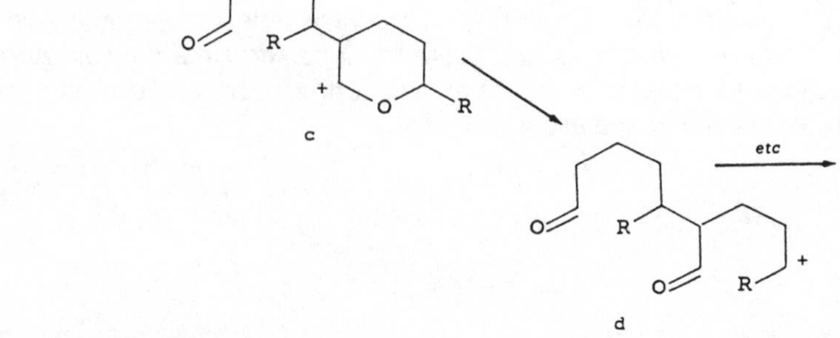

Scheme 3 Proposed Mechanism for the Polymerisation of Vinyl Ether I

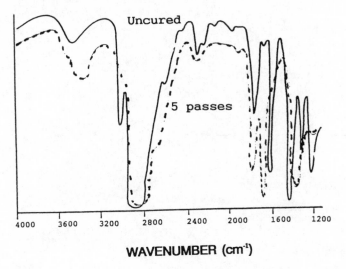

WAVENUMBER (cm⁻¹)

<u>Figure 7</u>   Spectra of Vinyl Ether II plus PI before and after cure

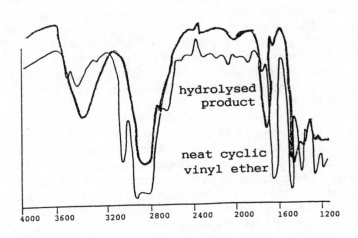

WAVENUMBER (cm⁻¹)

<u>Figure 8</u>   Spectra of Vinyl Ether II before and after acid hydrolysis

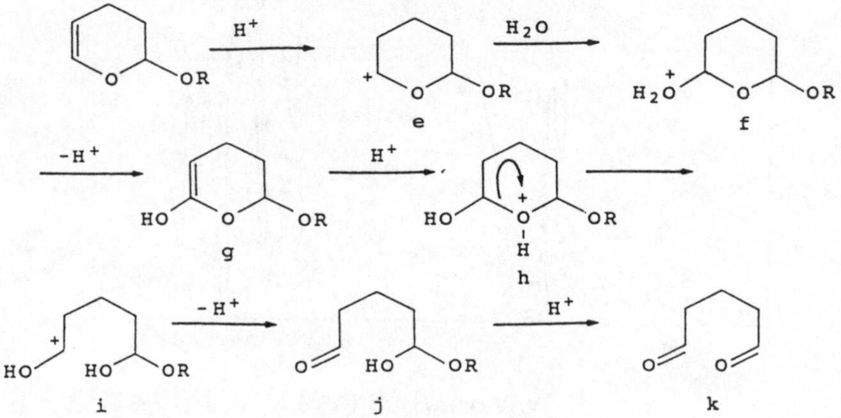

Scheme 4 Proposed Mechanism for Reaction of Vinyl Ether II

Vinyl ether II is essentially an unsaturated acetal and so it will readily undergo acid hydrolysis, the water may come from moisture in the atmosphere above the coating or it may be in the formulation to some extent. The mechanism proposed in scheme 4 relies on water for the first step, subsequent steps simply require the presence of an acid. The ultimate product is a smaller molecule than the starting material, containing a large amount of aldehyde functionality which would explain the results observed.

Scheme 5 proposes an alternative way of ring opening. However in this case more water is required and the final product contains a large amount of hydroxyl functionality. An answer to the question, "Which one is more likely?" can be obtained using IR spectroscopy. Figure 7 shows the spectra of vinyl ether II containing some photoinitiator before and after irradiation. The disappearance of vinyl functionality is quite apparent as is the appearance of aldehyde functionality. The bands caused by hydroxyl groups show only a modest increase which can be explained by both the final ether cleavage reaction and incomplete reaction i.e. it may stop at g or j.

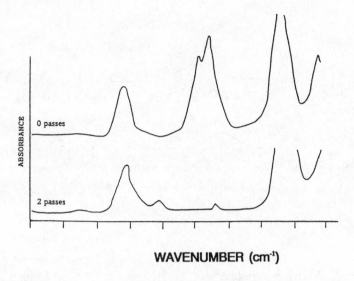

WAVENUMBER (cm⁻¹)

<u>Figure 9</u>   Spectra of Formulation D before and after cure

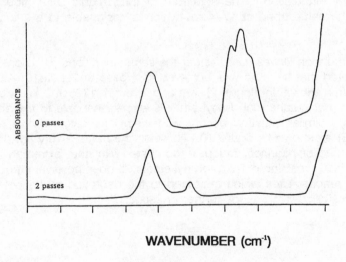

WAVENUMBER (cm⁻¹)

<u>Figure 10</u>   Spectra of Formulation E before and after cure

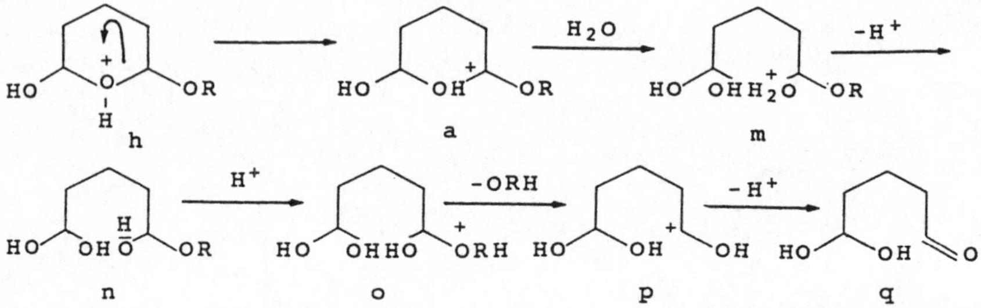

<u>Scheme 5</u> Alternative mechanism for the reaction of Vinyl Ether II

This theory was further tested by adding aqueous acid to the cyclic vinyl ether formulation. The spectrum of the organic layer (figure 8) yielded the same band at 1724cm^{-1} which is assignable to a saturated aldehyde.

Formulation D was cured under the same conditions as before and it was noted that full cure was achieved in 1 pass at 20m/min. After 2 passes (as shown in figure 9) a band around 1725cm^{-1} is evident, indicating that butane diol divinyl ether is also susceptible to the above reactions. Similar situation occurs with triethylene glycol divinyl ether (TEGDVE) as shown by figure 10. However the band is very small and so the reaction cannot seriously compete with the extremely fast polymerisation reaction of linear divinyl ethers. It does however introduce potential reactive sites which may lead to the degration of the polymer film a possibility which needs further investigation.

Conclusion

Cyclic divinyl ethers do not cure as fast as linear divinyl ethers or cycloaliphatic epoxides and they would appear to cure via a different mechanism. Acetal type cyclic vinyl ethers prefer to undergo cleavage reactions rather than polymerise. The acid hydrolysis of vinyl ethers can also occur in linear vinyl ethers but the rate of reaction is slow in comparison to the rate of polymerisation.

REFERENCES

1)   Crivello J.V., Lee J.L. and Conlon D.A., J. Rad. Curing 1983, 10(1), 6

2)   Lapin S.C., "Radcure 86: Conference Procs.", Association for Finishing Processes, Baltimore 1986.

3)   Dougherty J.A. Vara F.J. and Anderson L.R. "Radcure 87: Conference Procs." Association for Finishing Processes, Munich, 1987

4)   Lapin S.L. and Snyder J.R., "Radtech '90' Conference Procs.", Radtech North American, Chicago, 1990

5)   Eur. Pat. App. 0155 704 Union Carbide Corp., 1985

6)   Okada M. Sumitomo H and Ichiro T., Macromolecules 1977 10(3), 505

7)   Sumitomo H and Hastimoto J. Macromolecules 1977 10(6), 1327

Norbornene Resins as Substrate in Thiol-ene Polymerizations: Novel Non-acrylate Photopolymers for Adhesives, Sealants, and Coatings

Anthony F. Jacobine, David M. Glaser, Steven T. Nakos, Maria Masterson, Paul J. Grabek, Margaret A. Rakas, David Mancini, and John G. Woods

CHEMICAL AND MATERIALS SCIENCE GROUP, LOCTITE CORPORATION, 705 NORTH MOUNTAIN ROAD, NEWINGTON, CT 06111 USA

Introduction

Thiol-ene photopolymerizations occupy something of a unique position in the technical development of advanced radiation curing technology. Although this polymerization propagates by a free radical mechanism, its cure kinetics are quite different from the free radical chain mechanisms associated with acrylate and methacrylate photopolymerizations. The kinetics and mechanism of thiol-ene polymerizations are more akin to a step growth addition kinetic scheme even though radicals are involved in the propagation steps[1,2,3]. Unlike acrylate photopolymerization, thiol-ene polymerizations are not inhibited by oxygen[4]. This type of photopolymerization is further differentiated from conventional UV cationic technology in that it is not inhibited by ambient moisture[5].

Thiol-ene technology reached a fairly advanced stage of development early on when these conventional UV technologies were limited by some serious technical shortcomings. Later, concerns about the toxicology of low molecular weight acrylate species gave considerable impetus for the continuation of the development of this technology. This development is chronicled in over 200 United States patents as well as equivalent European and World Patent Office documents and many technical articles. Much of the pioneering work in this area was carried out by scientists at W. R. Grace and Company[6].

Initially, developmental efforts focussed on the addition of multifunctional thiol components to highly functionalized allylic resins (Scheme 1). This overcomes a major difficulty observed with many step-growth systems. This problem manifests itself in the observed difficulty to convert linear or difunctional enes and dithiols to high molecular weight linear polymer due to imperfections in functionalization and can be predicted from the Carothers Equation[2,3]. The use of enes and thiol components with functionalities greater than two ensures the rapid formation of a crosslinked network.

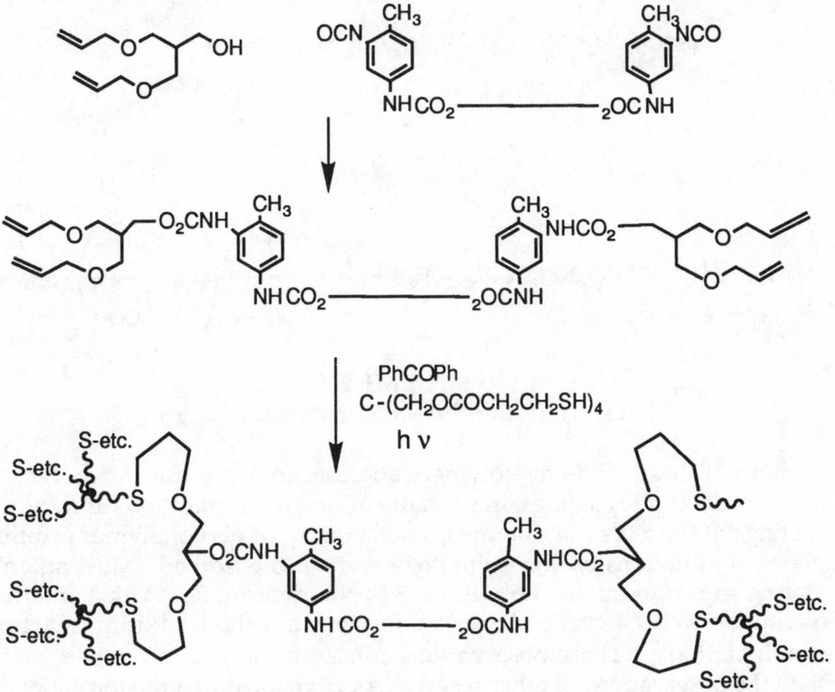

SCHEME 1
Preparation of Tetrafunctional Allylic Urethane and Photocrosslinking With Multifunctional Thiol

Additional technical advances were made when it was determined that many of the advantages of a thiol-ene polymerization could be obtained by using lower levels of multifunctional thiols as initiators for acrylate polymerizations[6] rather than crosslinkers (Scheme 2). No inhibition of polymerization was observed for these "hybrid" systems, polymerizations were fast, and end user acceptance was greater[6].

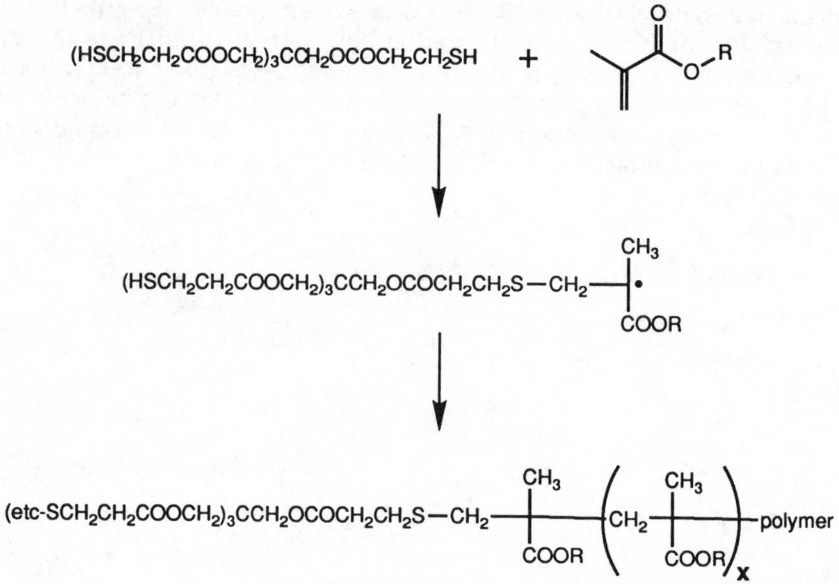

$$(HSCH_2CH_2COOCH_2)_3CCH_2OCOCH_2CH_2SH \quad + \quad \text{(acrylate)} \underset{O}{\overset{O}{\parallel}}\!\!-O-R$$

↓

$$(HSCH_2CH_2COOCH_2)_3CCH_2OCOCH_2CH_2S-CH_2-\underset{COOR}{\overset{CH_3}{\underset{|}{\overset{|}{C}}}}\!\!\cdot$$

↓

$$(etc\text{-}SCH_2CH_2COOCH_2)_3CCH_2OCOCH_2CH_2S-CH_2-\underset{COOR}{\overset{CH_3}{\underset{|}{\overset{|}{C}}}}\!\!\left(CH_2-\underset{COOR}{\overset{CH_3}{\underset{|}{\overset{|}{C}}}}\right)_x\!\!-polymer$$

SCHEME 2
Thiol "Initiation" of Acrylic Polymerization

Both of these thiol-ene polymerization technologies find wide acceptance in electronic coatings, in specialty adhesives applications and for the coatings of floor tiles, and in the manufacturing of photopolymer printing plates. Limitations of this technology were also observed. Most notable among these observed limitations was the determination that internal olefins as well as cyclic olefins underwent the thiol addition reaction slowly if at all[7]. These observations concerning cyclics were consistent with the observations of other researchers who noted that the formation of a highly stabilized, planar allylic radical species was more favorable than direct addition of thiol (Scheme 3).

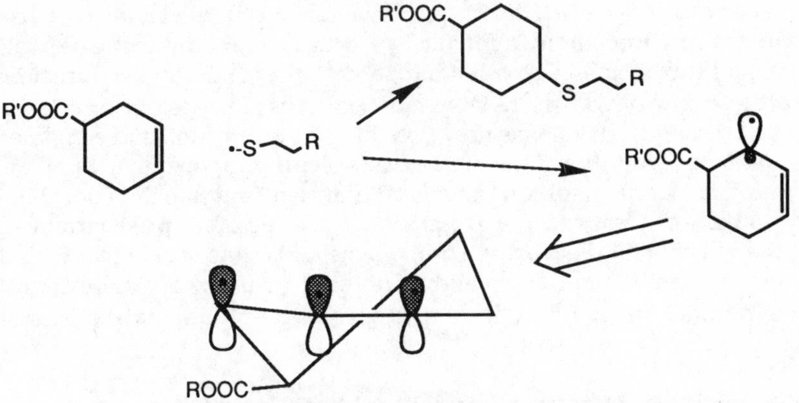

SCHEME 3
Allylic Abstraction *Versus* Addition In Cyclic Olefins

In 1986 we determined that certain cyclic olefins, most notably [2.2.1] bicycloheptene terminated oligomers would undergo rapid, exothermic, photoinduced crosslinking reactions with multifunctional thiols[8,9,10] (Scheme 4).

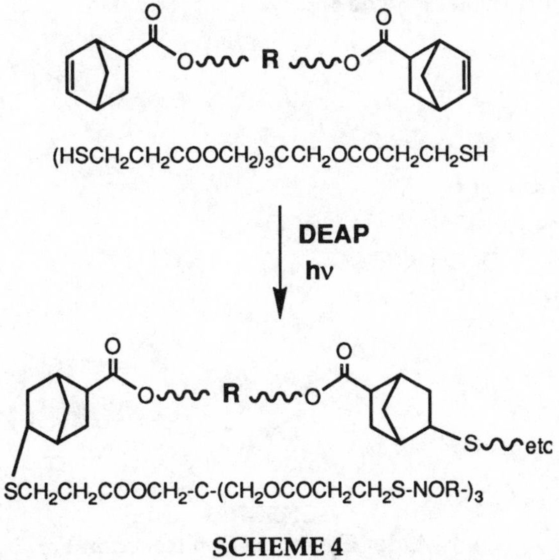

SCHEME 4
Photocrosslinking of Polyfunctional Norbornene Resin

The scope of this reaction is quite general. Norbornene functionalized aromatic hydrocarbons, aliphatic hydrocarbons, and urethane derivatives are all subject to this crosslinking process as are norbornene functionalized siloxane oligomers[11]. Design of networks of predetermined structure and crosslink density was now possible by the selection and combination of di and multifunctional norbornene olefin components with di- and multifunctional thiol components. Because neither the thiol nor the ene can homopolymerize, the overall network structure is predetermined by the selection of the reactive components. It is conceivable that with the development of our understanding of the structure-property relationship, the rational design of products with specific properties may be possible[12].

Preparation of Norbornene Functional Resins and Oligomers

The introduction of the norbornene nucleus into an organic molecule is straightforward and may be accomplished in a variety of ways. These methods include Diels-Alder reaction of acrylates with cyclopentadiene monomer (Scheme 5), reaction of the norbornene acid chloride with a suitable nucleophile, or the use of a number of reagents that contain the norbornene nucleus and are analogous to reagents used in acrylic chemistry (Scheme 6). The reaction can be carried out without solvent and yields are generally quite good (Scheme 5).

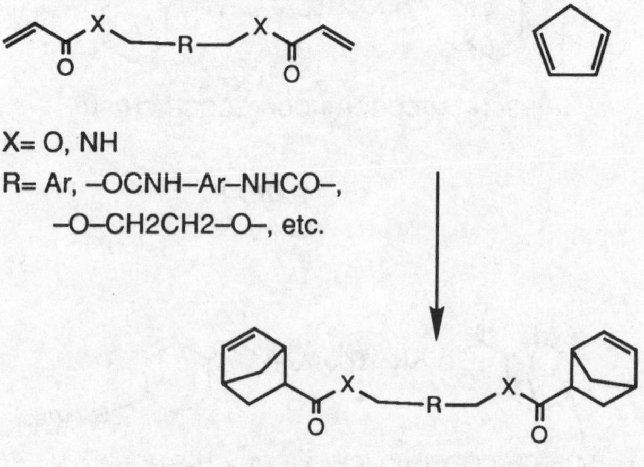

X= O, NH
R= Ar, –OCNH–Ar–NHCO–,
   –O–CH2CH2–O–, etc.

SCHEME 5
Diels-Alder Cycloaddition Reaction For
The Preparation of Norbornene Resins

Preparation of resins and oligomers via the reaction of monofunctional endcapping reagents such as hydroxyethyl norborn-2-ene-5-carboxylate with diisocyanates or norbornenemethyl chloroformate, norbornene isocyanate, nadic anhydride, or other norbornene isocyanate equivalents with a polyol or polyamine is also useful and efficient. Tischenko type dimerizations of aldehyde derivatives are also quite useful. These methodologies are outlined in Scheme 6 below and a partial listing of norbornene functional resins and oligomers synthesized by these methods is contained in Table I. The method of preparation is also listed in the table.

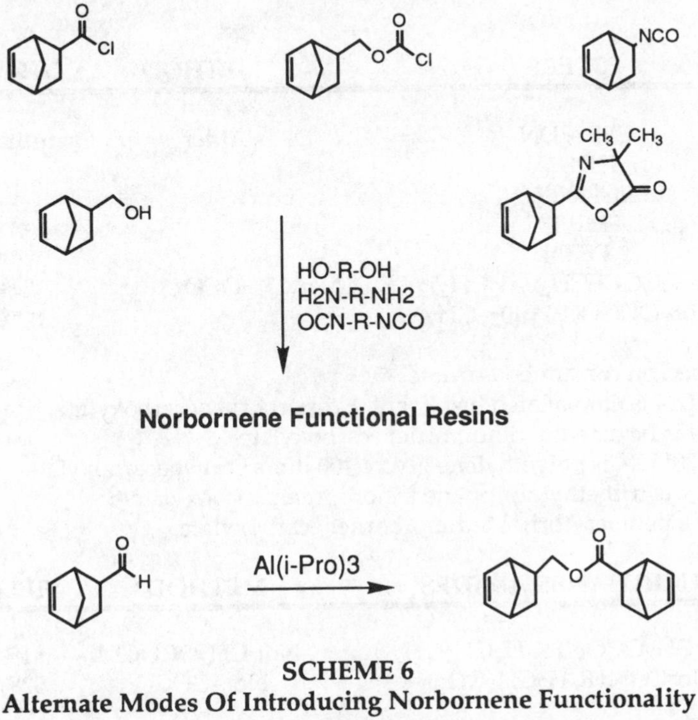

HO-R-OH
H2N-R-NH2
OCN-R-NCO

Norbornene Functional Resins

SCHEME 6
Alternate Modes Of Introducing Norbornene Functionality

By far the most accessible and general route to low to medium molecular weight multifunctional resins and oligomers is by the [4π+2π] cycloaddition reaction (Diels-Alder Reaction) of the corresponding acrylate ester described above in Scheme 5. The advantages of this synthetic strategy are based on the fact that there is a wide variety of acrylate precursors such as epoxy acrylate, polyester acrylates, urethane acrylates, and many more. Generally, the cost of these precursor materials is relatively low and they are widely available from a number of sources. In most cases the conversion of acrylate to norbornenecarboxylate via the Diels-Alder reaction is quantitative. The use of solvents is not necessary in the reaction scheme if the precursor is liquid at the reaction temperature (0 - 120°C).

In many cases structure-property relationships have been established for acrylic materials and this information base serves as a useful point of departure for further study.

TABLE I
NORBORNENE FUNCTIONAL RESINS AND OLIGOMERS

ESTERS	METHOD	YIELD
EBPA DN	Diels-Alder	Quantitative[†]
HDDN	"	"
PEG[400]DN	"	"
TMP TN	"	"
PETN	"	"
$(Nor-OCOO-CH_2CH_2CH_2)_2-$	Nor CH_2OCOCl	75%
$(Nor-OCOOC_6H_{10})_2-CH_2$	"	82%

† based on consumed acrylate
EBPA DN is ethoxylated bisphenol A dinorbornenecarboxylate
HDDN is hexanediol dinorbornenecarboxylate
PEG[400] DN is polyethylene glycol 400 dinorbornenecarboxylate
TMPTN is trimethylolpropane trinorbornenecarboxylate
PETN is pentaerythritol tetranorbornenecarboxylate

URETHANES/AMIDES	METHOD	YIELD
$(NorCH_2OCONHCH_2CH_2CH_2)_2$	Nor CH_2OCOCl	81%
$(NorCONHCH_2CH_2CH_2)_2-$	Nor$-COCl$	52%

The preparation of norbornene functional siloxane oligomers (silicones) requires the preparation of a special reagent for the introduction of the reactive functionality. This is accomplished in two ways. The first focuses on the selective reaction of 5-vinyl-2-norbornene with chlorodimethylsilane to yield 5-[(2-chlorodimethylsilyl)ethyl]-2-norbornene. An alternate reagent for the introduction of norbornene functionality is based on the Diels-Alder reaction of an acrylate based chlorosilane. Both of these methods are described in Scheme 7 below. Reaction of these materials with silanol terminated silicone fluids give either

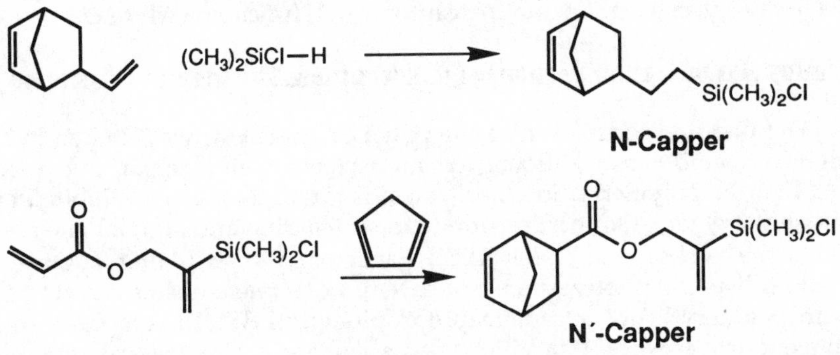

N-Capper

N'-Capper

SCHEME 7
Reagents For Preparing Norbornene Functionalized Siloxane Fluids

a telechelic norbornene (monoolefin terminated, 28N) or norbornene-propenyl (diolefin terminated, 28N') reactive silicone fluid (Scheme 8).

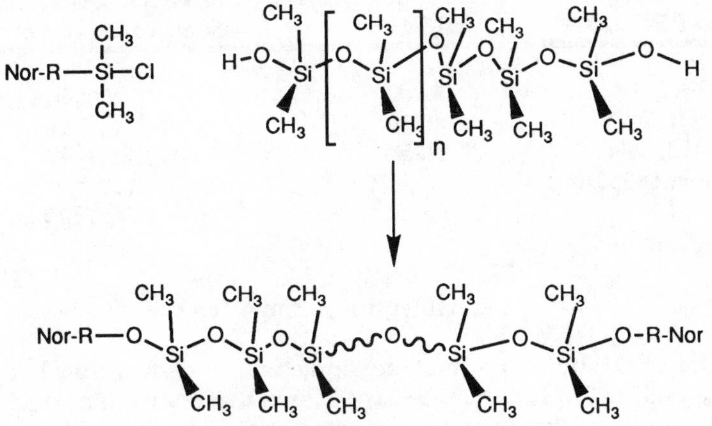

SCHEME 8
Routes to Norbornene Functionalized Siloxane Fluids

Formulation of the silicone norbornene fluids present an interesting problem as the solubility of hydrocarbon thiols in these silicone fluids is generally low. The empirical development of a crosslinking silicone thiol with the proper level of functionality in a molecular weight range to ensure compatibility is necessary to ensure complete crosslinking and optimum properties. For our purposes, optimum results were obtained with a 3000 molecular weight penta(mercaptopropyl) functional siloxane.

Photopolymerization Response Of Norbornene Resins and Oligomers

The photoresponse of thiol-ene polymer systems is well known to be high compared to conventional acrylate systems. This is in part due to the fact that the polymerization mechanism is for all intents not inhibited by ambient oxygen. Our earlier work had shown that among thiol-ene systems, the norbornene resins had a particularly high photoresponse[11]. Table II lists the photoresponse of several norbornene resins crosslinked with pentaerythritol tetramercaptopropionate (PETMP) compared to a conventional acrylic UV product (Loctite 350) and a conventional thiol-ene system, triallylisocyanurate (TAI) crosslinked with the same thiol (PETMP).

TABLE II
PHOTORESPONSE OF THIOL-NORBORNENE SYSTEMS

RESIN (w PETMP)	TACK FREE DOSE* (mJ/cm2)	DEPTH OF CURE (mm)** (50mJ/cm2, 600mJ/cm2)
EBPA DN	30-50	1.59, 4.46
HDDN	20-50	2.35, 7.11
TMP TN	30-50	2.35, 8.49
Loctite 350	1500-2000	0, 2.72
TAI	50-150	0.78, 1.679

*Light intensity of $10mW/cm^2$
**Light intensity of $25mW/cm^2$

Extractable studies on thiol crosslinked norbornene resins both with hexane and methylene chloride have shown that sol fraction levels drop rapidly with dose and fully cured materials have extractable levels, ca. 1% or less.

Silicone norbornene resins when crosslinked with a soluble silicone thiol also show high photoresponse when compared to an acrylated silicone of similar structure. In the case of a 28,000 molecular weight telechelic siloxane, the norbornene system will be fully cured to a depth of 2 mm with a dose of ca. 3 J/cm^2, while the silicone acrylate requires ca. 6J/cm^2 to cure to the same depth.

Properties of Crosslinked Norbornene-Thiol Polymers

The most extensively studied thiol crosslinked norbornene resin system is that based on ethoxylated bisphenol A (derived from the corresponding diacrylate). Studies on both the bulk polymer properties as well as adhesive properties indicate this resin has superior material and adhesive properties. A partial listing of these properties is listed below in Table III.

A more general study of this family of resins is also underway in our laboratories and is focussed on mechanical properties of crosslinked polymers derived from these novel resins. These results are listed in Table IV.

TABLE III
Typical Properties of PETMP Crosslinked Ethoxylated Bisphenol A Dinorbornenecarboxylate (EBPA DN PETMP)

STRESS-STRAIN PROPERTIES

Tensile Modulus	317,393 psi
Tensile Strength	7170 psi
% Elongation	3.94
T_{max} tan δ (T_g)	38-46°C

OUTGASSING

1.84 % Total Mass Loss   0.02 % Collected Volatiles-
                                            Condensable Materials

THERMAL CONDUCTIVITY

4.40 x 10^{-5} (kg-cal)-cm/m^2-°C-hr

WATER VAPOR TRANSMISSION

0.0364 gms./hr-m^2

ELECTRICAL PROPERTIES

Surface Resistivity	$2.86 \times 10^{14} \Omega\text{-cm}^2$
Volume Resistivity	$6.05 \times 10^{14} \ \Omega\text{-cm}$
Dielectric Constant	3.14 @ 100Hz
	3.12 @ 1 kHz
	3.08 @ 10 kHz
Dissipation Factor	0.01 @ 100Hz
	0.01 @ 1kHz
	0.01 @100kHz

THEMOGRAVIMETRIC ANALYSIS

$T_{1\%}$ wgt loss	196°C (nitrogen)
$T_{5\%}$ wgt loss	353°C (nitrogen)

TABLE IV
MECHANICAL PROPERTIES OF CROSSLINKED NORBORNENE RESINS

Norbornene Resin	Tensile Modulus (MPa)	Tensile Strength	Tg (max tan δ, °C)	TGA 1%, 5% (°C)
EBPA DN	1928	40.63	46	196, 353
TMPTN	2272	60.60	67	202, 349
HDDN	826	17.21	30	166, 356
PETN	1551	47.10	71	99, 322
HMDNorU	504	15.28	36	152, 305
HMNorAm	1722	52.68	62	129, 277

The effect on bulk polymer properties of both increasing the crosslink density in theses systems (by addition of increasing amounts of TMP TN monomer) and decreasing the crosslink density by chain extension (using dithiols glycol dimercaptopropionate, GDMP or polyethylene glycol[400] dimercaptopropionate, PEG[400] DMP) was also studied. These results are summarized in Table V and Table VI. In both cases the trends are clear in terms of modulus strength and Tg.

TABLE V
EFFECT OF INCREASED CROSSLINK DENSITY ON BULK
POLYMER PROPERTIES OF NORBORNENE RESIN

Resin*	Tg (°C)	Tensile Strength (MPa)	Elongation (%)	Modulus (MPa)
EBPA DN	38	30.35	3.83	1434
EBPA DN 2.5% TMPTN	39	33.10	4.04	1500
EBPA DN 5% TMPTN	41	33.58	3.85	1603
EBPA DN 10% TMPTN	42	42.6	3.88	1603
EBPA DN 50% TMPTN	52	44	2.20	2600
TMP TN	67	61	4.10	2300

*Crosslinked with Pentaerythritol Tetramercaptopropionate (PETMP)

TABLE VI
EFFECT OF CHAIN EXTENSION ON BULK POLYMER PROPERTIES
OF CROSSLINKED EBPA DN

DITHIOL*	TENSILE (MPa)	ELONGATION (%)	MODULUS (MPa)	Tg (°C)
none	30.35	3.83	1434	38
GDMP†	11.43	220	84	26
PEG[400]DMP†*	1.7	104	2.92	1.9

* Resin crosslinked with Pentaerythritol tetramercaptopropionate (50%)
  and chain extending dithiol (50%)
† GDMP Glycol dimercaptopropionate, $[HSCH_2CH_2COOCH_2]_2$
†*PEG[400] DMP Polyethylene glycol [400] dimercaptopropionate,
  $HSCH_2CH_2COO(CH_2CH_2O)_9-OCCH_2CH_2SH$

Adhesive Properties of Thiol (PETMP) Crosslinked Norbornene Resins

The adhesive properties of thiol crosslinked norbornene systems to a variety of substrates are generally good. This is probably due in part to two phenomena unique to this system. The first, shrinkage on cure, is generally low when compared to acrylic systems (3-5% vs. 5-10%) and forms adhesive bonds with less inherent strain. In addition, it is well known that thiols can act as efficient adhesion promoters to a variety of adherends, and while the exact mechanism of this interaction is open to speculation, the empirical observations are beyond dispute. Adhesive test results are summarized in Table VII.

TABLE VII
Adhesive Properties of Thiol Crosslinked Norbornene Resins

Resin (wPETMP)	Substrate (To Glass)	Tensile Strength (MPa)	Aged Strength 50°C, 100hr	60°C, 95% RH 10 days
EBPA DN	Steel Pin	6.34	8.62	15.9
TMP TN	Steel Pin	4.17	4.67	10.7
HDDN	Steel Pin	5.22	7.89	0
Loctite 350	Steel Pin	6.93	9.27	14.9
EBPA DN	Glass Lap	4.02	na	na
EBPA DN	Epoxy Glass	4.29		
EBPA DN	Phenolic	5.52		
EBPA DN	Steel	6.6		
Loctite 370	Steel	4.11		
Dymax 181	Steel	3.6		

In most of the cases reported above substrate failure and not adhesive or cohesive failure was the predominant mode of test specimen failure.

Properties Of Photocrosslinked Norbornene Functionalized Siloxane Resins

At the present time there is a tremendous amount of interest in the development of UV curable silicone materials[13]. The areas of interest and applications cover a wide range of mechanical, medical, and electronic uses where the unique properties of silicones (low Tg, good chemical and heat resistance, excellent elastomeric properties, and high gas permeability) outperform conventional materials. As noted above, norbornene functionalized silicones exhibit excellent photoresponse in addition to very good properties as UV curable elastomers. A summary of the properties of these materials is included below in Table VIII. Properties derived from and acrylated silicone of similar molecular weight (28A) is included for comparative purposes.

TABLE VIII
MECHANICAL PROPERTIES NORBORNENE SILICONES

Resin kMW	Strength MPa	Elongation %	Tear Strength pli	Extractables (% in Hexane)
28N[†]	0.43	399	low	4.55
28N *	5.02	466	68	4.6
28N' *	3.60	250	141	5.02
28A *	3.79	330	120	6.0

*Filled with 25% w/w High surface area reinforcing fumed silica
† 28N resin is norbornene terminated 28,000 molecular weight polydimethylsiloxane fluid crosslinked with a 3000 MW
penta(mercaptopropyl) functional polydimethylsiloxane.
28A resin is a 28,000 MW acrylate terminated siloxane fluid.

Toxicology of Norbornene Resins and Oligomers

Our initial experimental observations led us to conclude that thiol crosslinked norbornene resins exhibited high cure response and excellent polymer and adhesive properties. In addition to a superior technical profile, a major impetus for Loctite to explore the photopolymerization chemistry of norbornene resins and oligomers is based on the observation that existing acrylic monomer technology is being subjected to ever increasing regulatory scrutiny both in the United States and Europe. This scrutiny would be expected to be even greater for new acrylate based materials, especially low molecular weight monomers. The norbornene resins and oligomers that we have developed were until recently unknown, and accordingly, there is no literature concerning the toxicology of these materials.

We have completed a first round of toxicological testing for selected resins and intermediates. We anticipate additional testing in the near future. The results from the first tests indicate that all materials tested to date gave no toxic effects. In addition all materials tested gave negative Ames Test results (i. e. they are non-mutagenic) and were neither dermal nor ocular irritants under the test conditions.

Conclusions

The preceding study has demonstrated that di- and multifunctional olefins are efficient substrates for thiol-ene photopolymerizations. Norbornene resins exhibit an enhanced photoreactivity compared to the "normal" thiol-ene system. The polymers derived from this photocrosslinking process have very good bulk polymer properties as well as good mechanical electrical properties. The preparation of these resins and oligomers is generally straightforward. Preparation of norbornene resins and oligomers by the Diels-Alder reaction of acrylates with cyclopentadiene monomer gives the desired products in excellent yields, usually without solvent.

Preparation of UV curable silicone elastomers is also possible using norbornene functionalized silicone fluids photocrosslinked with an appropriate silicone thiol.

While the scope and utility of this photopolymerization process is still under investigation, initial results indicate that photoinitiated crosslinking of norbornene resins is an extremely useful method for the preparation UV curable adhesives, sealants and coatings.

Acknowledgements

The authors would like to take this opportunity to express their gratitude to J. Dowling and J. Dooley of Loctite (Ireland) Ltd. for their contributions to this study and Dr. Peter Pappas, Scientific Fellow, Loctite Corporation. We also thank Professor R. Danheiser of Massachusetts Institute of Technology for his timely advice and contributions.

References

1. C. S. Marvel and R. R. Chambers, *J. Am. Chem. Soc.*, **1948**, *70*, 993.

2. C. R. Morgan, F. Magnotta, and A. D. Ketley, *J. Poly. Sci. Poly. Chem. Ed.*, **1977**, 627.

3. R. W. Lenz, "Organic Chemistry of Synthetic High Polymers", Interscience, New York, 1967.

4. C. E. Hoyle, R. D. Hensel, and M. B. Grubb, *Polymer Photochemistry*, **1984**, *4*, 69.

5. B. L. Brann, Proceedings of RadTech Europe, 565, October 1989.

6. D. P. Gush and A. D. Ketley, Chemical Coatings Conference II, Radiation Coatings, National Paint and Coatings Association, Washington D. C., May 10, 1978.

7. O. Ito and M. Matsuda, *J. Org. Chem.*, **1984**, *49*, 17.

8. W. J. Steinkraus, J. G. Woods, J, M. Rooney, A. F. Jacobine, and D. M. Glaser, United States Patent 4,808,638 assigned to Loctite Corporation, 1989.

9. W. J. Steinkraus, J. G. Woods, J, M. Rooney, A. F. Jacobine, and D. M. Glaser, World Patent Office International Publication Number WO 88 02902 April 21, 1988.

10. A. F. Jacobine, D. M. Glaser, and S. T. Nakos, Polymeric Materials: Science and Engineering Preprints, 1989, 60, 211.

11. A. F. Jacobine, D. M. Glaser, and S. T. Nakos, in Radiation Curing of Polymeric Materials, ACS Symposium Series 417, American Chemical Society, Washington D. C. 1990.

12. J. G. Kloosterboer, *Adv. Poly. Sci.*, **1988**, *84*, 1.

13. A. F. Jacobine and S. T. Nakos, Photopolymerizable Silicone Monomers, Oligomers and Resins, in "UV Curing: Science and Technology, Volume III", S. P. Pappas, Editor, Plenum, New York, volume in preparation.

UV and γ-Irradiation of Polysilastyrene Films and Fibres

G.C. East[1], V. Kalyvas[1], J.E. McIntyre[1], B. Rand[2] and F.L. Riley[2]

[1] DEPARTMENT OF TEXTILE INDUSTRIES, THE UNIVERSITY OF LEEDS, LEEDS LS2 9JT, UK
[2] SCHOOL OF MATERIALS, DIVISION OF CERAMICS, THE UNIVERSITY OF LEEDS, LEEDS LS2 9JT, UK

ABSTRACT

Continuous polysilastyrene fibres, prepared by both melt and dry spinning, were subjected to UV irradiation and the structural changes were followed by FTIR measurements on equivalent film samples. Evidence was obtained for SiSi chain scission and for formation of SiOH, SiH, SiOSi, SiCSi and C=O groups. These changes occurred more rapidly in air than in argon and were generally sensitive to the duration of the irradiation. Crosslinking took place in all samples, along with formation of low molecular weight compounds. Whereas the former contributed to an increase in the ceramic yield on pyrolysis, the latter resulted in a lower Tg, increasing the difficulties of fibre processing. Unsaturated organosilicon compounds were prepared and incorporated into the fibres without significantly improving the crosslinking density. γ-irradiation did not produce any useful effects.

1 INTRODUCTION

Polysilastyrene (PSS)(I) is a member of the polysilane class of polymers whose backbone consists entirely of Si-Si bonds.

$$
\begin{array}{ccc}
CH_3 & CH_3 & \\
| & | & \\
(\ -Si-\ )x & (\ -Si-\ )y & (I) \\
| & | & \\
CH_3 & C_6H_5 &
\end{array}
$$

Tractable polysilanes were synthesised in the late 1970s[1] and since then they have attracted the interest of several research groups, whose findings have led to a better understanding of their properties and have revealed some potential applications[2]. These include

a. precursors to silicon carbide,

b. oxygen insensitive photo-initiators for vinyl polymerisation, and

c. photoresists in microlithography.

The most striking property of the polysilanes is their photochemical sensitivity. They absorb strongly in the UV region, due to the extensive sigma-bond delocalisation in the backbone. The values of λ_{max} and ε are sensitive to molecular weight, to the nature of the substituents and to the conformation of the backbone chain. The Si-Si bond, with energy of 280-350 kJ/mol, is the weakest of the bonds usually present in polysilanes and it is susceptible to homolytic fission induced thermally or by UV irradiation. Thermolysis of polydimethylsilane, through a complex and not completely understood mechanism involving redistribution reactions, has led to the production of SiC fibres. The process is illustrated in an oversimplified manner in Scheme 1[3].

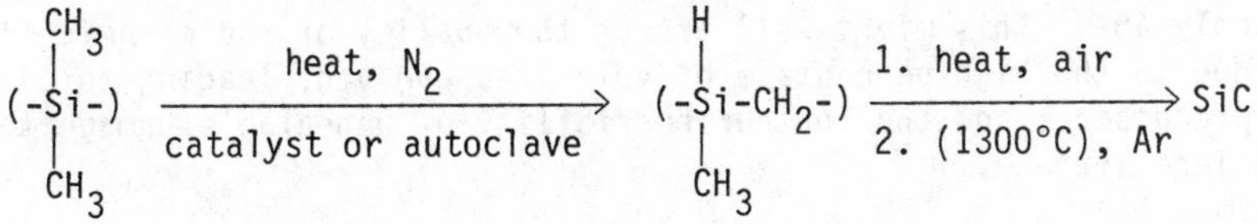

$$
\begin{array}{c} CH_3 \\ | \\ (-Si-) \\ | \\ CH_3 \end{array}
\xrightarrow[\text{catalyst or autoclave}]{\text{heat, } N_2}
\begin{array}{c} H \\ | \\ (-Si-CH_2-) \\ | \\ CH_3 \end{array}
\xrightarrow[\text{2. (1300°C), Ar}]{\text{1. heat, air}} SiC
$$

Scheme 1

UV irradiation of polysilanes results in degradation and crosslinking, the former being the dominant process in all cases, shown by a reduction in UV absorption intensity and by a drop in molecular weight. Crosslinking is noticeable in the cases where pendant unsaturation is present[4]. The reactive intermediates include silyl radicals and silylenes (Eq. 1).

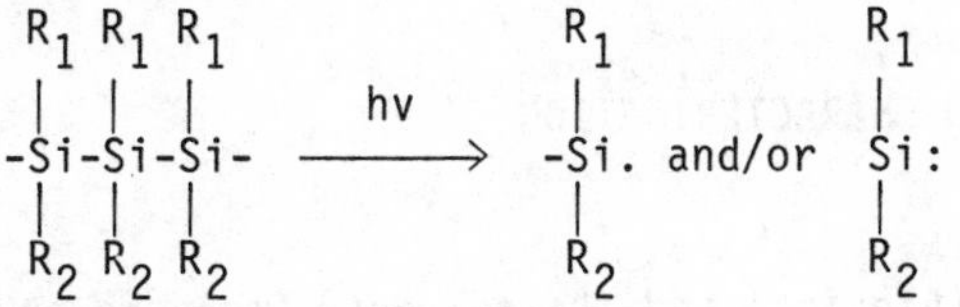

$$
\begin{array}{c} R_1 \; R_1 \; R_1 \\ | \;\; | \;\; | \\ -Si-Si-Si- \\ | \;\; | \;\; | \\ R_2 \; R_2 \; R_2 \end{array}
\xrightarrow{h\nu}
\begin{array}{c} R_1 \\ | \\ -Si. \\ | \\ R_2 \end{array}
\text{and/or}
\begin{array}{c} R_1 \\ | \\ Si: \\ | \\ R_2 \end{array}
\qquad (1)
$$

Initially it was thought[4,5] that polysilastyrene would give

rise to a better route to silicon carbide fibres than the one
established by Yajima's team (Scheme 1), following the stages outlined in Scheme 2.

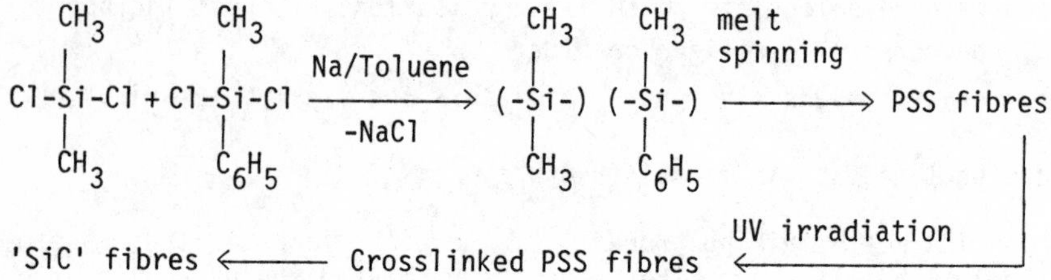

$$\underset{\substack{|\\CH_3}}{\overset{\substack{CH_3\\|}}{Cl-Si-Cl}} + \underset{\substack{|\\C_6H_5}}{\overset{\substack{CH_3\\|}}{Cl-Si-Cl}} \xrightarrow[-NaCl]{Na/Toluene} \underset{\substack{|\\CH_3}}{\overset{\substack{CH_3\\|}}{(-Si-)}} \underset{\substack{|\\C_6H_5}}{\overset{\substack{CH_3\\|}}{(-Si-)}} \xrightarrow[]{\substack{melt\\spinning}} \text{PSS fibres}$$

'SiC' fibres $\longleftarrow$ Crosslinked PSS fibres $\longleftarrow$ UV irradiation

Scheme 2

The advantages of this route are: (a) direct spinning of tractable PSS to precursor fibres without the thermolysis and fractionation steps needed in Yajima's process, (b) the exclusion in the
crosslinking process of oxygen, the presence of which is detrimental
to the high temperature mechanical properties of the ceramic fibre,
and (c) lower overall cost.

However, even under ideal conditions, there is an inherent
problem in this process in that the theoretical yield for SiC is
only 45%. This might well affect the quality of the ceramic fibres,
due to the high percentage of volatiles evolved, leading to porous
precursors, and the further possibility of unhealable damage to the
fibre surface.

Recently, high quality ceramic fibres have been produced, with
PSS as starting material. In this case, the route adopted is
similar to Yajima's. The polycarbosilane, resulting from a brief
thermolysis and fractionation of the PSS, is spun to fibres, UV-cured
and pyrolysed to ceramic (SiC) fibres[6].

No detailed information has been disclosed on the conditions of
the UV irradiation and subsequent pyrolysis of the resulting fibres.
The aim of the present study was to improve understanding of these
and other aspects of the processes.

2 EXPERIMENTAL AND CHARACTERISATION

Materials

The chlorosilanes (Aldrich) and the solvents (toluene and
tetrahydrofuran, HPLC grade from Aldrich) were purified by fractional
distillation. Tetravinylsilane and phenyltrivinylsilane were

synthesised from the appropriate chlorosilanes and vinyl magnesium chloride solution (Aldrich) and purified by double fractional distillation[14].

Instrumentation and Techniques

FTIR spectra were recorded in air on a 1725X Perkin-Elmer FTIR spectrometer. ^{1}H NMR spectra were recorded on a Jeol FX 90Q spectrometer at 90 MHz. UV spectra were recorded on a Perkin-Elmer model 402 UV/VIS or Lambda 2UV/VIS spectrometer. TGA and DSC were carried out on a Du Pont 951 Thermogravimetric Analyser and 910 DSC, equipped with a 2000 Thermal Analyser. UV irradiation of the films and fibres took place in a quartz tube fitted with a stopcock and septum, using a medium pressure mercury lamp of 50 W/cm power (total 125 W). The samples were rotated 180° half way through the specified irradiation time. Irradiation by γ-rays was performed in pyrex tubes, sealed under high vacuum or open to air, in a ^{60}Co unit. GPC analysis was performed on a system based on PSM 60 and PSM 1000 Zorbax columns in series and two UV detectors, set at 254 nm. It was calibrated with polystyrene standards. Reproducibility was~15% and accuracy was not expected to be better than 20%.

Synthesis of PSS

PSS was synthesised by reductive coupling of equimolar amounts of dichlorodimethylsilane and dichloromethylphenylsilane with sodium in refluxing toluene[1]. The reaction was quenched with isopropanol and after extensive extraction of the NaCl by-product with water, the high molecular weight PSS was separated from the insoluble and the low molecular weight fractions by centrifugation and fractional precipitation respectively. The highest yield thus attained was 55% along with 10% insoluble polymer and 35% low molecular weight polymer. The limiting viscosity number (THF at 25°C) was $[\eta]$ = 0.82-1.18 dl/g. On the hot stage microscope, softening of the polymer started at 60°C and flowing at 85-95°C. Tg, as determined by TMA on fibres, was 50°C. The DSC thermogram (10°C/min in N_2) revealed a glass transition at 65°C, a strong endothermic peak starting at 375°C with two maxima at 406 and 416°C, followed by an exotherm, all attributed to the decomposition of the polymer. From the TGA thermogram (10°C/min in N_2), the polymer initially lost up to 5% in mass, probably due to volatilisation of low molecular weight compounds and/or residual solvent, followed by a marked reduction in weight starting at 300°C and ending at 460°C. The ceramic residue (850°C in N_2) was 5-12% on various runs, which is lower than the reported values[5]. ^{1}H NMR (90 MHz)

showed two main broad peaks at 0.2 ppm ($-CH_3$) and 7.03 ppm ($-C_6H_5$).
Their relative areas were in an 8:5 ratio, which corresponds to the
chemical composition I, x = 0.45, y = 0.55. Elemental analysis, C
= 61.35, H = 7.7%, agreed fairly well with the structure proposed
above (x = 0.46, y = 0.54). Thus, the polymer had a slightly
higher content of $-Si(C_6H_5)(CH_3)-$ units than expected, indicating
a slightly higher incorporation of the $-Si(CH_3)_2-$ units in the insoluble and/or the low MW fractions. UV (THF, 20°C) revealed a
strong absorption at 331 nm (Si-Si) with a shoulder at 283 nm
($-C_6H_5$). FTIR absorption peaks (cm^{-1}) were assigned as follows:
3067, 3049, 1585, 1484, 1428 (C_6H_5), 3049, 2953, 1585, 1407, 959
($CH=CH_2$), 2953, 2894, 1407 (CH_3), 1246, 837 (CH_3 in $Si-CH_3$), 756,
731, 698 (CH_3, $CH=CH_2$, $Si-C_6H_5$), 1099 ($Si-C_6H_5$), 465 ($Si-Si-C_6H_5$),
2092 (Si-H), 1028, 1354 ($Si-CH_2-Si$), 1067 (Si-O-Si), 1055, 1138
($Si-CH_2-CH_2-Si$) and 3650 (free-OH). The presence of the otherwise
unexpected weak peaks at 3650, 2092, 1067 and 1028 cm^{-1} indicates
the involvement of free radicals during the polymerisation reactions,
as has been pointed out by other workers[2]. GPC analysis for the
sample with [η] = 1.18 dl/g gave a broad monomodal distribution
with $\overline{M}_n$ = 23 x 10^3 and $\overline{M}_w$ = 115 x 10^3.

Production of Fibres

PSS was spun into continuous fibres by both melt and dry
spinning. Melt spinning was performed at 220-260°C, through a
single hole spinneret. The fibres had a smooth surface and circular
cross section, with diameters ⩾ 60μm. Finer fibres with diameters
10-15 μm were obtained by drawing them over a hot plate. However,
due to breakages during drawing, continuous fibres were not
collected. PSS fibres were also obtained by dry spinning from
toluene or THF. Smooth, fine fibres, with diameters ⩾ 25 μm, were
obtained. Vinylsilanes were also incorporated into PSS fibres by
dissolving them in THF spinning dopes and dry-spinning. The relatively low spinning temperature of about 70°C permitted retention of
these additives by the fibres. Due to plasticising effects, the
maximum amount of the vinylsilanes it was possible to incorporate
in the PSS was 4%. With higher amounts the fibres appeared wet and
usually stuck together on the take-up device. All fibres were
collected in air and natural light, but were stored in the dark.

3 RESULTS AND DISCUSSION

The Effect of the Irradiation on Structural Changes

The FTIR spectra of PSS films, exposed to UV radiation in air or in an inert atmosphere (Ar or N$_2$), are shown in Figs. 1 and 2 respectively. The structural changes revealed are associated with the increase of the concentration of OH, SiH, C=O, SiOSi, SiC and SiCSi groups. Graphs of the ratios of the areas of the peaks arising from the above specified groups, along with the peak at 465 cm^{-1} (Si-Si-C$_6$H$_5$) are shown in Figs. 3 and 4. The ratio of the peak

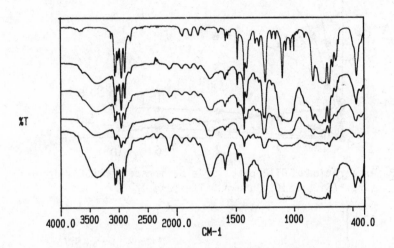

%T

Figure 1   FTIR spectra of a PSS film UV irradiated in air.

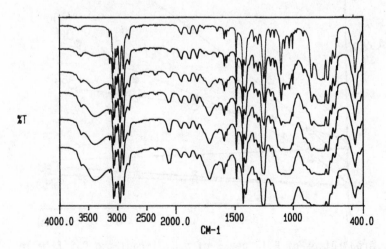

%T

Figure 2   FTIR spectra of a PSS film UV irradiated in argon.

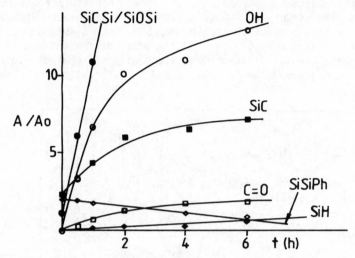

<u>Figure 3</u> Area ratios of FTIR peaks of a UV irradiated PSS film in
air. (SiCSi/SiOSi drawn half-scale.)

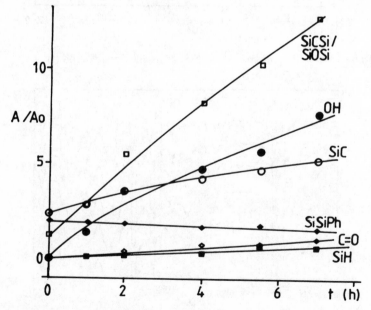

<u>Figure 4</u> Area ratios of FTIR peaks of a UV irradiated PSS film in
argon. (SiCSi/SiOSi drawn half-scale.)

areas for these different groups was calculated with reference to the mean area (A_0) of the peaks due to the C-H of the phenyl group at 3067, 1484 and 1428 cm^{-1}. All the curves are linear at low irradiation times. A decrease of the slopes is observed at longer irradiation times, especially of those bands correlated with faster structural changes (SiOSi, SiCSi and OH). This can be explained in terms of reduced concentrations of the reactive species on increasing the irradiation time and/or saturation of the FTIR peaks. All structural changes are faster in air than in argon, judged by the higher slopes of the corresponding curves. This indicates the positive role of atmospheric oxygen in these changes.

The structural changes are attributed to photochemical reactions, taking place during irradiation[2,4,8]. Generation of free radicals is expected to occur by scission of the weak Si-Si bond and, to a minor extent, of the SiC bond (Table 1). The presence of oxygen, water and traces of solvent affects the fate of these radicals.

Table 1 Bond dissociation energies $(E)^{15}$

Bond	E (kJ/mol)
Si-Si	280 - 350
Si-C	370 - 380
C-C	325 - 350
Si-H	360 - 420
C-H	320 - 440
Si-OH	475 - 535

Scission of the Si-Si bonds is indicated by the reduction of the intensity of the band due to Si-Si-Ph (Eq. 2).

$$\geqslant Si-Si\leqslant \quad \xrightarrow{\quad UV \quad} \quad 2\geqslant Si\cdot \qquad (2)$$

However, scission of the Si-Si bonds in the group $-Si(CH_3)_2-Si(CH_3)_2-$ can also take place, but this bond does not absorb in the region 4000-400 cm^{-1} and consequently cannot be detected by FTIR.

The silyl radicals may abstract hydrogen from various sources, generating Si-H bonds and new free radicals, react with oxygen or water to generate Si-O bonds (Eqs. 3-5) and finally react with other free radicals.

$$\gtrless Si\cdot + H\text{-}C\lessgtr \longrightarrow \gtrless SiH + \gtrless C\cdot \qquad (3)$$

$$\gtrless Si\cdot + O_2 \longrightarrow \gtrless SiOO\cdot \qquad (4)$$

$$\gtrless Si\cdot + H_2O \longrightarrow \gtrless SiOH + H\cdot \qquad (5)$$

Recombination of the silyl radicals is highly probable, due to the limited mobility of the chains in the solid phase. The disproportionation reaction observed in low molecular weight silanes is also possible, generating the highly reactive silene species (Eq. 6).

$$2 \;\; \overset{|}{-}Si(CH_3)\cdot \longrightarrow \overset{|}{-}Si(CH_3)H + CH_2=Si\lessgtr \qquad (6)$$

The silenes can react with a variety of groups. Reaction with oxygen yields formaldehyde and silanones (Eq. 7a). Formaldehyde could also be produced, according to Eq. 7b[9].

$$\gtrless Si=CH_2 + O_2 \longrightarrow \overset{|}{-}Si\text{-}CH_2 + \gtrless Si=O + HCHO \qquad (7a)$$
$$ \underset{O\text{-} \;\; O}{\phantom{-Si\text{-}CH_2}}$$

$$\gtrless Si\text{-}CH_2OOH \longrightarrow \gtrless Si\text{-}OH + HCHO \qquad (7b)$$

Formaldehyde is volatile. However, drying the irradiated film in vacuum at 40°C does not alter the relative intensity of the carbonyl band and consequently the contribution of trapped formaldehyde from these reactions to the carbonyl species must be small.

The increase of the concentration of Si-H groups is brought about by abstraction by the silyl radicals of hydrogen from various sources, including alkyl and phenyl groups from the polymer and/or the residual solvent (Eqs. 3, 8).

$$\gtrless Si\cdot + H\text{-}CH_2\text{-}Si\lessgtr \longrightarrow \gtrless Si\text{-}H + \gtrless SiCH_2\cdot \qquad (8)$$

The $\gtrless SiCH_2\cdot$ radicals can in turn combine with the same or another radical, creating new bonds. Two of these are of particular importance, because carbosilane structures are generated (Eqs. 9, 10).

$$\gtrless SiCH_2\cdot + \cdot H_2CSi\lessgtr \longrightarrow \gtrless SiCH_2CH_2Si\lessgtr \qquad (9)$$

$$\gtrless SiCH_2\cdot + \gtrless Si\cdot \longrightarrow \gtrless SiCH_2Si\lessgtr \qquad (10)$$

These desirable reactions are known to take place during the

thermolysis of PSS under argon[6] and also during the formation of
polycarbosilane by the thermolysis of polydimethylsilane (Scheme 1).
Thus 'crosslinking' bonds are formed, but at the expense of the
backbone Si-Si bonds. The increase in intensity of the band at
1000-1100 cm^{-1} is attributed, at least partially, to the formation
of these bonds, but the results cannot be quantified because the
SiOSi group absorbs in the same region.

The presence of oxygen and moisture is responsible for the
increase in concentration of the hydroxyl and carbonyl groups (Eqs.
4, 5, 11-16).

$$\geq SiOO\cdot + H\text{-}X \longrightarrow \geq SiOOH + X\cdot \quad (X=C, Si, OH) \quad (11)$$

$$\geq SiOOH \longrightarrow \geq SiO\cdot + \cdot OH \quad (12)$$

$$\geq SiO\cdot + R\cdot \longrightarrow \geq SiOR \quad (R=Si, CH_2, H) \quad (13)$$

$$\geq Si\cdot + \cdot OH \longrightarrow \geq SiOH \quad (14)$$

$$\geq SiCH_2\cdot + O_2 \longrightarrow \geq SiCH_2OO\cdot \quad (15)$$

$$2 \geq SiCH_2OO\cdot \longrightarrow \geq SiCH_2OH + \geq SiCHO + O_2 \quad (16)$$

Uncatalysed condensation of the Si-OH groups, a well known
reaction, is not likely to take place because it requires higher
temperatures (> 200°C). The formation of siloxane (SiOSi) and
carbosilane (SiCSi) groups results in an increase of the flexibility
of the chains. This facilitates the reactions of the free radicals,
but contributes greatly to the reduction of the Tg and thus the
softening of the films and fibres. The same effect is expected
from the formation of low molecular weight compounds, due to their
plasticising effect on the polymer.

The effect of UV irradiation on the structure of a PSS film,
loaded with 2% $C_6H_5Si(CH=CH_2)_3$, is shown in Fig. 5.

The inclusion of an unsaturated silicon compound was expected
to enhance the crosslinking process[4-5], through the following
reactions (Eqs. 17, 18).

$$\geq Si\cdot + CH_2=CH\text{-}Si\leq \longrightarrow \geq Si\text{-}CH_2\text{-}\dot{C}H\text{-}Si\leq \quad (17)$$

$$\geq Si\text{-}CH_2\text{-}\dot{C}H\text{-}Si\leq + R\cdot \longrightarrow \geq Si\text{-}CH_2\text{-}CH(R)\text{-}Si\leq \quad (18)$$

$$(R = Si, SiCH_2)$$

But if R = H or OH, as is highly probable due to the higher
mobility of these species, then the density of crosslinking will
be reduced. Hydrosilylation reactions between the -SiH and $CH_2=CH$-

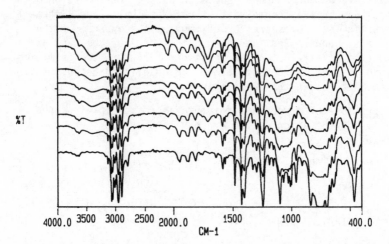

<u>Figure 5</u>   FTIR spectra of a PSS film, containing 2% PhSiVi$_3$,
          UV irradiated in argon.

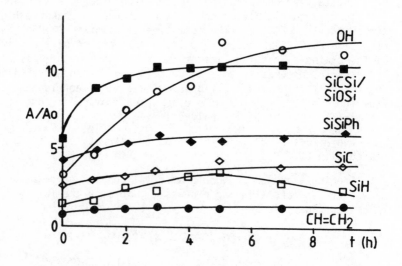

<u>Figure 6</u>   Area ratios of FTIR peaks of a PSS film, containing 2%
          PhSiVi$_3$, UV irradiated in argon.
          (SiCSi/SiOSi drawn half-scale.)

or =C=O, catalysed by the UV irradiation, can also yield crosslinking sites. From Fig. 6 it can be seen that the concentration
of the vinyl groups remains relatively constant. However crosslinking does take place, as shown by the lowering of the solubility
of the irradiated samples and the transformation of a solution of
PSS/PhSiVi$_3$/THF, but not those of THF/PhSiVi$_3$ or THF/PSS, into a
gel after standing for a few weeks exposed to natural light. Thus
the unaffected concentration of the vinyl groups can not be
explained, unless new ones are formed, possibly through the following reactions (Eqs. 19, 20), originally proposed for the colouration
of polystyrene exposed to UV radiation[10].

$$\geq Si-CH_2-\overset{\bullet}{C}H-Si\leq \; + \; R\cdot \; \longrightarrow \; \geq Si-\overset{\bullet}{C}H-\overset{\bullet}{C}H-Si\leq \; + \; RH \qquad (19)$$

$$\geq Si-\overset{\bullet}{C}H-\overset{\bullet}{C}H-Si\leq \; \longrightarrow \; \geq Si-CH=CH-Si\leq \qquad (20)$$

$$(II)$$

The structure II might contribute to the light yellow colour that
the film and fibres acquire after prolonged irradiation, in samples
both with and without added vinylsilanes.

   The intensity/time profiles of the Si-Si-Ph and Si-H absorptions
differ between the samples with and without added vinylsilane. In
the former the Si-Si-Ph peak area ratio increases slightly before
it levels off, instead of the continuous decrease observed with the
latter, and the Si-H ratio has a broad maximum, possibly due to
hydrosilylation reactions. Some further characterisation was
carried out, using UV spectroscopy and GPC. The films containing
vinylsilane were extracted with THF. The soluble fraction so
obtained became progressively smaller as the irradiation time increased. A shift of λ_{max} was observed, from 331.5 to 326 nm,
levelling off for the samples irradiated more than 3 h. Such a
shift can be explained by reduction of the Si-Si bond concentration,
as has been observed with other polysilanes[2,4,5,8]. The GPC data
showed that the molecular weights of the soluble fractions decreased
progressively from 115×10^3 to 7×10^3. In all UV irradiated
samples, broadening to higher wavenumbers of the two bands at 1246
(Si-CH$_3$) and 2098 cm^{-1} (Si-H) was observed, possibly due to an increase of the concentration of the O-Si-CH$_3$ and O-Si-H moieties,
as has been proposed by other workers[13].

   Finally, a comparison of the FTIR spectra of the irradiated
PSS films with that of a polycarbosilane, resulting from the
thermolysis of PSS (Fig. 7), shows significant differences.

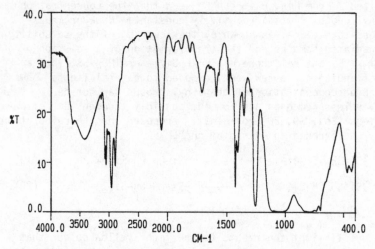

<u>Figure 7</u>   FTIR spectra of a polycarbosilane, resulting from
             thermolysis of PSS.

The intensity of the peak due to the Si-Si-Ph group is lower
in the polycarbosilane. On increasing the irradiation time, a new
band develops at 410-430 cm^{-1}, the intensity of which is higher in
the polycarbosilane. The intensities of the peaks due to hydroxyl
and Si-H groups are also higher. The weak band at 1354 cm^{-1} due
to Si-C-Si group can clearly be seen in the polycarbosilane but
not in the irradiated PSS. From this comparison it is concluded
that simple UV irradiation of PSS does convert it to a polycarbosilane, but its structure and consequently its properties are not
the same as those of the polycarbosilane resulting from the
thermolysis of PSS.

The Effect of Irradiation on the Ceramic Residue

Early experiments had shown that the ceramic residue of pyrolysed PSS fibres, UV irradiated in air or nitrogen, increased on
increasing the irradiation time. The maximum value attained was
40%. On increasing the irradiation time, the thick (diameter ~
200 μm) fibres became soft, light yellow and tended to stick together. Their solubility in THF decreased, although completely
insoluble fibres were not observed. During the subsequent pyrolysis (10°C/min up to 850°C in N_2), the shape of the fibres did not
survive[11].

The same fibres were irradiated by Y-rays in air or vacuum, and different results were obtained. Changes in the appearance of the fibres were not observed. The ceramic yields of these fibres on pyrolysis (10°C/min up to 850°C in N_2) are expressed as a function of duration of irradiation by curves, each with a single maximum, i.e. 20% in air and 16% in vacuo (Fig. 8). Again, the fibres lost their shape completely during pyrolysis. Irradiation in a nitrogen atmosphere gave similar results to vacuum irradiation, as judged by the treatment of selected samples. Additionally, FTIR spectra of all the fibres, irradiated by Y-rays in vacuum or in air, showed no apparent difference from the original sample. The spectrum of the latter revealed the presence of free but not of H-bonded hydroxyl groups and of Si-H and Si-C-Si/Si-O-Si groups, indicating that some oxidation took place during melt spinning, which had been performed in air at 220°C. The intensity of the peaks due to the above groups remained approximately the same for all the irradiated samples. The reactions taking place during irradiation by Y-rays and UV are different in that there is no evidence of structural changes on Y-irradiation.

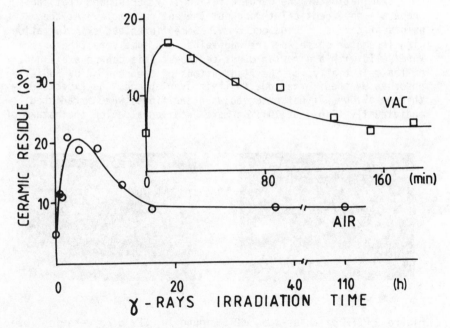

Figure 8   The effect of the duration of Y-irradiation on the ceramic residue of PSS fibres.

Experiments were carried out using fine (diameter ~ 30 μm) dry-spun fibres. These, produced with or without 1-4% tetravinyl-silane or phenyltrivinylsilane, were UV irradiated under a stream of nitrogen or in static argon for up to 36 h. The irradiated fibres became soft and during the subsequent pyrolysis (1°C/min up to 850°C in N_2), most of them were fused together. However, individual fibres maintained their shape and yielded ceramic fibres. The same results were obtained when the fibres were irradiated under a stream of nitrogen/tetravinylsilane vapour. The latter was included in order to promote crosslinking, at least at the surface of the fibres. This was expected to increase the Tg of the crosslinked polymer and subsequently reduce the fusing of the fibres during pyrolysis. This, however, was not the case, indicating that reaction between the SiH and CH_2=CH- groups did not occur to the required extent. Slower heating rates (0.1°C/min up to 80°C, isothermal for 3 h and heating up to 850°C at 1°C/min in N_2) of selected samples did not prevent fusion of the fibres but rather the reverse.

Examination of the ceramic residues by SEM showed that their shape was not identical throughout the entire sample. At the points of contact fusion occurred. Some fibres had skin formation with irregular shape, others had voids, and some very fine ones were solid with a circular cross-section. This can be explained, at least partially, by the fluctuations of the dose of UV-radiation absorbed by the fibres, due to their overlapping. As expected, this behaviour was not observed when the fibres were irradiated individually (Figs. 9, 10). A more detailed study on the nature of

   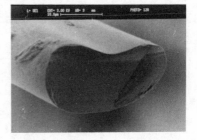

Figure 9 SEM of an as-spun PSS   Figure 10 SEM of a ceramic fibre
fibre, produced by dry spinning.   produced by pyrolysis (5°C/min up
                                     to 850°C in N_2) of a PSS dry-spun
                                     fibre UV irradiated in argon for 16h.

the ceramic residue as a function of the experimental conditions will be published at a later date.

Thus, UV irradiation of the PSS fibres results in some cross-linking and an increase of the concentration of reactive groups, whose concentration and/or reactivity is just enough to crosslink the fibres sufficiently to maintain the fibrous shape in subsequent pyrolysis. Surface softening results from formation of low molecular weight species and of siloxane and carbosilane groups. The latter groups, although thermally stable, increase the flexibility and so increase the back-biting reactions during pyrolysis and the loss of the polymer through the formation of cyclics.

In a recent patent, dicumyl peroxide has been shown to cross-link PSS, but under thermal conditions that are too extreme for fibres[12]. Some experiments were carried out in order to estimate the crosslinking efficiency of another typical thermal free radical initiator, asobisisobutyronitrile (AIBN) which has a lower decomposition temperature than dicumyl peroxide and does not contain oxygen. Continuous PSS fibres, containing 5% AIBN treated at 60°C for 12 h, did not show any structural changes, as judged by FTIR. Solution treatment of PSS in xylene with 10% AIBN at 100°C for 24 h gave no detectable change in the recovered PSS, either by FTIR or measurement of pyrolysis residue. Clearly, AIBN is not an effective crosslinking agent for PSS under these conditions.

4 CONCLUSIONS

PSS on UV irradiation undergoes structural changes, including the scission of Si-Si and the formation of Si-OH, Si-H, Si-O-Si, Si-C-Si and C=O bonds. Siloxane and carbosilane groups and low molecular weight photolysis products contribute to a reduction of Tg and consequently to a deterioration in the handling properties of the fibres. The inclusion of vinylsilanes does not seem significantly to improve the crosslink density. More uniform irradiation of finer fibres and individual handling of them during the pyrolysis step are needed to produce continuous ceramic fibres of improved quality. Y-irradiation of PSS fibres, unlike UV irradiation, produces no useful changes.

ACKNOWLEDGEMENTS

We would like to thank Dr. M.G. Dobb for SEM, Dr. J. Fisher and Mr. S.A. Barrett for NMR, Dr. P.G. Laye and Mr. F. Hongtu for TGA/DTA, Dr. K.D. Bartle and Mr. S. Pape for GPC, Mr. L. Johnson for some early work on DSC/TGA and Drs. A.H. Milburn and J.T. Guthrie for valuable comments. The work was supported by a grant from the SERC.

REFERENCES

1. (a) J.P. Wesson and T.C. Williams, J.Pol. Sci., Chem. Ed., 1980, 18, 959.
    (b) K.S. Mazdiyasni, R. West and L.D. David, J. Am. Ceram. Soc., 1978, 61, 504.
    (c) R.E. Trujillo, J. Organomet. Chem., 1980, 198, C27.
2. For a review on the properties and applications of polysilanes, see R.D. Miller and J. Michl, Chem. Rev., 1989, 89, 1359.
3. S. Yajima, Phil. Trans. R. Soc. Lond., 1980, A294, 419.
4. For a review on the work by West's team, see R. West, J. Organomet. Chem., 1986, 300, 327.
5. (a) R. West, L.D. David, P.I. Djurovich, H. Yu and R. Sinclair, Ceramics Bull., 1983, 62, 899.
    (b) R. West, J. Maxta, R. Sinclair and P.M. Cotts, Pol. Preprints, 1987, 28, 387.
    (c) R. Sinclair, in 'Ultrastructure Processing of Ceramics, Glass and Composites', L.L. Hench and D.R. Ulrich (Eds.), John Wiley, N.Y., 1984, Ch. 21.
6. E.P. A2,212,485 (Teijin Ltd.), 4.3.1987.
7. (a) A.L. Smith, Spectrochim. Acta, 1960, 16, 87.
    (b) T.R. Crompton, in 'The Chemistry of Organic Silicon Compounds', S. Patai and Z. Rappaport (Eds.), John Wiley, N.Y., Part 1, Chapter 6.
8. (a) W.P. Weber, in Ref. 5, Ch. 24.
    (b) J.M. Zeigler, L.A. Harrah and A.W. Johnson, Proc. SPIE, 1985, 539, 166.
    (c) R.D. Miller, G. Wallraff, N. Clecak, R. Sooriyakumaran, J. Michl, T. Karatsu, A.J. McInley, K.A. Inngensmith and J. Downing, Pol. Eng. Sci., 1989, 29, 882.
    (d) T. Karatsu, R.D. Miller, R. Sooriyakumaran and J. Michl, J. Am. Chem. Soc., 1989, 111, 1140.
    (e) R.D. Miller, in 'Silicon-based Polymer Science', J.M. Zeigler and F.W.G. Fearson (Eds.), Advances in Chemistry, Series 224, Am. Chem. Soc., 1990, Chapter 24.
9. A.D. Delman, M. Landy and B.B. Simms, J. Pol. Sci., 1969, A1, 7, 3375.
10. B. Ranby and J.F. Rabek, 'Photodegradation, Photo-oxidation and Photostabilisation of Polymers', John Wiley, N.Y., 1975, Chapter 4, page 165.
11. J.C. Ko, unpublished results.

12. US 4,804,731 (University of Florida Research Foundation Inc.), 14.2.1989.
13. (a) J. Devaux, J. Sledz, F. Schue, L. Giral and H. Naarman, J. Makromol. Chem., 1990, 191, 139.
    (b) P.T. Trefonas III and R. West, J. Pol. Sci., Pol. Lett. Ed., 1985, 23, 469.
14. S.D. Rosenberg, J.J. Walburn, T.D. Stankovich, A.E. Balint and H.E. Ramsden, J. Org. Chem., 1957, 22, 1200.
15. (a) Ref. 10, Chapter 2, page 45.
    (b) R. Walsh in Ref. 7b, Part 1, Chapter 5.

Analytical, Physical Chemical and Health and Safety Aspects

Experimental Techniques to Monitor the Degree of Cure in Radiation-cured Coatings

A.K. Davies

DEPARTMENT OF CHEMISTRY AND APPLIED CHEMISTRY, UNIVERSITY OF SALFORD, SALFORD M5 4WT, UK

1. INTRODUCTION

Selection of the optimum conditions for radiation curing is essential if the process is to be economical and the product is to have the desired physical properties. Excessive radiation doses whether ultraviolet (UV) or electron beam (EB), are not only wasteful in energy but may result in embrittlement and yellowing of the coating.

For the purpose of determining optimum radiation curing conditions and for quality control, various methods for monitoring curing have been developed.

With an acrylate resin the C = C double bonds of the acrylate groups disappear on radiation curing as the monomer molecules are converted to cross-linked polymer. The percentage conversion of an acrylate resin coating may be considered therefore to be the number of polymerisable double bonds which have reacted expressed as a percentage of the total number of such bonds present initially in the original uncured coating.

The most direct method of measuring the concentration of C = C double bonds is infra-red spectroscopy; the percentage conversion may be expressed as

$$\frac{(A_0 - A_t)}{A_0} \times 100\%$$

where A_t is the absorbance after exposure time t due to the C = C double bond (at the stretching frequency or the twisting frequency) and A_0 is the absorbance due to this

bond at the same frequency for the original uncured film.
Some workers prefer to use the area under the peaks
rather than the absorbance values in the above
expression.[1,2]

Alternatively, using a calorimetric method to
measure the enthalpy of polymerisation the degree of cure
is given by[3]

$$\frac{\Delta H_t}{\Delta H_o} \text{ x } 100\%$$

where ΔH_t is the enthalpy change of polymerisation after
a given period of irradiation (t) and ΔH_o is the enthalpy
change for complete conversion of the acrylate C = C
double bonds to C - C single bonds in the polymer.   The
theoretical value of ΔH_o is 86.1 kJ mol^{-1}.

In order to state in these exact terms whether a
coating is "well-cured", i.r. spectroscopy, calorimetry
or some other method which has been calibrated against
one or other of these two techniques is required. Often,
it is much quicker and more convenient to use simpler
techniques to enable one to decide when a coating is
"well-cured". In other words, the meaning of the term
"well-cured" is determined more by end-use performance
rather than simply by the disappearance of cross-linkable
moieties. Indeed, the term "functionally-cured" is
sometimes used to indicate the point at which the desired
performance characteristics of the coating have been
attained.[4] The attainment of these characteristics may
be judged by solvent double rubs, pendulum hardness and
adhesion tests for example.

Some workers equate "degree of cure" with percentage
conversion, but as Brann[45] has pointed out, a true
definition of "cure" is very difficult to formulate
because different applications may require varying
precentages of conversion. With a particular radiation
- curable resin it may be impossible to obtain one-
hundred percent conversion and indeed it may be quite
unnecessary. Moreover, it is important to remember that
"curing" imples cross-linking between multifunctional
monomers and oligomers and consequently, even when a
significant amount of unsaturation remains, the quantity
of extractable monomer may be exceedingly small. For
these reasons it is desirable to separate the terms
"degree of cure" and "percentage conversion". The degree

of cure may be considered to be the extent to which the properties of the coating approach those of the functionally cured state.

Measurements of the degree of cure and the kinetics of curing may be classified into two main types – (a) real-time and (b) intermittent. In real-time monitoring the measurement of some property of the coating is made continuously while curing takes place. In the intermittent type of measurement the coating is given a certain radiation dose after which measurements are made – the measurements being repeated after each subsequent exposure.

"Real-time" or continuous monitoring is possible with i.r. spectroscopy, i.r. radiometry, calorimetric methods, dilatometry, and with the UV-Curetester.

The ideal solution to the problem of measuring the degree of cure for the purposes of optimising curing conditions and for quality control, would be provided by a non-destructive, real-time quantitative method sensitive to small differences in the degree of cure as the coating entered the latter stages of the curing process and which could be used directly on a production line. Such a method would not make laboratory-based techniques redundant however for there will always be a role for these in the development of new products.

Certain kinds of measurement (for example dilatometric) are more suited to the initial or middle stages of curing while others (e.g. pencil hardness) are appropriate only for the latter stages of curing when the coating is dry and is approaching the fully-cured state. For this reason a more reliable assessment of the degree of cure of a coating is achieved by combining more than one technique. Frequently this means the combination of a quantitative technique (e.g. infra-red spectroscopy) with a simpler comparative technique which is at the same time convenient, rapid and suitable for in situ measurements.

There is another very good reason for not depending on the results of a single type of measurement. This is the fact that UV-cured coatings rarely possess uniformity of cure. For example, it is well known that oxygen-inhibition may cause incomplete surface cure, whereas high concentrations of photoinitiator may lead to poor light penetration resulting in the coating being well-

cured at the surface but under-cured in the deeper
layers[3,4]. The use of more than one technique is much
more likely to reveal such inhomogeneity of cure.

With electron beam curing there is the possibility
of chain scission and cross-linking occurring when this
form of radiation interacts with the oligomer and the
newly formed polymer[5]. Thus, it is particularly
important in assessing the degree of cure of an EB-cured
coating to use methods which are sensitive to further
change once the C = C double bonds have disappeared.

In principle, any physical or chemical property of a
radiation-curable system which changes as polymerisation
progresses can be used as a means of measuring the degree
of cure. Some methods are essentially qualitative and
rather subjective but with experience they serve as
useful guides to the degree of cure and, although they
lack exactness, they have the advantages of speed and
simplicity and they can be carried out in situ.

2.   EXPERIMENTAL TECHNIQUES

Finger-tip Test

In this simplest of tests, the surface of the
coating is first touched and then rubbed with the finger
tip to register any tackiness and then to determine how
readily the surface can be marked. A variation of this
test is the "thumb-twist" test.

Solvent Rub Test

This is the determination of the number of double
rubs (with a cloth soaked in a suitable solvent - usually
methyl ethyl ketone, MEK) which can be performed on the
coating before it is finally rubbed away to reveal the
substrate (normally a steel plate). A double rub is one
forward and backward motion of the solvent-impregnated
cloth. The test may be carried out by hand although some
mechanical means is preferable[4]. Using this test, useful
comparisons between different curing conditions can be
made.

Cotton Test

This is a purely qualitative test in which a cotton-
wool ball is touched on the coating. The tackiness of

the coating is assessed from the amount of cotton which adheres to the surface.

Permanganate Stain Test[4]

A quantity of potassium permanganate solution (1% w/v in water) is applied to the coating to cover an area of about 1/2" diameter. After five minutes the surface of the coating is rinsed with water and examined. The intensity of the brown stain is a measure of the unreacted material in the coating, i.e. an undercured coating will produce a deeper stain than a well-cured coating. The test can be quantified by colorimetric measurements. If, on repeating the test on the same coating some hours after curing, a darker stain is produced, migration of residuals to the surface from the lower regions of the film is indicated.

Calcium Carbonate/Carbon Black Test[4]

A mixture of calcium carbonate (99.5%) and carbon black (0.5%) is wiped over the surface of the coating with a tissue. After the residual powder has been wiped off, the degree of cure is judged by the amount of staining. The test is affected by any form of residual which migrates to the surface of the coating.

Hardness Tests

These are arbitrarily defined measures of the resistance of a material to indentation under static or dynamic load, to scratching, abrasion or wear or to cutting or drilling[6]. A number of hardness tests have been used to investigate the curing of radiation - polymerisable systems.

Microhardness[6]. This is the resistance to indentation over very small areas. One procedure employs a Tukon tester applying loads of 25 - 2600 g using a Knoop indenter which is a diamond cut to produce a diamond-shaped impression with a ratio of diagonal lengths of 7:1. The location of the indenter and measurement of the diagonals of the impression is accomplished with microscopes. The hardness number is the ratio of the applied load to the projected area of the impression (units: $kg\ mm^{-2}$). This method has been used, for example, to investigate the depth of cure in UV-cured dimethacrylate-based dental filling materials[7].

<u>Pencil Hardness</u>. In this test a sample strip is drawn under a pencil (for example one with changeable 0.5 mm flattened leads)[5] until a hardness grade is reached which will scratch the surface. The coating is then assigned a hardness value such as H, 2H, 3H etc signifying the hardest grade which will not penetrate the surface.

<u>Pendulum Hardness</u>. A more exact evaluation of hardness can be made by measuring the damping by the coating of the oscillations of a pendulum (Persoz hardness[8], König hardness[9,10]). The König pendulum hardness tester is convenient to use and is particularly useful for small samples of cured coatings. The pendulum is enclosed in a case to eliminate draughts and to facilitate control of humidity. The release mechanisms which lower the pendulum on to the specimen and begin its oscillations are operated from outside the case. The apparatus may be standardised by adjusting the centre of gravity of the pendulum (by movement of a weight up or down a threaded rod) so that a time of 250 seconds is obtained when the sample is plate glass. This is the time taken for the amplitude of the oscillations to decrease from 6° to 3° from the vertical which is measured on a scale at the lower end of the pendulum[10]. The higher the degree of cure, the longer the pendulum will oscillate. The pendulum suspension points rock on a small area of the coating and so it is easy to avoid surface imperfections; at the same time it is desirable to take the mean of a number of measurements.

In making the measurements it is important to realise that both the thickness of the coating and the hardness of the substrate have a marked influence on the results[3]. By adopting a standard procedure and a standard set of conditions, kinetic measurements are possible, once a dry coating has been formed, as further curing of the coating takes place[3,9].

<u>Sward Hardness Test</u>[10]. The device consists of a weighted rocker which rocks to and fro over the surface of the coating. The harder the surface the greater the number of rocking motions recorded. Because it 'samples' a much larger part of the coating than pendulum hardness testers it is more sensitive to surface imperfections.

Pencil, pendulum and rocker methods of hardness testing have an important role in curing studies because they are sensitive to small differences in the degree of cure in the latter stages of curing. More sophisticated techniques such as infra-red spectroscopy, tend to be rather insensitive to the subtle changes that occur as curing reaches completion. Indeed, this is especially so for EB curing[5] in which further changes in coating properties (possibly due to chain scission and cross-linking) may occur after i.r. spectroscopy has shown that all the polymerisable double bonds have disappeared.

Sutherland Rub Test[4]. In this test two samples are cut from the coated material and rubbed face to face under a certain load for a known period after which the surfaces are examined for scuffing.

Taber Abrasion Test[4]. In this test the coating is rotated under an abrasive wheel and either (a) the weight loss after a fixed number of cycles is measured or (b) the number of cycles required to just wear through the coating is recorded.

Some workers find it convenient to use their own simple hardness tests such as the '10p coin scratch test'[11]. These are rapid but of course, very subjective.

Dilatometry

Dilatometry is one of the most commonly used methods for following the course of a photopolymerisation reaction for the simple reason that such reactions are almost invariably accompanied by a change in volume. For practical reasons, dilatometry is most often used to measure photopolymerisation in solution or the early stages of polymerisation of the neat monomer. It is often the method of choice when detailed kinetic studies of photopolymerisation are undertaken. Frequently, the quantity of polymer formed, after a known period of irradiation in the dilatometer vessel, is determined by pouring the polymer/monomer mixture into a quantity of solvent in which the polymer is insoluble. The precipitated polymer is filtered off, dried and weighed. Characterisation by GPC of the resultant polymer is often carried out.

In a determination of polymer yields in highly polymerised samples[12] the sample was cooled to 77 K in liquid nitrogen and crushed to a fine powder. The powder was transferred to a high speed blender containing a solvent which dissolved the monomer but not the polymer.

The insoluble polymer was filtered, dried and weighed and the residual monomer determined by bromometry.

A very simple type of dilatometer, in the form of a graduated glass tube 23 cm long and 4 mm O.D. in which the monomer and photoinitiator were sealed, has been used successfully to measure the initial rate of UV-induced polymerisation[13]. Clearly, the measurements cannot be carried out beyond a certain stage.

The use of collapsable containers for the monomer sample have allowed photopolymerisation to be carried out to completion. Sample volumes as low as 100 μl have been employed and attempts have been made to produce polymer films within the dilatometer[14].

Point-by-point measurement of the level of the meniscus of the indicator liquid in a fine-bore capillary tube is a time-consuming and laborious process. Consequently a number of transducers have been developed which allow automatic recording of the volume change. For example, a thermomechanical analyser (TMA) was adapted by McGinniss and Ting[15] to provide a recording dilatometer in which the acrylate sample ($2cm^3$) contained in a collapsable polyethylene sample cell was placed in a 20 cm^3 glass syringe full of water. Contraction of the sample upon UV-irradiation was sensed by the linear variable differential transformer (LVDT) whose output to a mV recorder provided a continuous record of volume change against time. With a similar arrangement, some problems were encountered with the movement of the syringe plunger which was not as smooth as desired[16].

An alternative approach is to fix the syringe plunger in one position and to connect the luer end of the syringe barrel to a thin-walled precision bore glass capillary tube. Adjustment of the position of the water/air interface in the capillary and calibration are achieved by a micrometer syringe attached by a 'T'-piece to the apparatus[14]. Two types of transducer have been used to measure the movement of the water-air interface along the capillary in response to the contraction of the sample. One employs an optical method in which light from a tungsten lamp is passed through a green filter, through the side of the capillary and onto a selenium photocell which is connected directly, or for greater sensitivity via an amplifier, to an X-t recorder[17]. The capillary is situated behind a slit and the optical

properties of the arrangement are such that the ratio of
the signal obtained when the tube is full of water to the
signal obtained when it is full of air is about 10.
Thus, movement of the air-water interface in response to
the photo-curing of the sample produces a proportional
drop in the output signal.

An alternative transducer is a capacitance bridge in
which a glass capillary tube is sandwiched between fixed
metal plates. As the sample contracts and the water-air
interface moves along the capillary, there is a
proportional change in the capacitance of the assembly
which is registered by an X-t recorder[18].

Dilatometric techniques in which photocuring is
carried out to completion have a distinct advantage over
other techniques in that they allow the shrinkage of the
sample to be measured directly. A knowledge of the
percentage shrinkage is important in all applications of
photo-curable resins particularly in optical applications
where shape accuracy and absence of internal strain are
vital[19].

In another interesting variation of the dilatometric
technique the photo-curable sample is sandwiched between
a lower quartz plate and an upper thin glass cover-
slip[19]. A linear displacement transducer in contact with
the cover-slip registers the contraction of the sample
following its exposure to UV radiation from below. The
output of the transducer is recorded by a mV recorder or,
for high curing rates, by a memory scope. The total
shrinkage can be measured and the rate of photo-curing
can be expressed in terms of $t_{1/2}$ - the time taken for the
sample to undergo half its total shrinkage.

Infra-red Spectroscopy

Infra-red (i.r.) spectroscopy has gained great
prominence in the measurement of the kinetics of
radiation curing and the degree of cure. One advantage
of this technique over most others lies in the form of
the sample which is a thin film, so that the conditions
of irradiation can be made very similar to those
encountered in current commercial coating practice. It
is a quantitative technique and gives a direct measure of
the concentration of the carbon-carbon double bonds which
disappear on curing. This is done for acrylates by
measuring the fall in the absorbance of the sample at
~1630 cm^{-1} and/or ~810 cm^{-1} as photocuring progresses.

These frequencies correspond to the stretching and twisting vibrations of the C=C double bond respectively[3].

In one example[20] of the application of i.r. spectroscopy, a UV-sensitive acrylate ester composition was sandwiched between salt flats which had been pre-treated with a transparent fluorocarbon mould-release spray to prevent permanent adhesion of the cured film to the surfaces of the salt flats. A Teflon spacer (0.015 mm thick) allowed standard thicknesses of resin to be applied. The sandwiched sample was placed directly in the holder of the i.r. spectrometer and irradiated with UV radiation while the fall in intensity of the acrylate C=C double bond stretching band was recorded continuously.

With films applied to metal test panels similar data can be obtained using a specular reflectance attachment for the infra-red spectrometer[20]. With this arrangement it is possible to study the influence of various factors on the curing curve - in particular the effect of the atmosphere above the coating in relation to oxygen-inhibition. The percentage conversion can be evaluated directly from the chart.

Infra-red spectroscopy has been applied in an intermittent mode to the monitoring of the degree of cure in samples exposed to radiation from a standard laboratory UV curing unit[21]. The salt flats were drilled so that they could be held together with screws. A thin copper spacer allowed a reproducible quantity of the UV-curable resin to be sandwiched between the salt flats. A shutter device fitted with a 3 ft cable release was fixed over the sample and the whole assembly was placed on the conveyor belt of the curing unit. The sample was exposed to the UV source by releasing the shutter at the precise moment the assembly was in the optimum position under the UV lamp. After exposure in this manner the i.r. spectrum of the sample was obtained. The process was repeated a dozen or so times. Spectra were analysed in some detail - the calculation involving the input of six percentage transmission values so that, after each exposure, the percentage unsaturation corrected for variation in film thickness etc. was obtained.

Another study[22] showed that a percentage conversion of 97%, as measured by i.r. spectroscopy, corresponded to a hard film while a coating having almost as high a percentage conversion (95%) still had a tacky surface.

This difference was shown to be due to inhibition of surface cure by oxygen in the air. When air was excluded, a percentage conversion as low as 87% was sufficient to provide a tack-free surface as judged by the finger-tip method.

With pigmented systems or clear resins on opaque supports or where the coating thickness is > 50μm, multiple internal reflection i.r. spectroscopy (MIR) is another form of the technique which can be used[3], although, if difficulties arise in achieving uniform contact between the sample and the thallium bromide crystal, the results are likely to be at best qualitative[5].

Diffuse Reflectance Fourier Transform Infra-red spectroscopy (DRIFTS). This technique gives quantitative data but it is unsuitable for glossy coatings such as those given by clear acrylate resins[5].

Real-Time Infra-red (RTIR) Spectroscopy for Ultra-Fast Curing[23]. Infra-red spectroscopy has been used to monitor, in real-time, ultra-fast (< 1 second) photo-curing. The sample was coated on to a KBr salt flat and exposed at one and the same time to the polymerising UV beam and the i.r. monitoring beam (frequency 812 cm^{-1}). An infra-red spectrometer was used in conjunction with a transient memory recorder making it possible to follow polymerisation processes with half-cure times > 30 ms. In a study of the curing of a polyurethane-acrylate coating using this technique it was noted that even though most of the acrylic double bonds had reacted after a short exposure (0.1 s) it took much longer (0.5 s) for a tack-free surface to be formed and even longer (2 s) for the formation of a scratch-free surface. This was primarily due to oxygen-inhibition of curing in the top layers of the coating. This example emphasises the importance of applying more than one technique in assessing the degree of cure of a coating.

Photoacoustic Spectroscopy

Another spectroscopic technique which holds great promise in photocuring studies is photoacoustic spectroscopy (PAS). The chief reason for applying this technique is 'depth-profiling'[24] - the unique ability of PAS which, in principle, allows one to determine, for example, if the degree of cure at the surface of a coating differs from that of the underlying layers.

In photoacoustic spectroscopy the sample to be examined is contained in a gas-tight cell and is exposed to a modulated beam of light of variable wavelength. Because a portion of the absorbed light is degraded to heat, a thermal wave is produced in the sample. This thermal wave, which has the same frequency as the modulated light beam, travels to the sample boundary where it produces an acoustic wave in the gas. This wave propagates through the volume of the gas to a sensitive microphone where a signal is produced.

The thermal wave produced in the sample is heavily damped and may be considered to be fully damped within a distance $2\pi\mu_s$ where μ_s is the thermal diffusion length. It is normally assumed that only those waves originating from a depth $\leq \mu_s$ will make an appreciable contribution to the photoacoustic signal. The thermal diffusion length is a function of the thermal diffusivity (x) and the modulation frequency of the incident radiation (ω).

$$\mu_s = \sqrt{\frac{2x}{\omega}}$$

This relationship means that for a sample of given thermal diffusivity the depth examined may be varied by altering the modulation frequency.[24]

UV/visible PAS has been used to follow the curing of photopolymerisable resins but it has proved to be difficult to quantify the degree of cure as a function of depth[25]. The need for greater instrument sensitivity and higher modulation frequencies was emphasised. The amount of information relating to photocuring, which is available from UV-visible PAS appears to be rather limited at present.

Of much more immediate use is the closely related technique of Fourier Transform infra-red photoacoustic spectrosocpy (FTIR-PAS).[5,24] This technique has been used to examine the extent of oxygen-inhibition in the UV-induced curing of acrylates and also to investigate the degree of cure of acrylates polymerised by EB radiation.[5] In this latter study the acrylates were coated on to tracing paper. The intensity of the PAS spectrum was shown to depend not only on the concentration of the absorbing species but also on the thermal properties of the coating which change with increasing radiation dose. By ratioing the magnitude of the peak under investigation against an absorption band that exhibited little variation throughout the curing process, excellent kinetic curves were obtained. A significant finding of this study was that even when most of the available double bonds had been consumed, the properties of the film as judged by pencil hardness and solvent resistance, continued to change with increasing dose – presumably due to cross-linking between polymer chains.

Ultra-violet/visible Spectrophotometry

In certain cases this technique can be used to follow the curing process. In one example photo-crosslinkable polyesters of p-phenylene bis(acrylic acid) and related polyesters were studied in connection with laser imaging applications.[26] In these systems the spectral sensitivity extends into the near-UV region due to the conjugation of double bonds with the phenylene rings. For measurements of photo-crosslinking by UV-visible spectrophotometry the polyesters were coated as thin films on to quartz plates. Similar coatings on salt flats enabled the change in the i.r. spectrum to be examined. Upon irradiation with near UV radiation the coatings showed a marked fall in the absorbance of the π–π^* bands. Correspondingly, i.r. studies showed a decrease in absorbance at ~1610 cm^{-1} due to the carbon-carbon double bonds together with a shift of the 1710 cm^{-1} band to 1720 cm^{-1} indicating the formation of saturated esters. The experimental evidence was consistent with crosslinking due to photo-dimerisation of double bonds leading to the formation of cyclobutane rings.

Calorimetric Methods

Photopolymerisation reactions are generally exothermic since double bonds of monomers are converted into single bonds whenever the monomers add to polymer radicals. For this reason calorimetric studies of photopolymerisation have been carried out using a variety of calorimeter designs. For example, a highly sensitive thermal leak calorimeter equipped with a UV irradiation unit was used to measure photopolymerisation of lauryl acrylate initiated by benzoin methyl ether.[27] The heat of the photo-induced chemical reaction was determined by

detecting the temperature difference between the sample
and reference cell, using thermopiles (sensitivity 12.8
mV $^{\circ}C^{-1}$) calibrated against electrical heaters beneath
the sample holders.

Many examples of the application of calorimetry to
UV-curing have employed commercial differential scanning
calorimeters (DSC) adapted to allow the sample to be
irradiated with a UV lamp[28-30]. In one modification of a
DSC instrument a light guide was used to transmit the UV
radiation to the sample vessel which was a modified
transistor can.[28]

An important feature of modified DSC equipment is
that once the photo-reaction exotherm has been recorded,
measurement of 'post-cure' can be achieved by gradually
raising the sample temperature to complete the curing
process.

Differential Photocalorimetry (DPC)[1,31,32,33]. This
type of commercial instrumentation uses dual sample DSC
to measure the heat of reaction of one or two samples as
they are exposed to radiation - usually that from a high
pressure mercury vapour lamp, although other
interchangeable lamps are available. An internal
standard may be subjected to identical conditions of
light intensity, wavelength, temperature and atmosphere.
A photo-feedback system maintains the intensity of the
lamp constant over long periods.[31]

The dual-sample facility allows the photo-curing
behaviour of two samples to be compared under identical
irradiation conditions. Data analysis software provides
a plot of percentage conversion versus time. The
instrument also allows one to study the increase in the
glass transition temperature of samples with increasing
UV dose.

Infra-red Radiometry[34]

Infra-red radiometry is an in situ on-line technique
for monitoring photo-curing. The radiometer consists of
a tube containing a ceramic pyroelectric infra-red
detector, lenses and filters. It is shielded from light,
thermally isolated and mounted on a tray which carries
the coated sample. It detects the emitted thermal
radiation from an eliptical area of the sample
approximately 2 cm wide and 8 cm long. The output of the

radiometer is connected to a mV recorder or computer. The radiometer and sample are conveyed through the UV curing unit on the conveyor belt.

The recorded output of the radiometer is a plot of differential heating rate dT/dt against time. When a photocurable acrylate coating is passed, together with the radiometer, through the UV-curing unit two peaks are recorded - one corresponds to the heating effect of the UV lamp while the other is the exotherm due to the polymerisation process. By carrying out a second pass to measure the heating effect of the UV lamp alone, a second trace is obtained which can be subtracted from the first one to give the true exotherm of the polymerisation process.

The i.r. radiometer has considerable potential and represents a breakthrough in instrumentation. With it one can measure both the kinetics and the temperature profile of curing. It is a simple, rapid method for optimising curing variables under production conditions.

Dynamic Mechanical Analysis (DMA)[31]

Dynamic mechanical analysis is the measurement of the mechanical response of a material as it is deformed under periodic stress. The sample is forced to undergo oscillatory motions by the application of sinusoidal forces. Because the sample is viscoelastic, energy dissipation in the sample causes the resulting maximum strain to lag behind the maximum stress (phase shift). By measuring, as a function of temperature, the dynamic modulus (stress/strain) and the phase shift, subtle polymer transitions can be detected with greater sensitivity than is attainable with DSC. For example the change in Tg, as a polymer sample undergoes photo-crosslinking reactions, can be measured readily.

Evaporative Rate Analysis (ERA)[35,36]

The ERA technique involves the deposition of a 20 μl drop of a commercial pre-formulated test solution on to the surface of the coating. The test solution is a low boiling solvent (or mixture of solvents) containing a high boiling but completely volatile ^{14}C-labelled compound in the ratio ~1:1 x 10^{-5}. A thin-end-window Geiger Mueller detector is positioned immediately above the droplet on the surface and gaseous nitrogen is passed

at a controlled rate between the surface of the coating
and the detector.

The amount of swelling of the coating caused by the
low-boiling solvent decreases with increasing degree of
cure. Thus, the invasion of the labelled compound into
the coating decreases with increasing degree of cure.
This means that the rate of evaporation of the labelled
compound from a well-cured surface is faster than that
from an undercured surface. The retained levels of
radioactivity, shown by much flatter regions of the log
(count-rate) versus time curves, are inversely
proportional to the degrees of cure of the coatings.

Microwave Dielectrometry[37,38]

This method for measuring degree of cure is rapid
and accurate. It depends on the measurement of the
variation of the resonance frequency of a microwave
cavity containing the photo-polymerisable resin. The
frequency change is related to $\Delta\varepsilon'$, the change in the
real part of the dielectric constant. The relaxation
times of monomer dipole moments when inserted in the
macromolecules are several orders of magnitude longer
than those of the free monomers. The change of
dielectric relaxation is ascribed to the restricted
mobility of the dipoles in the polymer chains. The
polymerisation process is analysed by plotting the
normalised value of $\Delta\varepsilon'$ versus time. The curves obtained
correlate closely with those obtained using i.r.
spectroscopy.

Resistivity

Resistivity is another property of a radiation-
curable resin which changes with the degree of cure. In
one study[15] a Teflon electrical resistivity cell, having
stainless steel electrodes, was filled with a UV-curable
resin to give film thicknesses of between 3 and 45 μm.
The change in resistivity of the resin with increasing UV
dose was recorded continuously. It was concluded that
the increase in resistance was due to decreasing mobility
of the charge-carriers by the changing structure and
cross-linking of the resin during photo-curing.
Intermittent measurements were also made following a
series of exposures of the sample to the radiation from a
commercial UV curing unit.

Recent experiments have shown that measurement of the electrical conductivity of a commercial UV-curable acrylate resin, with a conductivity bridge operating at 1kHz can provide useful information about the curing process particularly with regard to oxygen-inhibition and through-cure[39]. Of particular interest is the change in conductivity (ΔC) in response to a small temperature rise ΔT of short duration. $\Delta C/\Delta T$ is sensitive to incomplete surface cure due to oxygen inhibition and/or lack of through-cure caused by attenuation of the UV radiation by the photoinitiator.

Surface Profilometry[40]

This technique has been applied with considerable success to the measurement of the thickness of the oxygen-inhibited layer of a UV-cured acrylate coating. The instrument was calibrated with several metal foils of known thickness. Following the exposure of a coating to UV-radiation any uncured resin on the surface was washed off with 1,1,1-trichloroethane. The difference in film thickness compared with that of a fully-cured portion of the coating was measured with the profilometer to give the thickness of the oxygen-inhibited layer. This thickness was measured as a function of prepolymer resin and monomer, photoinitiator type and concentration, pigment type, inhibitors and their concentration, light intensity and exposure time.

For a particular photosensitiser the thickness $\Delta\tau_{O2}$ of the inhibited layer was given by

$$\Delta\tau_{O2} = \frac{A}{I[PI]} + \frac{B}{E[PI]}$$

where A and B are constants, t, I and [PI] are exposure time, intensity and photoinitiator concentration respectively and E (=It) is the total exposure energy.

The UV Curetester[41,42]

This ingenious, commercially available instrument deserves special mention because it bridges the gap between simple tests on the one hand and the much more sophisticated instrumental methods on the other. It is equally useful for the optimisation of photocurable resin formulations and for quality control.

Unlike many instrumental techniques it is capable of measuring the cure properties of a coating applied to exactly the same specifications as those used on a commercial UV-curing line. Another advantage of the instrument is its virtually unique ability to show the difference in cure rate resulting from variation in substrate - a hitherto neglected area.

The UV Curetester measures and records in real time the "test force" required to move a stylus through or across a film of coating material whilst it is being irradiated by UV radiation. The stylus is a stainless steel sphere (1-3 mm diameter) mounted in one end of a rod-shaped holder. In the latest model of the instrument the stylus, which has a vertical force pressing it against the test surface, is stationary and the sample is attached to a conveyor-driven tray. The stylus is lowered automatically on to the test sample which is then driven forward at a pre-selected speed of between 1 and 9 cm per minute by a stepper motor.

The stylus holder is mounted on a pivot so that the horizontal test force creates a turning moment which is opposed by the action of a spring. The deflection of the stylus holder is measured by a Hall-effect sensor and magnetic arrangement. The change in the test force as the sample undergoes curing is recorded continuously by the integral chart recorder.

The instrument incorporates a means of monitoring the intensity of the UV lamps as well as a nitrogen gas-purge facility. Results are most commonly expressed in terms of t_{50} - the exposure time for the trace to reach half its total deflection. This appears to be a reliable way of reporting the speed with which a material cures under selected test conditions. The UV-curing characteristics of numerous inks and coatings have been evaluated successfully using this instrument.

Other Methods

Laser nephelometry[43], turbidimetry[44], ^{13}C NMR spectroscopy[45] and laser interferometry[46] are further techniques which have been used to study radiation curing.

3. CONCLUDING REMARKS

Infra-red spectroscopy in its various forms will continue to play an important role in radiation-curing studies and undoubtedly there will be increased use of RTIR spectroscopy and FTIR-PAS.

Infra-red radiometry is a promising technique and further refinements are anticipated which would make this method capable of monitoring cure in a truly 'on-line' fashion.

The versatility of differential photocalorimetry will assure that this technique will become more widely used, as will the simplicity of operation and wide application of the UV-Curetester.

All the techniques reviewed here have special individual features which make them useful; for example, dilatometry gives the percentage shrinkage - an important quantity which is not provided by other methods, while profilometry is extremely useful for measuring the thickness of the oxygen-inhibited layer. Similarly, hardness measurements and the ERA technique are capable of discriminating between the degrees of cure of coatings which cannot readily be distinguished by i.r. spectroscopy.

In short, methods for measuring the degree of cure of radiation-curable resins should not be considered simply as alternative means to achieve the same end. In order to gain a clearer understanding of radiation-curing, data from a number of techniques needs to be brought together to form a coherent picture of the curing process.

There is still room for further instrumental innovation in this important branch of surface-coating science and technology. Two areas in which innovation would be particularly valuable are in situ on-line monitoring and, in order to widen the range of applications of radiation curing, methods suited to measurement of the degree of cure of thick samples.

ACKNOWLEDGEMENTS

The author thanks Professor P. Pappas, Dr M.S. Salim, Dr P. Oldring and Dr J.T. Guthrie for helpful discussions and Harcros Chemicals (UK) Ltd for the use of library facilities.

REFERENCES

1. N S Allen, S J Hardy, A F Jacobine, D M Glaser, B Yang and D Wolf, Eur. Polym. J., 1990, 26, 1041.
2. J M Julian and A M Millon, J Coatings Technol., 1988, 60, 89.
3. C Decker, J Coatings Technol., 1987, 59, 97.
4. B L Brann, J.Radiation Curing, 1985, 12, 4.
5. R S Davidson, R J Ellis, S A Wilkinson and C A Summersgill, Eur.Polym.J., 1987, 23, 105.
6. McGraw-Hill Encyclopaedia of Science and Technology, McGraw-Hill Book Co., NY., 6th Edition, Vol.8.
7. W D Cook, J.Macromol.Sci., Chem., 1982, A17, 99.
8. C Decker and K Moussa, J.Polym.Sci., Part C: Polym. Lett., 1989, 27, 347.
9. J Ohngemach, M Koehler and G Wehner, Proc.RadTech Conf., 1989, 639.
10. I Llewellyn and M F Pearce, J. Oil Col.Chem.Assoc., 1966, 49, 1032.
11. N S Allen, C K Lo and M S Salim, Proc.RadTech Conf., 1989, 253.
12. J H O'Donnell and R D Sothman, J.Macromol.Sci.Chem., 1980, A14, 879.
13. V D McGinniss, J.Radiation Curing, 1975, 2, 3.
14. A K Davies, R B Cundall, N J Bate and L A Simpson, J.Radiation Curing, 1987, 14, 22.
15. V D McGinniss and V W Ting, J.Radiation Curing, 1975, 2, 14.
16. A K Davies, 1990, Unpublished results.
17. A K Davies, N J Bate and R B Cundall, Applied Optics, 1986, 25, 1245.
18. R B Cundall, Y M Dandiker, A K Davies and M S Salim, Polym.Paint Col. J., 1987, 177, 215.
19. J J M Lamberts and H C Meinders, Proc. Radcure Conf., Basel, Switzerland, 1985, SME-FC411.
20. G A Lee and G A Doorakian, J.Radiation Curing, 1977, 4, 2.
21. G L Collins, D A Young and J R Costanza, J. Coatings Technol., 1976, 48, 48.
22. A van Neerbos, J. Oil Col. Chem. Assoc., 1978, 61, 241.
23. C Decker and K Moussa, Proc. RadTech Conf. 1989, 231.
24. M W Urban, J. Coatings Tech., 1987, 59, 29.
25. R S Davidson and C Lowe, Eur.Polym.J., 1989, 25, 159.

26. K Iwata, T Hagiwara and H Matsuzawa, <u>J. Polym.Sci.</u> <u>Polym.Chem.Ed.</u>, 1985 <u>23</u>, 2361.
27. M Ikeda, Y Teramoto and M Yasutake, <u>J.Polym.Sci.</u>, <u>Polym.Chem.Ed.</u>, 1978, <u>16</u>, 1175.
28. A J Evans, C Armstrong and R J Tolman, <u>J.Oil</u> <u>Col.Chem.Assoc.</u>, 1978, <u>61</u>, 251.
29. C E Hoyle, R D Hensel and M B Grubb, <u>J.Radiation</u> <u>Curing</u>, 1984, <u>11</u>, 22.
30. D D (Ziem)Le, <u>J.Radiation Curing</u>, 1985, <u>12</u>, 2.
31. L C Thomas, <u>Internat. Laboratory</u>, March 1987, 30.
32. S R Sauerbrunn, Proc.RadTech Conf. 1988, 219.
33. E Fischer and W Kunze, Proc.RadTech Conf., 1989, 669.
34. G B Tanny, A Lubelsky, Z Rav-Noy and E Shchori, Proc. Radcure Conf. May 1985, Basel, Switzerland, SME-FC 440.
35. J L Anderson, R F Russell and C C Slover, <u>J.Radiation</u> <u>Curing</u>, 1985, <u>12</u>, 10.
36. J L Anderson and C C Slover, Proc.RadTech Conf., 1988, 225.
37. C Carlini, M Martinelli, P A Rolla and E Tombari, <u>J.Polym.Sci., Polym.Lett.Ed.</u>, 1985, <u>23</u>, 5.
38. C Carlini, F Ciardelli, P A Rolla, E Tombari, G Li Bassi and C Nicora, Proc.RadTech Conf., 1989, 369.
39. A K Davies, 1990, to be published.
40. F R Wight and I M Nunez, <u>J.Radiation Curing</u>, 1989, 16, 3.
41. J H Nobbs, D Deurden and P K T Oldring, Proc. Radcure Conf., Munich, May 1987.
42. J H Nobbs, J T Guthrie, D Duerden and P K T Oldring, Proc. RadTech Conf., Tokyo, October 1988.
43. M Fizet, C Decker and J Faure, <u>Eur.Polym.J</u>, 1985, <u>21</u>, 427.
44. A K Davies, N J Bate and R B Cundall, unpublished results.
45. V P Thalacker and T F Boettcher, Proc. Radcure Conf., May 1985, Basel, Switzerland.
46. C Decker in 'Radiation Curing of Polymers', D R Randell Ed., Royal Soc.Chem., London, 1987.

Use of Infra-red Spectroscopy to Monitor the Degree of Cure in Acrylate and Epoxy Curable Systems

R.S. Davidson, K.S. Tranter, and S.A. Wilkinson

DEPARTMENT OF CHEMISTRY, CITY UNIVERSITY, NORTHAMPTON SQUARE, LONDON EC1V 0HB, UK

1 INTRODUCTION

The use of uv or visible radiation to transform a liquid into a solid film requires that a photochemical reaction occurs in which new bonds are formed. Normally the liquid films consist of a reactive diluent of low molecular weight to which a functionalised oligomer has been added. The performance of the resulting film is very dependent upon the nature of the oligomer. On many occasions liquid films are subjected to radiation, u.v. visible or e.b. and the resulting film is either still wet or "tacky" and the conclusion drawn that cure has not occurred. However, it is important to distinguish between those cases where the radiation has led to the desired chemistry but the nature of groups and arrangement of substituents has not allowed a hard film to be formed and the other case where no or little radiation initiated reaction occurred. The value of employing i.r. spectroscopy to examine films following or during radiation (as in FTIR spectroscopy[1]) is that one can assess how much of the desired chemical reaction has occurred. If reaction does occur but formation of a hard film does not, then one has to look at the nature of the oligomer etc to determine which features are responsible for this effect. Another valuable feature of the application of i.r. spectroscopy is that it allows one to change parameters and determine how these changes affect the rate of the polymerisation process. This methodology was employed to determine the mechanism of e.b. curing of acrylates and epoxides[2].

2 CURING OF ACRYLATES

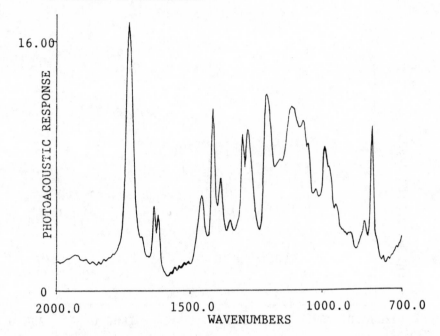

Figure 1   FTIR - PA Spectrum of Uncured TMPTA

Figure 1 shows an i.r. spectrum of an acrylate commonly used in u.v. curing formulations. The absorption at 810 cm^{-1} is associated with the C-H deformation of double bond associated with the acrylate group. The intensity of the absorption is a function of the number of acrylate groups in the molecule and the thickness of the film. However, since film thickness and hence effective path lengths cannot be accurately reproduced, absorption measurements cannot be compared directly to infer a quantitative change. The use of a reference peak in the spectrum, 'an internal standard', removes this problem enabling spectra to be compared directly and related and we chose C-H deformation absorption at 1380 cm^{-1} associated with the C-CH$_3$ since on cure none of these bonds is formed or destroyed.

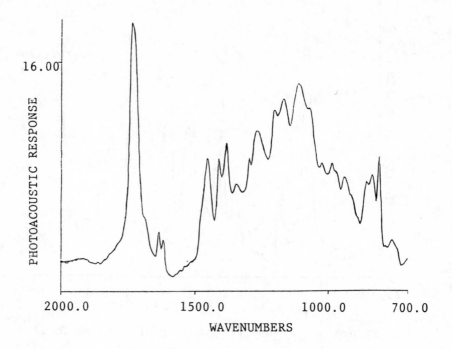

Figure 2   FTIR - PA Spectrum of TMPTA Cured at 5 kGy

From Figures 2 and 3 it can be seen that submitting
TMPTA to e.b. raidation leads to a decrease in
intensity of the 810 cm^{-1} absorption showing that
acrylate groups are being consumed.

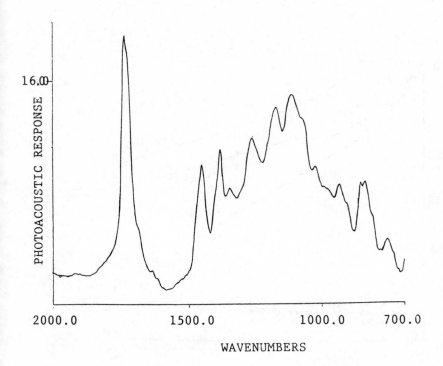

Figure 3　　FTIR - PA Spectrum of TMPTA Cured at 40 kGy

Measurements of this type were used to quantify the
effects of additives upon the e.b. curing of
isodecyl acrylate.

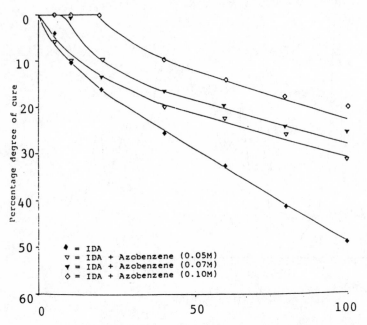

Figure 4   Effect of Added Azobenzene Upon
             the Rate of Cure of Isodecyl
             Acrylate

In one case (Figure 4) the effect of added azobenzene
was studied since in many radiation curable coatings
azo dyes are added. Figure 4 shows that azobenzene
retards cure and this is ascribed to it being a good
electron acceptor and thereby scavenging the slow
electrons which promote the cure of the acrylate.

By way of contrast, 1,3,5-trichlorobenzene (Figure 5)
promoted cure due to its ability to undergo
dissociative electron capture thereby producing aryl
radicals which are good initiators.

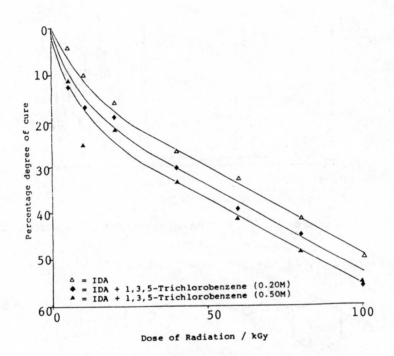

Figure 5   Effect of Added 1,3,5-trichlorobenzene
Upon the Rate of Cure of Isodecyl
Acrylate

3 CURE OF EPOXIDES

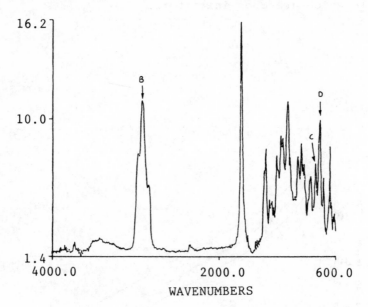

Figure 6   FTIR - PA Spectrum of Diepoxide (1)

Figure 6 shows the i.r. spectrum of (1) which is

(1)

commonly used in radiation curable epoxy formulation.
The absorption band at 794 cm^{-1} is associated with the
epoxy function.

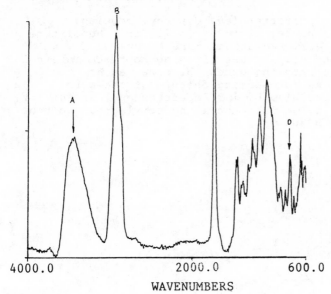

WAVENUMBERS

Figure 7   FTIR - PA Spectrum of Diepoxide (1)
             Cured at 40 kGy

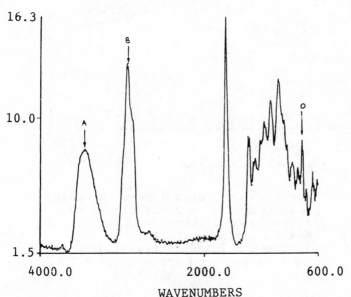

WAVENUMBERS

Figure 8   FTIR - PA Spectrum of Diepoxide (1)
             Cured at 100 kGy

If a cationic initiator (eg. diphenyliodonium
hexafluorophosphate) is added to (1) and the mixture
subjected to e.b. radiation, cure occurs
(Figures 7 and 8). By use of an appropriate internal
standard, the rate and extent of cure can be
determined. An interesting feature of the cure
process, as revealed by i.r. spectroscopy was that it
was attended by an increase in hydroxyl group content
(Figures 6,7,8 and 9).

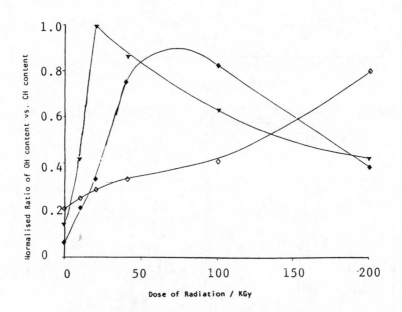

Figure 9   Increase in Hydroxyl Group Content
            Upon Curing of Epoxide (1) in the
            Presence of Diphenyliodonium
            Hexafluorophosphate ▼,
            Triphenylsulphonium Hexafluorophosphate
            (◆) and an Iron Arene Complex (◊)

We attribute this to water participating as a chain transfer agent, the source of the water being in the atmosphere above the film.

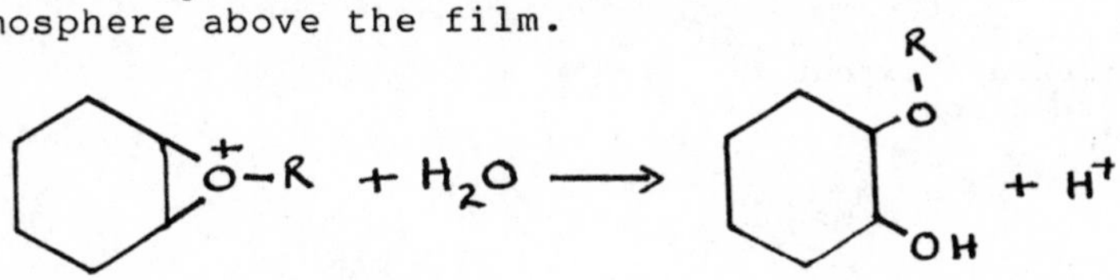

The hydroxyl groups formed in this way can also participate as chain transfer agents leading to enhanced crosslinking[3]. It will also be noted, from an inspection of Figures 6,7 and 8, that during cure the carbonyl group absorption at 1731 cm^{-1} shows an increasing amount of resolution as cure proceeds. It is possible that this is due to hydrogen bonding between the ester carbonyl and hydroxyl groups which will increase as film formation progresses. Another most valuable feature of the cure process as monitored by i.r. spectroscopy was that the hexafluorophosphate anion was progressively destroyed during cure. To date the mechanistic aspect of cure in cationic systems have concentrated upon the fate of the cation. Our rationale for decomposition of the anion is that the following reactions occur[3].

$$HPF_6 \rightleftharpoons HF + PF_5$$
$$PF_5 + 4H_2O \longrightarrow (HO)_3P=O + 5HF$$

Clearly the extent of hydrolysis is related to the concentration of water etc. It should also be noted that phosphorus pentafluoride is an initiator for epoxide curing in its own right[4].

These findings clearly impinge upon the role of curing epoxides using cationic initiators. If our mechanism is correct, increasing the amount of water in the fomulation should help by virture of it aiding crosslinking in the epoxy and increasing the rate of production of hydrogen fluoride.

As can be seen from Figure 10 this proved to be the case.

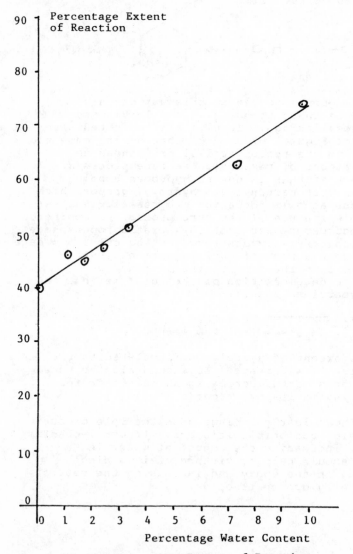

Figure 10   Graph to Show Extent of Reaction
Versus Water Content in the Curing
of an Epoxy Resin Formulation After
25 Minutes of Irradiation

4 SAMPLING TECHNIQUES

A wide variety of sampling techniques is available each offering some particular advantage or limitation. The Table lists some of the features of the techniques we have employed for studying degree of cure. Figure 11 shows the main optical features of the devices used with the various techniques. For a variety of reasons the spectra recorded for any one particular compound by all the techniques vary. Of course it is true that each spectrum shows all the absorptions characteristic of the compound but the relative intensities of one peak to another may vary between spectra recorded using different techniques. The underlying reasons for this phenomenon lie in the optical processes which are controlled by the film and the way it is presented to the incident radiation. Thus, to follow a particular cure process it is necessary to stay with one technique and even this is not totally foolproof. With the chosen technique one has to be sure that the sample is presented to the interrogating beam in a uniform and reproducible manner. As cure occurs, there is a change in refractive index and this in turn will affect the extent of surface reflection and the refractive process, and consequently sampling cannot be kept uniform throughout the cure process. Fortunately, it appears that these optical effects do not lead to large errors and one can make valid mechanistic deductions based on the raw spectral data.

We have found specular reflectance particularly valuable for studying the cure of epoxies and, in addition, have tried to separate the specular reflectance from the reflectance-absorption spectrum. Reference to Figure 12 shows how the two processes arise. The use of a highly reflective backing plate enhances the contribution from the reflection-absorption process. A formulation containing an epoxide was u.v. cured on a reflective substrate and then another was cured on a non reflective substrate. The spectra are shown in Figures 13 a,b and c. Figure 13 a is a spectrum containing both the specular and reflection-absorption components, whereas Figure 13 b only contains the specular component which is characterised by its apparent first derivative peaks. By subtracting the specular reflection spectrum (Figure 13 b) from the composite spectrum (Figure 13 a) the pure reflection-absorption spectrum was obtained

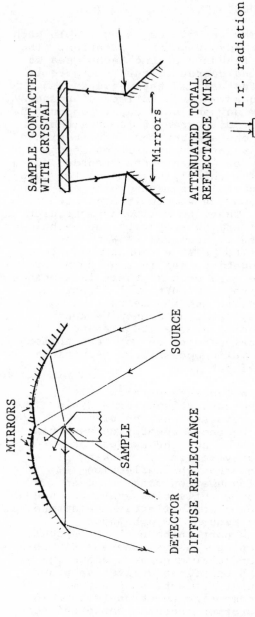

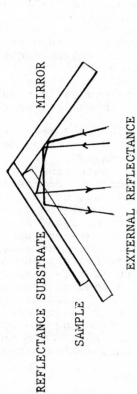

SAMPLE CONTACTED
WITH CRYSTAL

← Mirrors →

ATTENUATED TOTAL
REFLECTANCE (MIR)

I.r. radiation

Sample

Microphone

PHOTOACOUSTIC CELL

MIRRORS

SOURCE

SAMPLE

DETECTOR

DIFFUSE REFLECTANCE

MIRROR

REFLECTANCE SUBSTRATE

SAMPLE

EXTERNAL REFLECTANCE

Figure 11   Sampling Attachments for Recording I.R. Spectra

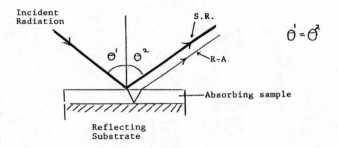

S.R. - Specular Reflectance Ray

R-A - Reflection-Absorption Ray

Figure 12   Ray Diagram for Typical Specular Reflection
Accessory

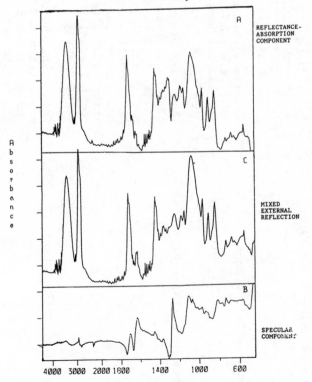

Figure 13   EXTERNAL REFLECTANCE SPECTRA FOR A
CURED EPOXY RESIN FORMULATION

TABLE 1 Comparison of Sampling Techniques for Studying Surface Coatings

TECHNIQUE	REMARKS
1. Transmission Spectroscopy	Clear films can be coated onto NaCl or KBr discs and cured. This method does not work for epoxies. It is difficult to control film thickness.
2. Diffuse Reflectance	Excellent powders but specular reflectance must be eliminated. Films can be studied but samples have to be cut out from cured sheet.
3. Attenuated Total Reflectance	When used with a clamping device which enables an even and reproducible pressure to films, this is an excellent technique. It is particularly valuable with thin films which have a yellow absorbancy since one effectively has a very long pathlength which the interrogating beam has to follow. Samples do not necessarily have to be cut out of the cured coating.
4. Specular Reflectance	This technique requires litle sample preparation and can give excellent spectra. The spectra can be a composite of specular and reflection absorption components. Highly viscous to solid films can be studied. Sometimes difficult to get a good signal to noise ratio, can be very good for studying thin hard and soft films. Thermal as well as optical properties of film determine quality of spectra.

(Figure 13 c). This is characterised by having a more uniform baseline thereby aiding quantitative treatment of the data. Using this system, we found that the subtraction factor for the uncured film was 0.35 whereas that for the cured material was 0.82. This change in subtraction factor is due to the specular reflection component becoming more important as the cure takes place which in turn is a measure of the change in dielectric properties (and hence refractive index) caused by polymerisation. Further work is in progress to determine the extent to which these phenomena can be used to understand the curing process.

ACKNOWLEDGEMENTS

We gratefully acknowledge financial support from the SERC, The British Technology Group and Wiggins Teape and Research and Development Ltd. Also we have had most fruitful discussions with Mr Alan Strawn of Spectra Tech Ltd.

REFERENCES

1. C. Decker and K. Moussa, Macromolecules, 1989, 22, 4455.

2. R.J. Batten. R.S. Davidson and S.A. Wilkinson, Polymer Paint Col.Jnl., 1989, 179, 176.

    R.S. Davidson, S.A. Wilkinson and A.K. Webb, Radtech Europe (Florence), 1989, 513.

3. R.S. Davidson and S.A. Wilkinson, Manuscript in preparation.

4. R. Hoene and K.H.W. Reuchert, Makromol.Chem., 1976, 177, 3545.

    F. Andruzzi, A. Prescia and G. Ceccarelli, Makromol. Chem., 1975, 176, 997.

5. A. Sing, Radtech 1988, New Orleans, 84.

Measurement of Dissolution Kinetics of Thin Polymer Films

Gareth J. Price and Jill M. Buley

SCHOOL OF CHEMISTRY, UNIVERSITY OF BATH, BATH, BA2 7AY, UK

1. INTRODUCTION

The other papers presented at this symposium have given ample evidence of the current interest in the changes that can be induced in polymers using light or other radiations. Perhaps one of the areas of greatest current activity[1] is the development of new materials, for photo-, electron beam and X-ray resists.

There are two major parts of the resist process; the production of an image by irradiation followed by its development[2]. Although dry etching and plasma development processes can be used, solvent developing is still most common. This depends on solubility differences between the irradiated and unirradiated regions of the polymer film, usually due to crosslinking or chain scission to lower molecular weights. Thus, measurement of these solubility characteristics is important to the production of new and improved resist materials. In this paper, we describe our development of a method to study the solubility of thin polymer films that can easily be applied to systems such as those used as resists.

The principle of our apparatus is the same as that of the so-called "Quartz Crystal Microbalance", QCM[3]. This utilises the piezoelectric properties of quartz to give a very sensitive mass detector. Certain cuts of quartz exhibit this effect of oscillating at a certain characteristic frequency when placed in a suitable electric field. Sauerbrey[4] showed that this resonant frequency changed in a linear manner dependent on the mass of any material loading the crystal.

Hence, the mass can be related to the frequency change by

$$\Delta F = -(F_O^2/A \ N \ \rho_q) \ M_p \tag{1}$$

where F_O is the fundamental resonant frequency and A the active area of the uncoated crystal, ρ_q is the density of quartz and N is the "frequency constant" of the particular crystal of quartz from which the mass of coating, M_p can be calculated. In our work, we have used crystals oscillating at approximately 10 MHz giving a theoretical sensitivity of approx. 10^{-9} g Hz^{-1}. Sensitivities close to this have been achieved in gases or vapours and exploited in a range of chemical sensors where sub-nanogram detection limits have been claimed[5,6].

Operation of the crystals immersed in liquids is more difficult due to the damping effect of the liquid[7]. However, these problems have been overcome and applications in analysis[8] and electrochemistry[9] have been published. The technique has also been applied to the study of photoresists by Hinsberg *et al.*[10,11] who found that it produced accurate, reproducible results and was applicable to a wide range of systems.

Other methods for investigating solubility in resist type materials have been reviewed by Rodriguez and co-workers[12] and include the so-called "dip-and-dry" method[13] where a polymer film is immersed in solvent for known lengths of time, removed and the film thickness determined using a stylus instrument. These measurements have several sources of error since dissolution does not cease immediately on removal from the solvent and due to uncertainties in the thickness measurements. Uberreiter *et al.*[14] used a microscope to observe pellets of polymer in a flowing solvent. Other "in-situ" methods include moitoring of changes in the capacitance of the system[15] or the ellipsometric properties of the surface[16]. However, these have limitations as to the accuracy of the results and also in the types of system to which they can be applied. Most recent studies have involved laser interferometry[12] where the reflection of laser light from the surface of a thin film is compared to the incident beam. The accuracy of this apparatus is high but the method is expensive and quite complex to set up.

Thus, although several methods are available for this work, none offers all of the desirable properties: speed, convenience, economy and wide applicability. This paper will demonstrate that our QCM method has these properties

and should provide an excellent method for studying dissolution and solubility in photoresist systems.

2. EXPERIMENTAL

As an introduction to our work and to serve to illustrate
our methods, we chose to study poly(methyl methacrylate),
PMMA. This polymer has been studied by other workers and
so gave us a good basis to compare our technique but it is
also in commercial use as a positive photoresist so that
the results are of commercial significance. Most of the
work illustrated here used a polymer supplied by BDH Ltd.
having a number average molecular weight of 56000 and a
polydispersity of 2.0 as measured by GPC. NMR spectroscopy showed it to be primarily atactic. For the molecular weight studies, narrow (1.02-1.06) polydispersity
samples from Polymer Laboratories Ltd. covering a range
from 6100 to 1400000 were used. The molecular weights
quoted are those supplied by the manufacturers for GPC
calibration purposes. The solvents used were all of reagent grade.

The quartz crystals (International Crystal Manufacturers Inc., USA) used in this work were 0.017 mm in thickness and 12.0 mm in diameter with a gold plated (i.e. active) area of approximately 19.6 mm^2. They were cleaned
with acetone, chloroform and the solvent to be used and
coated with a thin film (0.5 - 5 μm) of polymer. For this
initial work, both sides of the crystal were used so that
coating was performed by dipping the crystal into a 1-5 %
solution of PMMA in chloroform, depending on the desired
film thickness, and allowing the solvent to evaporate.
The crystals were then heated for 1 hr at 160 °C to remove
remaining solvent and anneal the film.

A schematic of our apparatus is shown in Figure 1. A
coated crystal was placed in a holder which could be rapidly immersed in a stirred container of solvent thermostatted to ± 0.1 °C and its resonant frequency recorded.
When this was stable, it was immersed and the frequency
change followed with time until no further change was
noted. To give signals of the required stability, the
apparatus was contained in an aluminium box to act as a
Faraday cage.

The electrical circuit required to operate the crystals was based upon that of Bruckenstein and Shay[7], modified to function with our crystals. Details will be pub-

lished elsewhere[17]. The frequencies were recorded on a
Racal Dana 1030 Frequency Counter measuring to ± 0.1 Hz
and read into an IBM compatible microcomputer via a
Metrabyte IEEE interface card where subsequent data treatment was performed. Measurements were made at 1 s intervals, each being the average of ten separate readings.

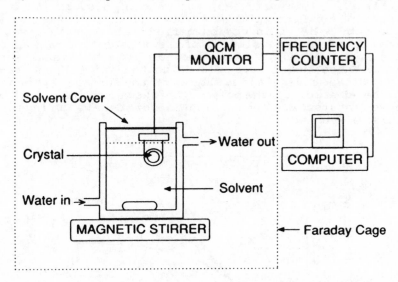

Figure 1.   Schematic Diagram of QCM Dissolution Apparatus.

3. RESULTS AND DISCUSSION

Our experiments yield a series of frequency-time curves.
Using Equation (1) and introducing the density of the polymer, ρ_p, the frequency can be transformed into the film
thickness, T_f, giving;

$$T_f = [(N\ \rho_q)\ /\ (F_o^2\ \rho_p)]\ \Delta F \tag{2}$$

It was assumed that the density of the film was given by
that of the bulk polymer, 1.170 g cm^{-3}.

As an illustration of our results, Figure 2 shows
results for the dissolution of the wide distribution PMMA
at three different initial film thicknesses in a mixture
of 2-butanone (MEK) and propan-2-ol (IPA) at 25 °C, plot-

ted as functions of both film thickness and frequency
change. For the initially 0.42 μm film, the total change
in frequency for complete dissolution is 11.7 kHz. Since
the uncertainty in measuring the frequency is less than ±
10 Hz, we can measure the film thicknesses with a resolution of approximately 0.1 nm. The results also show that,
within the limits studied, the rate of dissolution is independent of film thickness. Two other features are also
apparent: a brief induction period before rapid dissolution starts and the slowing of the dissolution rate as the
last layers of polymer are removed from the crystal surface. These factors are related to the initial penetration of the solvent and swelling of the surface layer, and
to the adhesion of the polymer to the substrate. Both of
these features will be studied in more detail during the
course of our work.

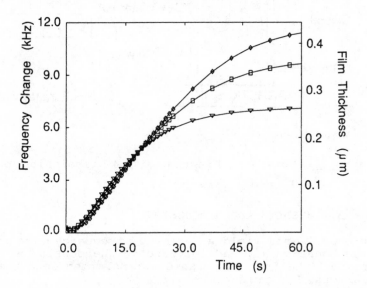

Figure 2. Change in Frequency and Film Thickness *versus*
Time for PMMA Dissolution in 60:40 v:v MEK/IPA at 25 °C

For the purposes of this work, the rate of dissolution has been taken as the initial slope of the thickness-
time plot at the end of any induction or swelling period.
To gain an idea of the reproducibility of our measurements, Figure 3 shows four separate experiments performed
on different days for the dissolution in MEK at 25 °C of

films approximately 1 μm in thickness. The results are
normalised to the same initial thickness and are displayed
as the percentage of the film removed with development
time. The curves are very similar and calculation of the
rates gave 2.422 ± 0.017 μm min^{-1}, the four runs agreeing
to within ± 1 %. Precise comparison of our results with
those of other workers is not possible since the dissolution rate in a particular solvent depends on a large
number of factors such as molecular weight (as will be
shown below), polydispersity, the microstructure of the
polymer and its thermal history. Thus it is impossible to
work on identical samples to those used for the Literature
studies. However, all our results show good agreement
with published work for similar systems. In this paper,
we will show that the main factors affecting dissolution
may be accurately measured using our QCM technique.

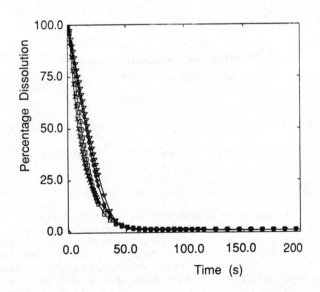

Figure 3. Illustration of the Reproducibility of Dissolution Experiments for PMMA in MEK at 25 °C.

The effect of the different solvents on the rate of
dissolution is clearly shown in Figure 4 which plots the
results for a homologous series of acetates where faster
rates were found for the lower members of the series. This
also demonstrates the range of dissolution rate that is

amenable to measurement by our method; the process being
essentially complete in less than 20 s with methyl acetate
while for amyl acetate, in excess of 30 min is required.
The same trend of dissolution rate for these solvents was
found by Ouano[14].

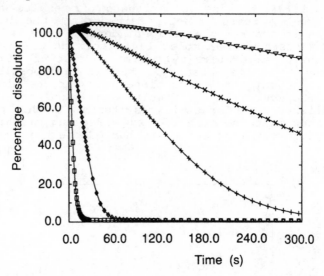

Figure 4.   Dissolution of PMMA in acetate solvents at 25°C.

□   methyl;   ◇   ethyl;   +   isopropyl;   ×   *n*-butyl;   ∇   *n*-pentyl.

Effect of Solvent Composition on Dissolution

    Most photoresist processing is not carried out in pure
solvents but in a mixture of a good solvent with a moder-
ating non-solvent and this has been the focus of study of
several workers.   Two such systems involving MEK have been
studied here.   Figure 5 shows results at 30 °C for the wide
polydispersity PMMA in pure MEK, pure IPA and six mixtures
of varied composition. MEK is a good solvent for PMMA and
gives rapid dissolution while IPA is a non-solvent and
displays only a small degree of swelling.   As might be expected on first consideration, the mixtures give dissolution rates intermediate between those of the single components as has been demonstrated by Cooper *et al*[18] using the
laser interferometric technique.

    Many workers have assumed that the properties of mix-
ed solvent systems would be the average of those of the

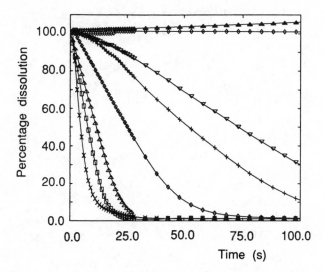

Figure 5. Dissolution of PMMA in mixed MEK/IPA solvents at 25 °C.

<u>Vol. % IPA:</u>   × 0;   □ 20;   Δ 40;   ◇ 50;   + 60;   ∇ 70;   ◇ 80;   ▲ 100.

single components. However, Cooper and co-workers[18] demonstrated that this is in fact not so and their findings are confirmed by the results shown in Figure 6. Methanol is also a non-solvent for PMMA but addition of up to approximately 20 percent by volume to MEK actually accelerates the rate by as much as a factor of two. Similar effects were found on the addition of water. This effect has been ascribed to the small, mobile methanol molecules diffusing rapidly into the polymer structure and plasticizing it so as to allow better dissolution.

Effect of Polymer Molecular Weight on Dissolution.

One of the main parameters that affects polymer solubility is the molecular weight. This is clearly shown in Figures 7 and 8 for the narrow polydispersity PMMA standards between 6100 and 1400000 at 25 °C. Clearly, polymers having lower molecular weights dissolve at a considerably faster rate than those with higher values. Another significant point is that the polymers with molecular weights over 100000 show swelling prior to dissolution and that the extent also increases with increasing molecular weight.

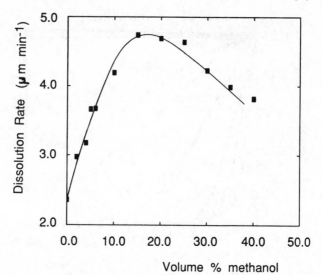

Figure 6. The effect of added methanol on the dissolution rate of PMMA in MEK at 25 °C.

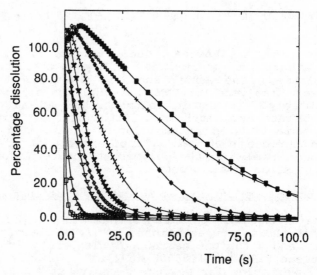

Figure 7. Dissolution rates for PMMA of varying molecular weight in MEK at 25 °C.

Molecular weights: □ 6100; △ 10300; ▽ 22200; ◇ 34500; ▼ 67000; × 107000; ◆ 330000; + 820000; ■ 1400000.

Ouano has suggested a relation between dissolution rate, DR, and polymer molecular weight, M, of the form

$$DR = a\ M^{b}$$

where a and b are constants for a particular system. Figure 8 shows our results plotted in double logarithmic form.

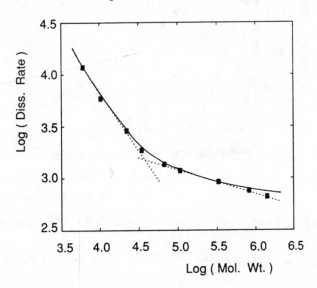

Figure 8. Variation of dissolution rate of PMMA in MEK at 25 °C with polymer molecular weight.

However, the plot is not linear as predicted by this relation. The curve could though, be the resultant of two linear portions as shown by the dashed lines, these corresponding to molecular weights where swelling is, or is not, significant. However, the evidence is tenuous and much further work needs to be done to confirm these effects.

Effect of Temperature on Dissolution

Also of great importance to the dissolution process is the solvent temperature with dissolution being faster at higher temperatures. Figure 9 indicates this for PMMA with molecular weight 107000 and polydispersity 1.03 in MEK over the range 16 - 45 °C. The swelling of the polymer also appears to be temperature dependent, this being negligible above 30 °C.

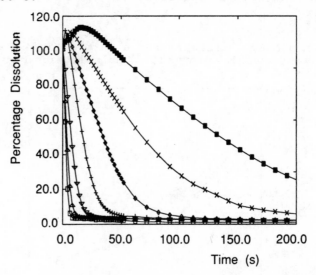

Figure 9. Dissolution curves for PMMA (M = 107000) in MEK for a series of temperatures.

Temp. (°C): □ 16.4; △ 20.6; ▽ 25.3; + 29.9; ◇ 35.1; × 40.0; ■ 45.0.

When the results are plotted in the Arrhenius form, a linear relation is obtained as shown in Figure 10. From this plot, the following relation was derived:

$$DR = 2.04 \times 10^9 \ exp - [5239.5 \ / \ T]$$

where DR is the dissolution rate in $\mu m \ min^{-1}$ and T is the absolute temperature. This leads to an apparent activation energy for the dissolution of 43.5 $kJ \ mol^{-1}$.

Application to Photochemical Reactions

To illustrate the potential application to photoresist systems, crystals were coated with PMMA and exposed to U.V. irradiation from a medium pressure mercury lamp and their dissolution rates measured. PMMA is a positive photoresist which undergoes chain scission on irradiation. Thus, the rates should be faster in irradiated samples and this is demonstrated in Figure 11 where increasing the dose also increases the dissolution rate. This clearly indic-

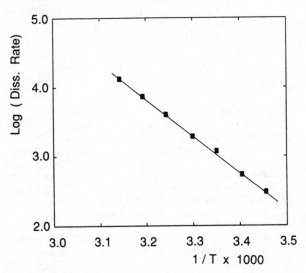

Figure 10. Arrhenius plot for the dissolution of PMMA (M 107000) in MEK.

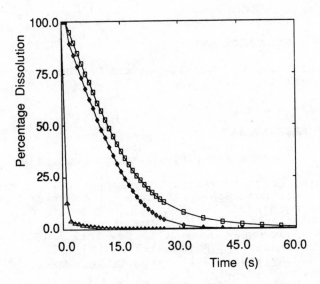

Figure 11. Dissolution rates of PMMA in MEK at 25 °C for polymers exposed to U.V. irradiation.

Irradiation time (min): Δ 30; ◇ 10; □ 0.

ates that our methods are suitable for application to resist type materials under conditions very similar to those employed commercially.

4. CONCLUSION

The work described in this paper demonstrates that our method for studying dissolution processes in thin polymer films is potentially extremely useful for application to resist systems. We have produced results comparable to those of other workers and have demonstrated that we are able to study a range of effects of fundamental importance to the dissolution process in polymers. There are few, if any, restrictions on the types of polymer that can be studied and a wide range of solvents, temperatures and dissolution rates can be examined. The crystals are small and may be conveniently introduced into irradiation systems under conditions very close to those used for commercial resist development. Perhaps the major advantages are that results may be obtained very quickly with little "set-up" time and (from an academic point of view) that the apparatus is inexpensive.

5. ACKNOWLEDGEMENT

We are grateful to the Science and Engineering Research Council for financial support of this work.

6. REFERENCES

1. E. Reichmanis and L.F. Thompson Chem. Rev. 1989, **89**, 1273.
2. Introduction to Microlithography, L.F. Thompson and C.G. Willson (Eds), ACS Symposium Series 219, ACS, Washington, D.C., 1983.
3. G. Guilbault in *Methods and Phenomena - 7* C. Lu and A.W. Czanderna Elsevier, New York, 1984.
4. G. Sauerbrey Z. Phys. 1959, **155**, 206.
5. J. Hlavay and G. Guilbault Anal. Chem. 1977, **49**, 1890.
6. J.F. Alder and J.J. McCallum Analyst 1983, **108**, 1169.
7. S. Bruckenstein and M. Shay Electrochim. Acta. 1985, **30**, 1295.

8. M. Thompson, C.L. Arthur and G.K. Dhaliwal <u>Anal. Chem.</u> 1986, **58**, 1206.
9. M.R. Deakin and D.A.Buttry <u>Anal. Chem.</u> 1989, **61**, 1147.
10. W.D. Hinsberg, C.G. Willson and K.K. Kanazawa <u>J. Electrochem. Soc.</u> 1986, **133**, 1448.
11. W.D. Hinsberg and K.K. Kanazawa <u>Rev. Sci. Inst.</u> 1989, **60**, 489.
12. F.Rodriguez, P.D.Krasicky and R.J.Groele <u>Solid State Technol.</u> 1985, 125.
13. A.C. Ouano <u>Polym. Eng. Sci.</u> 1978, **18**, 306.
14. K. Ueberreiter and F. Asmussen <u>Makromol. Chem.</u> 1961, **43**, 324.
15. W. Oldham <u>Opt. Eng.</u> 1979, **18**, 59.
16. J.S. Papanu, D.W.Hess A.T. Bell and D.S.Soane <u>J. Elecrochem. Soc.</u> 1989, **136**, 1195.
17. G.J. Price and J.M. Buley Manuscript in preparation.
18. W.J. Cooper, P.D. Krasicky and F. Rodriguez <u>J. Appl. Polym. Sci.</u> 1986, **31**, 65.

Health and Safety Aspects of Radiation Curing

B.C. Ross

HEALTH & SAFETY EXECUTIVE, BOOTLE, MERSEYSIDE, L20 3QZ, UK

Introduction

Radiation curing of materials such as coatings and adhesives is a well established and expanding industrial process. The technique is used in the production of paper, plastics, packaging materials and many other consumer items. The main radiation sources employed are ultra-violet light and electron beams although gamma radiation is used in some applications. Exposure to radiation from these sources can be hazardous to the health of persons associated with the process. Therefore it is essential that radiation processing equipment is designed to adequately protect persons during both normal production and maintenance operations.

The effect of radiation on the human body is the subject of continuing research. Historically such research has led to a general increase in the standard of worker protection. There is evidence to suggest that this trend will continue (1). Therefore, it is advisable for designers and users of equipment to establish a forward looking approach to the subject so that exposures are kept as low as reasonably practicable.

Current Legislation

In the UK safety regulations for employees are made under the Health and Safety at Work Act 1974 (HSW Act)(2). This act applies to all sections of employment. In 1986 The Ionising Radiations Regulations (3) and a supporting Approved Code of Practice were introduced. The regulations include requirements which are binding on the UK by virtue of its membership of the European Community. The HSW Act also applies to work with non-ionising radiation such as ultra-violet, but so far no regulations have been made covering this topic. One of the most important parts of the HSW Act is Section 6 which places certain obligations on designers, manufacturers, suppliers, importers and installers of plant and equipment. The obligations require that these persons do what is reasonably practicable to make articles and substances safe and without risks to health in respect of risks which are reasonably foreseeable. The provisions of Section 6 apply equally to British and Foreign products. In addition some categories of equipment have or will become subject to European Community Directives.

These directives establish a European Common Market in respect of products certified as complying with safety requirements laid down in the relevant directive. Products which fully conform to such provisions including those made to harmonised or international standards related to the directives are deemed to comply with S6 of the HSW Act.

Designers of radiation curing equipment will be particularly interested in the Machinery Directive which sets safety requirements for all functioning machines (4). These requirements relate to mechanical, and electrical safety as well as noise, vibration, emission of dust and gasses, ionising and non-ionising radiation. An important feature of the directive is the requirement for a risk assessment to estimate the probability and degree of injury or damage to health. Guidance on the risk assessment procedure, and appropriate control systems is given in the directive. In the UK at the present time emphasis is placed on the use of British Standard BS5304 1988 Safety of Machinery which covers many of the requirements in the directive (5).

Supporting the directive is a programme of harmonised standards. When a harmonised standard is adopted each member state has to replace any existing standard with the European standard. A working group has been convened to produce a standard for ionising and non-ionising radiation covering low frequencies, radio frequencies, microwaves, infra-red, visible, ultra-violet, x, α, β and γ rays and electron and ion beams (6). The standard will deal with:

     (i)   Radiation measurement;
     (ii)   Methods for designing radiation shielding;
     (iii)   Selection of shielding materials;

The target date for implementation of the complete standard is March 1993. All existing national and international standards and regulations will be considered in the production of the standard.

Ultra-violet Radiation Curing

For radiation protection purposes the UV spectrum is divided into three wavelength regions:

UV-A (400 - 315nm), UV-B (315 - 280nm) and UV-C (280 - 100nm). These regions have significantly different potential for causing damage which is confined mainly to the surface of the eye and skin. The effects include inflammation of the cornea and the possibility of cataract induction. Exposure of the skin can lead to erythema (burning), photosensitisation and skin cancer.

The Health and Safety Executive is guided by the Threshold Limit Values (TLV's) recommended by the American Conference of Government Industrial Hygienists (ACGIH) (7). For UV-A the TLV is $10Wm^{-2}$ ($1mW\ cm^{-2}$) for periods greater than 10^3 seconds while for shorter duration exposure the total dose should not exceed 10^4 Joule .m^{-2}. In the bands B and C the maximum permissible exposure varies markedly with wavelength being a minimum at 270nm. The maximum permissible exposure times in any 8 hour period are given in Appendix 1.

It should be noted that these values do not represent a fine dividing line between safe and dangerous levels. Also the values do not apply to exposure of photosensitive persons or to persons exposed continually to photosensitizing agents.

Protection against exposure of UV radiation can best be achieved by properly engineered equipment which reduces emission of UV to the workplace. Designers should consider both direct and reflected radiation as well as any need to remove protective panels for maintenance or adjustment purposes while the source is energised. All safety information should be identified clearly in the manufacturers instructions which should also include details of UV levels at any accessible position. This will enable the user to assess the exposure of employees.

Electron Beam Curing (EBC)

EBC machines are used to irradiate surfaces of materials with beams of electrons which have been accelerated through a voltage of typically 150kV. The acceleration takes place in a vacuum vessel and the electrons pass through a cooled thin metal window and are absorbed in the machine structure. This process produces X-rays but these pose a smaller radiation hazard than the electrons. The electron beam is absorbed in the surface layer of material. This, coupled with the high currents used would lead to extremely high dose rates giving severe instantaneous injury to anyone exposed to the direct beam.

Effective shielding must be provided to reduce doses to acceptable levels. It should be possible to achieve dose rates near to the general level of natural background radiation. Machines which are not self contained should be housed in a shielded enclosure which excludes access. The safety control system for both self contained and enclosed machines is required to be of high integrity because of the extreme nature of the radiation hazard. The system requirement is for two independent control channels each of which should be capable of interrupting the high voltage supply:

(a) before a beam escapes from the machine;

(b) before the creation of any aperture which allows access to the internal radiation fields in the machine.

This can be achieved for example by mechanical interlock switches on relevant removable panels or moving structural assemblies and a series of external radiation monitors. These systems should be immune from common mode failure, and not susceptible to earth faults (8). In addition it is important for designers to give consideration to:

(a) Separation of the safety related system from the overall system. This would ensure that any modification to the overall system would not compromise the operation of the safety controls;

(b) Inclusion of comprehensive diagrams and descriptions of the structure and operation of the safety system in the users handbook;

(c) The safety integrity of programmable electronic systems including hardware and software (9)

The Way Ahead

The radiation curing business is a worldwide activity, therefore international cooperation is necessary to achieve acceptable equipment safety standards. This can be done through the creation of standards under the auspices of the International Standards Organisation. This is a long term process which should be encouraged. However, short term needs can be served by cooperation between regulatory authorities and organisations such as the International Association for the Advancement of Radiation Curing by UV, EB and Laser Beams. The Health and Safety Executive would be interested in exploring this avenue for the development of practical guidance material. This approach has been used successfully in the production of guidance on safety at gamma irradiation plants in conjunction with the UK Panel on Gamma and Electron Irradiation (10).

Technical guidance covering all aspects of radiation safety at Gamma and Electron Irradiation facilities is being finalised at the present time by the International Atomic Energy Agency (IAEA). HSE has been involved in the preparation of this document which is due for publication later this year (11).

Eeff(WM⁻²)	MAXIMUM PERMISSIBLE EXPOSURE
1×10^{-3}	8 HOURS
8×10^{-3}	1 HOUR
5×10^{-2}	10 MINUTES
5×10^{-1}	1 MINUTE
3	10 SECONDS
30	1 SECOND
3×10^2	0.1 SECOND

APPENDIX 1 Where Eff in the Effective Irradiance

REFERENCES

1. INTERNATIONAL COMMISSION ON RADIOLOGICAL PROTECTION - RECOMMENDATIONS OF THE COMMISSION - DRAFT FEBRUARY 1990. ICRP/90/G-01

2. HEALTH AND SAFETY AT WORK ETC ACT 1974 HER MAJESTY'S STATIONARY OFFICE LONDON

3. THE IONISING RADIATIONS REGULATIONS 1985, STATUTORY INSTRUMENT NO 1333 HER MAJESTY'S STATIONERY OFFICE, LONDON 1985

4. SAFETY OF MACHINERY, BASIC CONCEPTS, GENERAL PRINCIPLES FOR DESIGN PREN 292 1989 CEN BRUSSELS

5. BRITISH STANDARD BS5304 1988 CODE OF PRACTICE FOR SAFETY OF MACHINERY

6. SAFETY OF MACHINERY - RADIATION CEN/TC 114/WG13 MARCH 1990 CEN BRUSSELS

7. ACGIH THRESHOLD LIMIT VALUES AND BIOLOGICAL EXPOSURE INDICES. AMERICAN CONFERENCE OF GOVERNMENTAL INDUSTRIAL HYGIENISTS (1989).

8. BRITISH STANDARD 2771 1986 ELECTRICAL EQUIPMENT OF INDUSTRIAL MACHINERY

9. HEALTH AND SAFETY EXECUTIVE PROGRAMMABLE ELECTRONIC SYSTEMS IN SAFETY RELATED APPLICATIONS HER MAJESTY'S STATIONARY OFFICE LONDON 1987

10. HEALTH AND SAFETY EXECUTIVE AND UK PANEL ON GAMMA AND ELECTRON IRRADIATION. RADIATION SAFETY FOR OPERATORS OF GAMMA IRRADIATION PLANTS. HSE TECHNOLOGY DIVISION RADIATION AND POWER GROUP MAGDALEN HOUSE, STANLEY PRECINCT, BOOTLE, MERSEYSIDE L20 3QZ

11. IAEA TECHNICAL SERIES DOCUMENT ON THE RADIATION PROTECTION AND SAFETY ASPECTS OF IRRADIATION FACILITIES. IAEA VIENNA (1990)

Subject Index

Aberchrome 540 actinometry, 184
Absorption maxima, 186, 191, 193
Absorption of
 UV radiation, 216
Acetonitrile, 190, 191, 193-196
Acetophenone photoinitiators
 Influence of substituents, 163-165
Acid catalysed curing, 199
Acrylate and epoxy curable systems
 Degree of cure by IR spectroscopy, 400-415
Acrylate resins, 269-283
Acrylated monomers
 Hybrid systems containing, 285
Acrylated oligomers
 Hybrid systems containing, 284, 285, 287
Acrylates, 46, 47, 56, 57, 76, 77, 111, 130, 131,
140, 145, 326
Acrylates(s) resin(s), 379, 387, 389, 390, 393, 395
 Cross-linked polymer from, 379
 double bonds of, 379
Acrylic acid reaction product with
 Adipic acid/hexane diol-based polyester, 8
 BISPHENOL A epoxide, 8
Acrylic monomers
 Dual and hybrid curing systems containing, 247, 304
Acrylic polymerisation
 Thiol "initiation" of, 344
Acylphosphine oxides use in
 UV cured paints, 109, 110, 113-119
Adhesion promoter
 Acrylated phosphate (EBECRYL 169), 303, 319, 320,
 322-324
Adhesives
 Multi-layer laminating, 150
 Radiation curable, 147-159
 UV-Aerobic cure, 155
 UV-anaerobic cure, 156
 UV-curable for opaque substrates. 157
 UV-moisture cure, 153-154
 UV-thermal cure, 149-153
Adhesives, sealants and coatings
 Non-acrylate photopolymers for, 342-357
Aliphatic urethane acrylates
 Application areas, 11
Alkoxylated aliphatic monomers, 273
Alkoxylated trifunctional monomers, 270
Alkylamino radicals, 189

Allylic resins
   Addition of multifuctional thiol components, 343
alpha-Amino alkylradicals, 59
alpha-Aminoketones
   UV absorption spectra, 172
alpha-Dimethylamino acetophenones in UV curable
systems
   Conclusion, 180
alpha-Ether radicals, 204, 209-212
alpha-Hydroxy radicals, 211
alpha-Hydroxlcyclohexyl radical, 210, 211
alpha-Methylbenzoinmethyl ether, MBME, 201, 208-211
alpha-Oxygen radicals, 204
American Conference of Government Industrial
Hygienists
   Threshold Limit Values, 432
Amine acrylate (accelerator)
   In UV cured paints, 111
Amine derived radical, 52, 53
Amine synergist(s)
   Acrylated amines, 126
   Aminobenzoates, 126, 128, 129
   4-(2-Hydroxyethyl)morpholine, 223, 225
   N-methyl diethanolamine, 293
   Screen print varnish containing, 282
   Tertiary, 125-127
   UVECRYL P104, 303, 309, 320
Aminoketones
   Photocleavage reaction of, 175, 176
   Stable conformations of, 177, 178
   Triplet state of, 178
Ammonium thiocyanate, 141, 142
Aromatic epoxy resins, 303
Aromatic peroxy radical, 198
Aromatic pinacols
   Thermal initiators, 153
Aromatic urethane acrylates
   Application areas, 12
Attenuated total reflectance (ATR), 412, 414
Benzil dimethyl ketal, 16
Benzoinmethyl ether (BME), 201, 208, 209
Benzophenone (BP)
   UV spectrum in DEGDVE, 201, 208-212
Benzophenone-ketyl radical, 52
1-Benzoylcyclohexane-1-ol (BCHO), 201, 208-213
Benzoyl radical, 210, 211
Benzoyl radicals, 48
Benzoyloxy radicals, 182, 188, 189, 191, 193, 194,
   197, 198
Bimolecular
   Process, 49
Bimolecular hydrogen-abtraction, 47
Bis-benzoyloxy radicals, 189
Bisphenol A type epoxy-acrylates, 247-268
   CRAYNOR CN 104A80, 248

Blue offset printing ink
  (4-morpholino) acetophenone derivs, in, 167-168
  Alpha-amino ketones in, 167-169, 172-173
  Curing by medium pressure mercury lamps, 167-169
  Curing of, 167-169
  Formulation, 165
British Standard BS5304 (1988)
  Safety of Machinery, 431
Bromometry
  Determination of residual monomer by, 386
Butane diol diacrylate (BDDA), 272
Butanedioldivinyl ether (BDDVE), 201, 204, 333
C^{13} NMR spectroscopy, 396
Calcium carbonate/carbon black test, 383
Calcium hydride, 201
Calomel electrode, 202, 203
Calorimeters, 391-392
  Differential photo (DPC), 392
  Differential scanning (DSC), 392
Calorimetric methods
  Real-time, 381, 383
Calorimetry, 380
  Photo, 397
CAROTHERS equation, 343
Cationic photocuring of
  Coatings, 200
Cationic photoinitiation, 302, 304
Cationic photoinitiators, 40
  Sulphonium salts, 40
Cationic photopolymerisation, 199
Cationic polymerisation, 199, 200
  Electron beam induced, 204
  Photoinduced, 207
  Redox induced, 208
Cationic polymerisation of
  DEGDVE, 206-209
Cationic polymerisation with
  Iodonium salts, 200, 202
  Sulphonium salts, 200, 202
Cationic propagation, 199
Cationic species, 199
Cationic UV curable coatings, 284-301
  Acrylate oligomer contribution to film properties,
  284-301
Cationic UV curable ink jet ink, 138, 146
  Difunctional acrylate monomer use in, 141, 145
Cationic UV initiated polymerisation
  Cycloaliphatic diepoxide, 284
Ceramic decal market, 74
Ceramic residue from pyrolysed PSS fibres, 370-373
  Effect of duration of γ-irradiation on, 371
  Effect of irradiation on, 370-373
Chain growth polymerisation, 147
Chloroform, 183, 185-188, 191
Chlorosilanes, 360
Chromium comples dyes
  Ink jet ink use, 132, 142-145

Clear coatings
  (4-morpholino)acetophenone derivatives in, 173, 174
  Curing of, 173, 174
Coatings
  "Degree of cure", 380, 381
  "Functionally cured", 380
  "Well cured", 380
  Adhesion tests on, 380
  Pendulum hardness of, 380
  Solvent double rubs on, 380
Coatings and inks
  Curing of, 302
Cobalt carboxylates, 155
Cobalt complex dyes
  Ink jet ink use, 132, 142, 143, 145
Comparison of UV and EB curing methods, 41, 42
Computer control to
  Monitor cooling water and air flow, 66
  Monitor lamp power and age, 66
  Monitor temperature, 66
  Monitor UV radiation level, 66
  Program lamp power, 66
Conformal coated PCB, 155
Continuous ink jet printing (CIJ), 132, 134
Cotton test, 382, 383
Cross-linked covercoat, 77
Crosslinked networks, 46
Crosslinked polymer networks, 147
Crosslinking
  Processes, 46, 47
Crosslinking of
  Multifunctional monomers, 46
  Multifunctional oligomers, 46
Cumene hydroperoxide, 157
Cure of epoxides, 406-410
  FTIR spectrum of cured diepoxide, 407
  FTIR spectrum of diepoxide, 406
  Hydroxyl group formation, 409
  Increase in hydroxyl group content, 406-408
  Reaction versus water content, 409, 410
  Specular reflectance in study of, 411-415
  Specular reflectance spectra, 411-413
  Use of cationic initiator, 407-409
  Use of diphenyliodonium hexafluorophosphate, 407-409
  Use of iron arene complex, 408
  Use of triphenylsulphonium hexafluorophosphate, 408
  Water as chain transfer agent, 409
Cure speed
  On acrylates, 247, 254, 256, 258, 267
Curing agents
  Thermally activated, 150
Curing of acrylates, 401-405
  Aryl radicals as initiators, 405
  Comparison of sampling techniques, 414
  Effect of 1,3,5-trichlorobenzene on cure rate, 405
  Effect of azobenzene on cure rate, 404

FTIR spectrum of cured TMPTA, 402, 403
FTIR spectrum of uncured acrylate, 401
Sampling techniques, 411-415
Curing speed, 54
Cyclic vinyl ethers
  Acid hydrolysis of, 332
  1,4-Bis(3',4'-dihydro-2'H-pyran-2'H-oxy)butane, 328,329
  Comparison of cure rate with cyclic epoxides, 335
  Cure response, 326
  Curing conditions, 333
  Curing results, 333-335
  2-(3,4-Dihydro-2H-pyranyl) methyl-3,4-dihydro-2H-pyran-2-carboxylate, 328
  Discussion of cure results, 335-340
  Effects of light on, 326-341
  Evaluation of cure, 332
  Formulation used, 332-333
Cyclic vinyl ethers cont.
  FTIR to follow cure of, 326, 334-340
  Intrumentation for MS and IR characterisation, 332
  IR for characterisation, 329-331
  Mechanism of polymerisation, 336, 338, 340
  MS for characterisation, 329, 330, 332
  Preparation from acrolein, 327-329
  Solvent resistance of final film after curing, 333
  Structure comparison with cyclic epoxy resin, 327, 328
Cycloaliphatic diepoxide polymers
  Specific characteristics of, 284
  Use in can coatings, 284
  Use in graphic arts, 284
  Use in packaging applications, 284
Cycloaliphatic diepoxide resin
  UVR 6110, 287-289
Cycloaliphatic diepoxide/polycaprolactone polyols
  Hybrid systems containing, 284, 285
Cycloaliphatic diepoxide/TMPTA
  Hybrid system, 285
Cycloaliphatic epoxide/acrylate monomers, 286
  Polymerisation of, 286
Cycloaliphatic epoxy resins
  Dual and hybrid curing systems containing, 303, 304
Cyclobutane rings by
  Photodimerisation of double bonds, 391
Cyclohexanone, 210
Cyclopentadiene, 346
  Reaction with acrylates, 346
CYRACURE UVI-6974, 201,287
Daylight Simulator, 89, 92
  Storage modulus/temperature profiles by, 92, 95-97
Decals
  Applications, 73
  Composition, 73, 74
  Covercoat composition, 81
  Ink medium composition, 81
  Lithographic, 73, 74, 82

Mitographic (or silk-screened), 74
Prepolymer characterisation by GPC, 79, 80
Processors, 73
Types, 73
UV ink development, 82
Degree of cure, 380, 381
"Thumb-twist test", 382
Calcium carbonate/carbon black test, 383
Cotton test, 382, 383
Finger-tip test, 382
Hardness tests, 383
Intermittent, 381
Pencil hardness, 381
Permanganate stain test, 383
Real-time, 381
Solvent rub test, 382
DERAKANE XZ-86799, 287
Diacrylate
Use in blue printing ink, 165, 166
Dialkylphenylacylsulfonium salts, 206
Diaryliodine radical, 202
Diaryliodonium salts, 201, 202
Reduction potential of, 206
Dicyclopentyl acrylate (DCPA), 303, 312, 314
Dielectric constant, 204
Diethyl ether, 204
Diethylene glycoldivinyl ether (DEGDVE), 201, 204,
206, 210-212
Diffuse reflectance, 412, 414
Diglycidyl ether of Bisphenol A (DGEBA)
Cure response of, 304, 305, 307-309
Diacrylate ester of (EBECRYL 3700), 303, 304, 307-
309, 322
Half-acrylate ester of (EBECRYL 3605), 304, 307-325
Influence of different photoinitiators on cure,
304, 308, 309
Influence of diluting monomers on cure, 304, 312-
318
Influence of inert atmosphere on cure, 304, 310-311
Influence of thermal treatment on cure, 304, 318-
319
MEK resistance of cured, 306, 310
Metal adhesion of cured, 304, 319-323
Reactivity, 304
Reactivity with initiators, 306
Surface hardness, 304, 309, 311, 312
Diglycidyl ether of Bisphenol A(EPI-REZ 510), 304,
305, 319, 321, 322
Dilatometer, 385, 386
Dilatometry
Real-time, 381, 385-387, 397
3,4-Dimethoxybenzoyl radical, 176
2,2-Dimethoxy-2-phenylacetophenone (DMPA), 201, 208-
213
Dipentaerythritol penta/hexaacrylate (DPETPA), 249,
251, 254, 256-258, 261-268
2,7-Ditert-butylperoxycarbonyl-9-fluorenone, 183-198

Ditrimethylolpropane tetraacrylate (DTMPTTA), 249,
    250, 251, 255, 256, 260-268
Divinyl ethers, 204, 326,
    Cationic polymerisation of, 199-215
    Dual and hybrid curing systems containing, 304
    Preparation, 326-329
Double or Dual-Cure process, 107
Dropping mercury electrode, 202
DSC, 361
Dual cure system, 304
    Acrylic and epoxy functionality, 304
DYMAX 181, 354
Dynamic Mechanical Analysis (DMA), 84-102, 393
    Assessment of degree of cure, 87, 89
    Definition of technique, 84
    Glass transition by, 85
    Glass transition temperature (Tg), 85
    Loss modulus by, 84
    Loss modulus curve, 85-87, 89
    Storage modulus by, 84-87, 92-95
    Thermogram of primary coat system, 86
    Thermogram of single coat system, 92-94
    Viscoelastic properties by, 85
EB coatings, 5, 19-21, 43-45
EB curing of acrylates and epoxides
    Mechanism by IR spectroscopy, 400
EBC equipment, 22-30
    Broad beam type, 27, 28
    Filament or cathode, 22, 23
    High vacuum chamber, 23
    Linear cathode type, 25-27
    Scanned beam type, 23-24
    Wire-ion-plasma (WIPR) system, 27-30
Electrochemical reduction of
    Onium salts, 201
Electrocurtain design, 25
Electrocurtain processor-depth-dose profiles, 32
Electrocurtain type electron processor, 26
Electromagnetic deflection system, 23
Electron beam curing, 204, 381, 385, 390
    Accelerating voltage, 31, 32
    As competitor to UV curing, 68
    Beam current, 31
    Capital cost, 68
    Cationic mechanism, 40
    Clear coatings, 68
    Delivered dose, 33, 34
    Equipment for safe use, 433, 434
    Hazards, 433
    Inert atmosphere, 35
    Initiation, 38
    Laminating applications, 68
    Machine safety, 34, 35
    Megarad, 33
    Ozone production, 35
    Physical size, 68
    Pigmented coatings, 68

  Primary process, 38
  Radical mechanism, 38
  Secondary process, 39
  Shielding needed, 433
  Terminology, 31
  Worldwide installations, 44, 45
Electron donating free radicals, 202
Electron transfer, 200
Electronic ground states, 176
Elongation at break
  On cured acrylates, 247, 250, 252, 253, 255, 258,
  263
End-of-pulse transient absorption spectra of
  9-Fluorenone(s), 193-195
Epoxides, 46, 47, 145
Epoxidized dicyclopentadiene monoacrylate (EDA), 303,
  312, 314, 316, 320, 321
Epoxies
  Polymerisation rate of, 200
Epoxy acrylate
  Hybrid systems containing, 284, 285
Epoxy acrylate oligomer, 223
Epoxy acrylate oligomers, 287, 288
  Aromatic epoxy diacrylate, 287
  Bisphenol A based, 247
  Cycloaliphatic diepoxide/polycaprolactone triol,
  290-296
  DERAKANE XZ-86799, 287
  PHOTOMER 3104, 287
  SETACURE AP 570, 287
Epoxy acrylate resins, 274, 278, 283
Epoxy acrylates, 269, 347
  Dual and hybrid curing systems containing, 304
  European market, 270
  Offset printing inks containing, 230
  Properties, 275, 276
  Resins containing, 278
Epoxy acrylates - use in
  Clear coatings, 173, 174
  Litho inks and varnishes, 8, 282
  Nonpigmented coatings, 173, 174
  Varnishes and inks, 8
  White lacquer, 165
  Wood coatings, 8
3,4-Epoxycyclohexyl methyl-3,4-epoxycyclohexane
carboxylate (ECC), 285, 287-289
  Use with Polycaprolactone polyol hybrid system,
  285, 287, 288
Epoxy monomers, 200
Epoxy resin ER 510, 303
Epoxy-acrylate oligomer, 248, 250, 258
ESR spectroscopy, 204
  Low temperature, 50
2-(2-Ethoxyethoxy)ethyl acrylate (EOEOEA), 249, 251,
253, 254, 256-258, 261-268
Ethoxy ethoxy ethyl acrylate (EOEOEA)
  Toxicity problems, 272

Ethoxylated Bisphenol A diacrylate (BPADA), 249, 251,
  253, 256-258, 261-268
Ethoxylated Bisphenol A dinorbornenecarboxylate
(EBPADN), 348, 350, 352-354
Ethoxylated trimethylol propane triacrylate
(TMPEOTA), 249, 251, 261-268
Ethylhexyl acrylate (EHA)
  Toxicity problems, 272
European standard
  Date for implementation, 432
  Ionising and non-ionising radiation, 432
Evaporation rate analysis (ERA),393, 394
Excitation, 200
Excitation wavelength, 190
Excited singlet state, 47, 55, 195
Excited triplet state, 47, 54, 56, 189, 197
Excited triplet state of
  9-Fluorenone chromophore, 189, 197
External reflectance, 412
Extinction coefficients, 185, 186, 188, 197
Ferrocene, 157
Film application
  In UV cured paints, 111
Film curing
  In UV cured pints, 111
Film hardness
  In UV cured paints, 112
Finger tip test, 382
9-Fluorenone(s), 182-198
  Absorption properties of, 183
  Absorption spectrum of, 190, 191, 197
  Chromophore, 190, 194
  End-of-pulse transient absorption spectra of, 193
  Perester derivatives, 182-198
  Photoreduction in presence of tert. amines, 182
  Transient absorption profiles for, 196
Fluorescence
  Quantum yields, 188
Fluorescence emission maxima, 185, 193
Fluorescence properties of
  9-Fluorenones, 183, 187
Fluorescence spectra, 184
Fluorescent lamps
  Spectral energy distribution of, 108
Fluorimeter, 184
Foil printing, 61
Fourier Transform Infra Red (FTIR), 358, 361, 362,
  400
  ATR techninque, 89
  Measurement of degree of acrylate cure, 89
Free radical
  Electron donating, 214
  Photogenerated, 214
Free radical curing, 199
Free radical polymerisation, 195
Frequency tripled neodymium laser, 184
Functional oligomers/monomers, 147, 149 152, 153, 155

Functionalised monomer, 400
Furan, 183, 185-187
Gallium doped lamp
  Emission spectrum of, 109, 110
GEIGER MUELLER detector, 393, 394
Gel Permeation Chromatography (GPC), 206, 361, 369, 385
  Ink jet ink analysis by, 142-145
  Use in examining ink jet ink dyes and salts, 132, 135
GENOMER RCX-88 990 LV, 289
Glass transition temperature (Tg), 148, 150
Glass transition temperatures
  On cured acrylates, 247, 252, 255, 257-259, 268
Glycidyl ethers
  Dual and hybrid curing systems containing, 304
Glycol dimercaptopropionate (GDMP), 353
Ground state-projection of for
  Aminoketones, 179
H-donor (s)
  Amine synergists, 47, 56, 125
  Amino alcohols, 48, 51, 57
  N,N-dialkylamino aromatics, 48, 51
Half acrylates
  Hybrid cure of by radiation, 302-325
Half-wave potentials, 202, 203
HALL-effect sensor, 396
Hardness test, 383
Health and Safety at Work Act 1974, 431, 435
Health and Safety Executive, 432, 434, 435
Heated spiral point cathode, 23
Hexafluorobenzene, 193, 195
Hexanediol diacrylate (HDDA), 223, 249, 251, 256, 257, 261-268, 272
  Use in white lacquer, 165, 270
Hexanediol dinorbornenecarboxylate (HDDN), 348, 350, 352, 354
Hiding power
  In UV cured paints, 112-114, 116, 119
  Measurement with ACS SpectroSensor-2, 112
High irradiance monochromator, 184
High speed penetrating electrons, 22
High-energy radiation, 199
Highly ethoxylated trimethylolpropane triacrylate (HEOTMPTA), 249, 251, 254, 255, 258, 261-268
Highly pigmented systems, 54
Highly propoxylated glyceryl triacrylate (HPOGTA), 249, 251, 253, 261-268
Homolysis of
  Perester groups, 188, 191
Hybrid/dual cure systems
  Acrylic and epoxy containing, 302-323
Hybrid curing systems, 302-325
Hybrid systems
  Acrylated oligomer containing, 284, 285
  Acrylated monomer based, 285
  Acrylated oligomer based, 285, 287, 288

Hybrid systems cont.
  Acrylic and epoxy functionality, 302, 304
  Cycloaliphatic diepoxide/polycaprolactone polyols
  containing, 284, 285
  Cycloaliphatic diepoxide/TMPTA based, 285
  Epoxy acrylate containing, 284, 285
  Optimum properties using, 284
  Urethane acrylate containing, 284, 285
  Vinyl ether containing, 285
Hybrid systems (epoxy acrylate/cycloaliphatic
epoxide)
  Adhesion characteristics, 295
  Crosshatch adhesion, 291
  Cure speed, 290, 291, 293, 294
  Curing conditions, 290
  Differential scanning calorimetry, 295-297
  Impact resistance, 291
  MEK double rubs, 291, 292, 294
  Pencil hardness, 291
  PERSOZ hardness, 291
  Results and discussion, 290-298
  Shelf life stability, 291
  Solvent resistance, 290-292, 295
  Viscosity, 291, 294
Hybrid systems (urethane acrylate/cycloaliphatic
epoxide), 296-300
  Crosshatch adhesion, 298
  Cure speed, 296, 298, 299
  Impact resistance, 296, 298
  MEK double rubs, 296, 298, 300
  Pencil hardness, 298
  PERSOZ hardness, 298
  Shelf life stability, 298
  Surface hardness, 296
  Viscosity, 298
Hybrid UV curable ink jet ink, 183
Hydrogen atom abstraction, 189
Hydrogen fluoride
  Production in curing of epoxides, 409
Hydrogen transfer, 56
2-Hydroxy-2-methyl-1-phenyl-propane-1-one
  Norrish I type cleavage on, 112
  UV absorption spectrum of, 110
Hydroxyl functional caprolactone monoacrylate (TONE
M100), 303, 312, 316
Hydroxyphenones
  Effect of introduction of groups, 54
In-line coating, 62
Incident dose, 207
Induction period, 56
Infra-red radiometer, 393
Infra-red radiometry, 392, 393, 397
  Real-time, 381
Infra-red spectrometer, 388, 389
Infra-red spectroscopy, 379, 380, 385, 387-389, 397
  Absorbance, 379, 380

Acrylate and epoxy systems-degree of cure by, 400-415
Diffuse relectance Fourier transform(DRIFTS), 389
Double bond concentration by, 379
Multiple internal reflection (MIR), 389
Real-time (RTIR), 381, 389, 397
Stretching frequency, 379
To assess degree of chemical reaction, 400
Use to examine films, 400
Inhibition period, 209
Initiation, 200
Initiator Chemistry, 161
Ink jet ink(s)
  Solvent based, 132-146
Inter-coater drying, 62
Inter-colour applications, 62
Inter-colour drying, 62
Intermix process, 107
Intermolecular hydrogen-abtraction, 56
International Atomic Energy Agency (IAEA), 434
Intersystem crossing (i.s.c.), 47, 49, 175
Iodonium salts, 200, 202
Ionising and non-ionising radiation
  European standard, 432
IRGACURE 184, 201, 281, 282, 303
IRGACURE 500, 303, 309-312, 317-324
IRGACURE 651, 16, 80, 201, 237, 288-290, 293, 294, 303, 306-309, 313, 316, 320, 322
Irgalith Blue GLB
  Use in blue printing ink, 166, 173
Isobornyl acrylate (IBA), 249, 251, 253, 255-258, 261-268, 272, 303, 317, 318, 320-324
Isodecyl acrylate (IDA), 249, 251, 255, 257, 261-268, 272, 403
Isopropyl thioxanthone (ITX), 201, 208-212, 223, 230, 232, 233
  Mechanism with aromatic amines, 233
  Use in colour-pigmented formulations, 230, 233
Isopropyl thioxanthone (ITX) cont.
  Use with aromatic amines, 233, 235, 236
  UV spectrum in DEGDVE, 211
Ketone derived radical, 52
Ketyl radical, 52, 210,
KNOOP indenter, 383
KONIG hardness, 384
  On cured acrylates, 247, 250, 252, 254, 256, 258, 264
KRATOS GM252 monochromator, 184
LAMBERT-BEER equation, 217
Laser flash photolysis, 183, 184, 190-197
Laser interferometry, 396
Laser-nephelometry, 55, 396
Latent acidity, 199
Lauryl acrylate (LA)
  Photopolymerisation of, 249, 251, 255, 257, 258, 261-268, 391, 392

Light source
    Phillips HOK-6 lamp, 57
Linear variable differential transformer (LVDT), 386
Liquid films
    Cured by UV or EB radiation, 400
    Formation of "tacky" films, 400
LOCTITE 350, 350, 354
Long-lived catalysts, 149, 150
Low viscosity acrylate resins
    Commercially available, 278
    Commercially available water thinnable, 280, 283
    Conclusions, 283
    Developable etch and solder resists, 275
    Historical development, 272
    Likely areas of interest, 283
    Logic, 270
    Options, 271
    Problems with monomer diluents, 273
    Radiation cured formulations for, 269-283
    Roller coating varnish formulation, 281
    Screen print varnish application, 282
    Toxicity considerations, 271-273, 283
    Trends in for radiation cured formulations, 269-283
    Volatile organic solvent use, 274
    Water thinnable, 279, 280, 283
Low viscosity oligomers, 269, 271
    Development, 277
Machinery Directive
    Mechanical and electrical safety, 431
    Risk assessment, 431
Magnesium carboxylates, 155
Magnesium titanate
    Absortion spectrum of, 108, 116
Mandrel
    Absortion spectrum of, 108, 116
Medium pressure mercury lamp
    Emission spectra, 106
    Infra-red radiation, 107
MEK rubs, 289, 291
Melamine resins, 46, 47
(Meth)acrylated resins, 151-153, 155, 156
Methacrylates, 47, 76, 77
    In decal production, 80
    In UV cover coats, 76
Methyl ethyl ketone (MEK), 274, 280, 281, 382
    QCM Dissolution Method use in, 419-425
Methyldiethanolamine, 183, 184, 188, 189, 191, 192
Methylmethacrylate, 183, 184, 187, 188, 197
Microhardness, 383
Microlithography
    Introduction to, 428
Microwave dielectrometry, 394
Molecular modelling, 174-180
    of aminoketones, 174-180
    of photoinitiators, 174-180
Molecular modelling package, 174
Molecular photochemistry, 60

"Mono-cure" process in
  UV pigmented paint systems, 109
Monomer radicals, 190
Monomeric vinyl ethers, 326
Monomers
  Cross linking of, 380
  European market, 270
  Functionality, 5, 7
  Low Draize reading, 6
  Odour/taint and volatility, 5
  PHOTOMER 3000 series, 8
  PHOTOMER 4000 series, 7
  Properties, 5
  Skin irritancy, 6
  Trimethylol propane ethoxylate acrylate, 6
  Tripropylene glycol diacrylate, 6
  Viscosity and solubility, 5
Multi-colour coating, 62
Multifunctional norbornene olefin components with
  Multifunctional thiol components, 346
Multifunctional thiol, 150, 151
Multifunctional thiols as
  Acrylate polymerisation initiators, 344
n-Hexane, 201
N-Vinylpyrrolidone (NVP), 303, 312
  Use in white lacquer, 165, 166
Neopentyl glycol diacrylate (NPGDA), 272
Nitrous oxide, 205, 206
Nonpigmented coatings
  Alpha-aminoketones in, 173, 174
  Curing of, 173, 174
Norbornene functional resins and oligomers, 347-348
  Diels-Alder reaction in preparation of, 346
  Preparation of, 346-350
Norbornene functionalised
  Aliphatic hydrocarbons, 346
  Aromatic hydrocarbons, 346
  Siloxane oligomers, 346, 348-350
Norbornene functionalised derivatives, 346
Norbornene functionalised siloxane fluids
  Reagents for preparing, 349
  Routes to, 349
Norbornene functionalised siloxane resins
  Mechnical properties of, 355
  Properties of, 355
Norbornene functionality
  Alternate modes of introducing, 347, 348
Norbornene resins and oligomers
  Photopolymerisation response of, 350
  Toxicology of, 355, 356
Norbornene resins as substrate in
  Thiol-ene polymerisations, 342-357
Norrish type I cleavage, 15, 16, 112, 286, 288
Norrish type II reactions, 56
Novel t-butylperester derivative of
  9-Fluorenone, 182-198
Novolac epoxy, 150, 151

O_2 inhibition, 47
Offset lithography, 124
Offset printing, 43
Offset printing inks
  Formulation, 230
Oligomeric vinyl ethers, 326
Oligomers
  Amine synergists, 14, 15
  Cross linking of, 380
  Emulsions and water thinnable, 7, 10
  Epoxy acrylates, 7, 8
  Polyester acrylates, 7, 8, 9
  Specialised, 7
  Speciality acrylates, 14
  Urethane acrylates, 7, 9-14
Onium salts, 199-207, 211, 214
Optical fibre coatings
  Acrylate based systems, 85
  Advantages, 84
  DMA in establishing degree of cure, 84
  DMA in investigating post cure effects, 84
  DMA in resin design, 84
  Effect of post cure on physical properties, 89-91
  Effect of UV dose on physical properties, 87, 88
  Electron beam cured, 97, 98
  Elongation at break, 87, 89
  Polyurethane acrylate resins, 85, 100
  Primary, secondary or single types, 85, 87
  Properties, 84
  Properties of films, 99
  Radiation curable, 84-102
  Tensile strength, 87, 89
  Tensile testing on INSTRON machine, 89, 92
Optical fibre technology, 84
Optical fibre telecommunications, 101
Organicsilicon compounds
  Unsaturated, 358
Overprint varnishes, 43
Oxonium salts
  BTSPF$_6$, 202
  BTSSbF$_6$, 202, 203
  DBMSPF$_6$, 203, 205, 206
  p-CH$_3$OBTSSbF$_6$, 203, 205, 206
  Ph$_2$IPF$_6$, 201-206, 209-213
  Ph$_3$SSbF$_6$, 202, 203, 205, 206, 212
  PTPDS, 202, 203, 212, 213
  Reduction potential of, 203
Oxygen complexes, 195
Oxygen quenching, 57
Oxygen radical cation, 194
Oxygen scavenging, 58
Oxygen-inhibition in
  Free radical polymerisation, 47, 48, 56-59
p-Phenylene bis(acrylic acid)
  Photocrosslinkable polyesters of, 391
p-Toluene sulphonic acid, 149

Pencil hardness, 291, 298, 381, 384, 391
  On cured acrylates, 247, 250, 252, 254, 256, 258,
  265
Pendulum hardness, 170, 171, 173, 224, 225, 380, 384
  Pigmented coatings, 224-226, 228, 229, 234, 239,
  242, 243
Pendulum hardness by Konig method, 112
Pendulum hardness of cured films, 112, 114, 116, 119
Pentaerythritol tetraacrylate (PETTA), 249, 251, 257,
  261-268
Pentaerythritol tetramercaptopropionate (PETMP), 353
  Norbornene resins crosslinked with, 350
  Triallylisocyanurate (TAI) comparison within
  resins, 350
Pentaerythritol tetranorbornenecarboxylate (PETN),
  348, 352
Pentaerythritol tri and tetra acrylate esters (PETA),
  273
Pentaerythritol triacrylate (PETIA), 249, 251, 254,
  258, 261-268
Perfecting applications, 62
Perkin-Elmer Model Lambda 7 spectrometer, 184
Perkin-Elmer Model LS-5 fluorimeter, 184
Permanganate stain test, 383
PERSOZ hardness, 291, 298, 384
PF-6 counterion, 204
Phenolic diene, 150, 151
Phenothiazine, 201
2-Phenoxyethyl acrylate (2PEA), 249, 251, 253, 261-
268
Phenoxy ethyl acrylate (PEA)
  Toxicity problems, 272
Phenyl type radical, 194
Phenyltrivinylsilane, 360
Photo-DSC measurements, 201
Photoacoustic cell, 412
Photoacoustic spectroscopy (PAS), 389-391, 414
  Depth profile by, 389
  Fourier transform infra-red (FTIR), 390, 397
  UV/visible, 390
Photoactivity, 187
Photocalorimetry, 200
Photocatalyst(s), 46
  Polymerisation and crosslinking processes, 46-60
Photochemical processes, 48, 56
Photochemistry of
  Fluorenone and derivatives, 182
  Perester derivs. of fluorenone, 182-198
Photocleavage reaction of
  Aminoketones, 175
Photocrosslinking, 147
Photoelectron beam and x-ray resists
  New materials for, 416
Photogenerated
  Organic free radicals, 200
Photogenerated initiators/catalysts, 149, 150, 153
Photoimaging, 60

Photoinduced reaction, 190
Photoinitiated polymerisation, 147, 153
Photoinitiation, 188, 190
Photoinitiator(s), 5, 15, 46, 147, 150, 386, 395
Photoinitiator(s) for, 201
  Absorption curve, 104
  Acetophenone derivatives, 126, 127
  Acetophenone type, 163-165, 220
  Acetophenones, 220
  Acylphosphine oxides, 50, 53, 109, 110, 113-118,
  220, 236, 237
  alpha, alpha-Dialkoxy acetophenones, 50
  alpha, alpha-Dialkoxy deoxybenzoins, 50, 53
  alpha-Alkoxy deoxybenzoins, 50
  alpha-Cleavage type, 230
  alpha-dimethylamino acetophenones, 164-181
  alpha-Hydroxy alkylphenones, 50, 54
  alpha-Keto coumarins, 51
  Aminoacetophenones, 164-165
  Aromatic carbonyl compounds, 47
  Behaviour in pigmented formulations, 163-165
  Benzil ketals, 16, 126-130, 220, 237
  Benzildimethyl ketal, 288-290, 293, 294
  Benzils, 51, 53
  Benzoin and benzoin ethers, 126, 127, 201
  Benzoin butyl ether, 15, 220
  Benzoin ethers, 220
  Benzoin methyl ether, 391, 392
  Benzophenone(s), 14, 17, 51, 53, 57, 125, 127-131,
  140, 201, 220, 223, 225, 231, 240, 248, 281, 282,
  293, 303, 309, 320
  Benzoyldialkylphosphonates, 51, 53
  2-Benzyl-2-dimethylamino-1-(4-morpholinophenyl)-
  butanone-2, 240
Photoinitiator(s) for cont.
  Cationic, 302
  CG24-61, 303
  Characteristics required, 126
  CYRACURE UVI 6974, 201, 287-290, 296
  DAROCUR 1173, 248
  Decal production in, 76, 77, 79, 80
  Effect of concentration on rate of curing, 217
  Effect on light penetration, 381
  Efficiency of, 17, 207, 210, 213
  ESACURE TZT, 248
  Excited, 57
  9-Fluorenone as, 188, 197
  Fragmentation, 49
  Free radical, 200, 201, 302
  Free radical generators, 126
  Free radical polymerisation for, 150, 153, 155, 156
  Free radical processes, 47-49
  FX 512, 303, 306-313, 316, 318-324
  Ground-state, 57
  H-abstraction, 49
  Hydrogen abstraction, 17, 125

2-Hydroxy-2-methyl-1-phenyl-propane-1-one(HMPP),
112, 223, 225, 237, 238, 240, 242
Improvement for litho inks, 130
In UV cover coats, 76
IRGACURE 184, 201, 281, 282, 303
IRGACURE 500,   303, 309-312, 317-324
IRGACURE 651,   16, 201, 237, 288-290, 293, 294,
303, 306-309, 313, 316, 320, 322
Isopropyl benzoin ether, 220, 237
Isopropylthioxanthone(ITX), 201, 208-212, 223, 225,
230
Ketals, 220
Litho inks, 124-131
Low taint and odour, 127
Methyl[4-(methylthio)phenyl]-morpholino-propane,
130
Michler's Ketone, 220, 234, 235
Molecular modelling of, 174-180
New for use in pigmented systems, 107
O-Acyl- oximinoketones, 50
Onium salts, 199
Performance in clear coatings, 217
Performance in pigmented coatings, 217, 231-243
Performance with aromatic amines, 233-236
Substituent effects, 172-174
Photoinitiator(s) for cont.
Technology-drawbacks, 302
(Thio)xanthones, 51, 53, 125, 128, 130, 201
Trarylsulfonium hexafluoro antimonate salt, 286-290
2,4,6-Trimethyl benzoyl diphenyl oxide, 113, 114
(2,4,6-Trimethyl-benzoyl)diphenylphosphine oxide
(TMDPO), 170, 201, 236-238, 240
2,4,6-Trimethylbenzoylethoxyphenylphosphinec oxide,
201
Type I, 48-50, 52-55, 59
Type II, 48-53, 56, 59
Use in clear coatings, 173-174
UV absorption spectra of, 110, 128
UV in pigmented systems, 216-243.
UVE 1014, 303
UVI 6974, 212, 213
Photoinitiator(s) for
Polymerisation and crosslinking processes, 46-60
Photolysis, 200
Quantum yield, 187
Photolysis of
t-Perester groups, 187
PHOTOMER, 4
PHOTOMER 3005, 8
PHOTOMER 3016, 8
PHOTOMER 3038, 8
PHOTOMER 3104, 287
PHOTOMER 4028, 7
PHOTOMER 4039, 7
PHOTOMER 4061, 6, 7
PHOTOMER 4094, 7
PHOTOMER 4116, 15, 17

PHOTOMER 4127, 7
PHOTOMER 4149, 6, 7
PHOTOMER 4182, 15, 17
PHOTOMER 4193, 7
PHOTOMER 4215, 15
PHOTOMER 6052, 12
PHOTOMER 6118, 13
PHOTOMER 6129, 11
PHOTOMER 6140, 11
PHOTOMER 6162, 12
PHOTOMER 6202, 12
PHOTOMER 6250, 11, 289
PHOTOMER 6261, 12
PHOTOMER 6305, 13
PHOTOMER 6316, 289
PHOTOMER 7020, 14
PHOTOMER 7031, 14
PHOTOMER EL1303, 10
Photomultiplier (RCA IP28A), 184
Photon absorption, 175
Photophysical processes, 48, 56
Photopolymerisation, 60, 184, 187-190, 195, 197, 198,
    385, 386
  Efficiencies, 184
Photopolymerisation activity
  Perester derivs. of fluorenone, 182-198
Photopolymerisation efficiency, 198
Photopolymerisation of
  Coatings, 206
  t-Butylperester derivs. of benzophenone, 182
  Vinyl monomers, 182
Photopolymerisation through
  Homolytic scission, 182
Photoreduction
  Quantum yields, 184, 187, 188
Photoreduction of
  Fluorenone, 182
Photoreduction properties of
  9-Fluorenones, 183
Photoredux induced cationic polymerisation, 199-215
  Principles of, 200
Photoredux induced cationic polymerisation of
  Divinyl ethers, 199-215
Photosensitisers, 200, 209, 214
  Excited state of, 200, 202, 204
Phthalocyanine pigment
  Use in blue printing ink, 165
Pigment characteristics, 118
Pigmented coatings
  Acrylic, 219
  Passage of light through, 217-219
  Polyester-styrene, 219
  Selection of UV-photoinitiator, 219
  Types of pigments used, 219
Pigments
  Absorption curves, 104
  Carbon (black), 115, 117, 231

Titanium dioxide, 105, 113-116
Transmission spectra, 129
Pigments for coatings
  Basic lead carbonate, 221
  Bismuth oxychloride, 221
  Black, 219
  Chromium oxides, 221
  Coloured inorganic, 219
  Coloured organic, 219
  Cyan, 231
  Extender, 219
  Iron oxide, 221
  Magenta, 231
  Magnesium titanate, 236, 237
  Metal effect/(aluminium), 220, 221, 223-230
  Metal oxide/mica, 221, 223
  Metallic, 219-221
  Nacreous and interference, 219
  Orange, 231
  Pearl lustre, 218, 220-222, 224-230
  Titanium dioxide/(mica), 217, 221-229, 235-237
  White, 219
  Yellow, 231
  Zinc sulphide, 236
Polarographic measurements, 202
Poly(methyl methacrylate)
  Molecular weights, 418
  Polydispersivity, 418
  Resist materials use in, 418
Polycaprolactone polyols
  Coatings from, 285
Polycaprolactone triol
  TONE 0301, 287-289
Polycaprolactone triol 300, 287-289
Use in UV curing cationic system, 287-289
Polycarbosilane
  Fibres from, 360, 362
  Formation by thermolysis of polydimethylsilane, 367
  FTIR spectra material from PSS thermolysis, 370
Polydimethylsilane
  Production of SiC fibres from, 359
  Thermolysis of, 359, 367
Polyester acrylates, 278, 347
  European market, 270
  Offset printing inks containing, 230
  Properties, 275, 276
  Resins containing, 278
Polyester acrylates - use in
  Varnishes, 9
  White lacquer, 165, 167
  Wood sealers, 9
Polyethylene glycol 200 diacrylate (PEG200DA), 249-251, 261-268
Polyethylene glycol 400 diacrylate (PEG400DA), 249, 251, 253-255, 258, 260-268
Polyethylene glycol 400 dinorbornenecarboxylate (PEG400DN), 348, 353

Polyethylene glycol 400 dimercatopropionate
(PEG400DMP), 353
Polyfunctional norbornene resin
  Photocrosslinking of, 345
Polymerisation
  Efficiency, 207
  Enthalpy change of, 380
  Thermogram, 207
Polymethacrylates, 76
Polymethylmethacrylate, 189, 191, 192
Polysilanes, 359
  Absorbance in UV region, 359
  As photoinitiators for vinyl polymerisation, 359
  As photoresists in microlithography, 359
  As precursors to silicon carbide, 359
  Degradation and crosslinking of, 359
  Photochemical sensitivity, 359
  Si-Si bond in, 359
  UV irradiation of, 359
Polysilastyrene films and fibres, 358-375
  Conclusions, 373
  UV and alpha-irradiation of, 358-375
Polysilastyrene (PSS), 358
  Advantages and disadvantages of use in fibre
  production, 360
  As route to silicon carbon fibres, 359, 360
  Bond dissociation energies, 365
  Carbosilane generation from on irradiation, 366
  Dichlorodimethylphenylsilane in synthesis of, 361
  Dichlorodimethylsilane in synthesis of, 361
  Formaldehyde formation from on irradiation, 366
  Polycarbosilane from, 360
  Synthesis of, 361, 362
Polysilastyrene (PSS) fibres
  Ceramic residue of after UV irradiation, 370-373
  FTIR spectra after alpha-irradiation, 371
  SEM of, 372, 373
Polysilastyrene (PSS) films
  Effect of irradiation on structural changes, 362-
  370
  Effect of presence of unsaturated silicon compounds
  on structure, 367, 368
  FTIR peaks-area ratios of after UV irradiation in
  air, 364, 365
  FTIR peaks-area ratios of after UV irradiation in
  argon, 364, 365
  FTIR spectra of, 362-365
  FTIR spectra of after irradiation in presence of
  unsaturated Si compounds, 368, 369
  FTIR spectra of after UV irradiation in air, 363
  FTIR spectra of after UV irradiation in argon, 363
  Silene species formation in production of, 366
Polysilastyrene (PSS) UV crosslinking
  Instrumentation and techniques, 361
  Materials, 360, 361
Polyurethane acrylate
  Use in blue printing ink, 165, 166

Polyurethane-acrylate coating
  Study of curing of, 389
Potassium permanganate, 383
Pre-coating applications, 62
Primary initiating radicals
  Scavenging of, 56
Printed ink films, 124
Printing presses, 62
Problems of pigmentation
  Industrial solutions, 105
Propagating macro radicals
  Scavenging of, 56
2-Propanol, 183-188
Propanol-2-ol (IPA)
  QCM Dissolution Method use in, 419, 420, 422, 423
Propenyl ether oligomers, 200
Propoxylated glyceryl triacrylate (GPOTA), 249, 251, 261-268
Propoxylated neopentyl glycol diacrylate (NPGPODA), 249, 251, 253, 261-268
Propylene carbonate
  Solvent for photoinitiator, 286
Proton transfer, 56
Pulsed xenon lamp, 184
QCM Dissolution Apparatus, 418, 419
  Quartz used, 418
QCM Dissolution Method
  Conclusion, 428
  Effect of different solvents, 421, 422
  Effect of PMMA molecular weight, 423-425
  Effect of solvent composition, 422-424
  Effect of temperature, 425, 426
  Effect of UV irradiation, 426-428
  Electrical circuit, 418, 419
  Frequency and film thickness v time, 419-421
  Frequency-time curves, 419
  Poly(methyl methacrylate) used, 418
  Reproducibility of dissolution experiments, 420, 421
  Results and discussion, 419-428
Quantum mechanical calculations, 174, 176
Quantum mechanical methods, 174
Quantum yield of
  Radical formation, 58
Quantum yields, 185, 186
Quartz
  Damping effect of liquids, 417
  Mass to frequency change relationship, 417
  Piezoelectric properties, 416
  Resonant frequency change, 416
  Used in QCM method for resist materials examination, 418
Quartz Crystal Microbalance (QCM), 416
Quenching equation, 57, 58
Quinine sulphate, 184
Radiation curable adhesives
  Auxiliary cure mechanisms for, 147-159

Radiation curable systems, 199
Radiation cured coatings
  Background, 4
  Techniques to monitor degree of cure, 379-399
Radiation cured formulations
  Monomers, 5
Radiation cured products
  European market, 3
Radiation curing, 430-435
  Advantage over thermosetting curing, 199
  Current legislation, 431, 432
  Health and Safety aspects, 269, 272
  Safety standards - the way ahead, 434
Radiation curing equipment, 430
  Machinery Directive, 431
  UV curing, 17
Radiation curing technology
  Applications, 1
  Equipment, 3
  New applications, 3
  Raw materials, 3
Radiation sources, 430
Radical cations, 195
Radical processes, 48
Redox initiation
  High energy radiation induced, 204
  Photochemically induced, 206
Reduction potential of
  Oxonium salts, 203
Reflection of
  UV radiation, 216
Reflection-absorption ray, 413
Reflector unit, 61
Reflector/lamp units, 62
Refraction of
  UV radiation, 216
REHM-WELLER equation, 202, 204
Resist materials
  QCM method for study, 417, 418
  Solubility investigation methods, 417
  Solubility investigation-methods limitations, 417
Resist process
  Development of image, 416
  Dry etching, 416
  Plasma development, 416
  Production of image, 416
  Solvent development, 416
  Study of thin polymer films, 416
Resistivity, 394, 395
Rivelling effect, 104
Rocker methods, 384, 385
S decay, 48
Scanner type electron processor, 23
SETACURE AP 570, 287
Silk screen, 43
Silk-screen decal production
  Drawbacks, 75

SILWET L720
  Surfactant, 287-289, 293
Silyl radicals
  As reactive intermediates, 359
Silylenes
  As reactive intermediates, 359
Singlet complex, 194
Singlet ground states, 175
Singlet states, 185-187
Skin irritancy, 4, 6
Solvent containing oligomers, 269
Solvent rub test, 382
Speciality monomers
  Supplier, 4
Speciality oligomers
  Supplier, 4
Spectroscopic measurements, 184-187
Specular reflectance ray, 413
Specular reflection accessory, 413
Storage oscilloscope, 184
Substituent effect, 56
Sulfonium salts, 200
Sulfur radical, 202
Sulphonium salts, 201, 202
Surface profilometry, 395, 397
SUTHERLAND rub test, 385
Sward hardness test, 384, 385
T decay, 48
T quenching, 48
t-Butyloxy radicals, 182
TABER abrasion test, 385
Tableware decoration
  Decalcomanias(decals), 73
  Direct screening of solvent based inks, 73
  Offset printing processes, 73
Tensile strength
  On cured acrylates, 247, 250, 252
Tensile strength at break
  On cured acrylates, 253, 255, 257, 262
Tert-amines, 48, 51, 53, 56, 58, 59
4-Tertbutylperoxycarbonyl-7-nitro-9-fluorenone, 183-198
2-Tertbutylperoxycarbonyl-9-fluorenone, 183-198
Tetraethylene glycol diacrylate (TTEGDA), 249, 251, 253, 261-268, 303, 313, 314
Tetrafuctional allylic urethane
  Photocrosslinking with multifunctional thiol, 343
  Preparation of, 343
Tetrahydrofuran, 204
Tetravinylsilane, 360
TGA, 361
Thermomechanical analyser (TMA), 386
Thick pigmented coatings
  Chipboard on, 120-122
  Double or Dual-Cure process, 107
  Future trends, 123
  Intermix process, 107

Magnesium titanate in, 107, 113
New photoinitiators for, 107
Production by UV technology, 103-123
Zinc sulphide in, 107, 113, 114, 116
Thilo-ene photopolymerisations
Difference from (meth)acrylate polymerisations, 342
Thin polymer films
Measurement of dissolution kinetics, 416-429
Thiol crosslinked norbornene resins
Adhesive properties of, 354
Effect of chain on bulk polymer properties, 353
Effect of increased crosslink on bulk density, 353
Electrical properties, 352
Mechanical properties, 352
Outgassing, 351
Properties of, 351-356
Properties of-PETMP crosslinked, 351, 352
Stress-strain properties, 351
Thermal conductivity, 351
Therogravimetric analysis, 352
Water vapor transmission, 351
Thiol-ene polymerisations, 342-357
Allylic abtraction v addition in cyclic olefins, 345
Limitations of, 344
Norbornene resins as substrate in, 342-357
Use in coatings of floor tiles, 344
Use in electronic coatings, 344
Use in photopolymer printing plates, 344
Use in speciality adhesives, 344
[2,2,1]Bicycloheptene terminated oligomers in, 345
Thiol-ene system, 151, 153
Three dimenstional networks, 46
"Thumb-twist test", 382
Titanium dioxide
Absorption spectra, 106, 112
Titanium dioxide pigments
Al_2O_3 coated, 113
Anatase, 105, 113-116
By chloride process, 113
By sulphate process, 113
Rutile, 105, 113-116, 237
SiO_2 coated, 113
UV radiation absorption, 105
Titanium dioxide (rutile)
Coatings use, 217, 219, 235-238
Use in white lacquer, 165-167, 173
TONE 0301, 287
Transient absorption changes, 184
Transient absorption profiles for
9-Fluorenone(s), 196
Transmission of
UV radiation, 216
Transmission spectroscopy, 414
Trapping agents
Phosphorus (III) derivatives, 58
Tert-amines, 58
Thiols, 58

Triarylsulfonium salt, 157
1,1,1-Trichloroethane, 395
Triethylamine, 183, 187
Triethylene glycol diacrylate (TIEGDA), 249, 251,
    261-268
Triethylene glycol divinyl ether (RAPIDCURE DVE-3),
    304, 313, 314, 320, 333, 340
2,4,6-Trimethylbenzoyl diphenylphosphine oxide
(TMPDO)
    UV absorption spectrum of, 110
2,4,6-Trimethylbenzoylethoxyphenylphosphine oxide
(TMPEO), 201, 208, 209
Trimethylolpropane triacrylate (TMPTA), 110, 251,
    256, 270, 273, 293
    Crosslinking of, 55
    In UV cured paints, 111
    Use in white lacquer, 165-167, 249
Trimethyolpropane trinorbornenecarboxylate (TMPTN),
    348, 350, 352-354
Triphenylsulfonium salts, 201, 202
Triplet absorbances, 197, 198
Triplet absorption maxima, 197
Triplet complex, 194
Triplet excited state, 182
Triplet state(s), 176, 178, 191, 193, 194
    Aminoketones, 178
Tripropylene glycol diacrylate (TPGDA), 248-251, 253,
    257-259, 261-268, 270, 272, 274, 281, 293
Tris(2-hydroxyethyl) isocyanurate tracrylate
    (THEICTA), 249, 251, 253-255, 257-259, 261-268
TUKON tester, 383
Turbidimetry, 396
Unimolecular
    Fragmentation, 47
    Process, 49
Unsaturated polyester
    Use in clear coatings, 173, 174
    Use in nonpigmented coatings, 173, 174
Unsaturated polyester resins, 46, 47
Urethane acrylate
    Hybrid systems containing, 284, 285
Urethane acrylate oligomers, 288, 289
    Aliphatic polyester urethane triacrylate, 289
    Aliphatic urethane hexafunctional acrylate, 289
    GENOMER RCX-88 990 LV, 289
    Pentafunctional aliphatic urethane acrylate, 289
    PHOTOMER 6250, 289
    PHOTOMER 6316, 289
Urethane acrylates, 269, 347
    Application areas, 11-13
    Dual and hybrid curing systems containing, 304
    European market, 270
    Offset printing inks containing, 230
    Preparation, 9
    Properties, 275, 276
    Typical properties, 14
    Variables, 9

Urethane acrylates in
  UV water based coatings, 10
UV absorption spectroscopy, 189, 191, 192, 197
UV and alpha-irradiation of
  Polysilastyrene films and fibres, 358-375
UV and EB coatings
  Applications, 284, 285
UV and EB cured formulations, 5
UV and EB curing, 270
  Benefits, 4
  Current applications, 43
  European market, 270
  Formulations, 76
UV based coatings from
  Water emulsions, 10
  Water thinnable oligomers, 10
UV coating of
  P.V.C., 64
UV covercoat developments, 76
UV covercoat from
  Polymethacrylate, 77
UV covercoats
  Firing conditions, 83
  Future in decal production, 83
  Photoinitiator/photosensitiser combinations, 83
  Pilot scale preparation, 77, 78, 80
  Requirements, 76
  Small scale preparation, 76, 77
  Two stage cure process, 76
UV curable adhesives
  Applications, 147, 148
  Attractive features, 148
  Auxiliary cure mechanisms for, 147-159
  Challenges, 149
  Compositions, 149
  Cure thickness, 148
  General characteristics, 148
  Limitations, 148
  Opportunities, 149
  UV-aerobic, 149, 155
  UV-anaerobic dual cure, 149, 151, 156
  UV-moisture, 149, 153, 154
  UV-thermal, 149-153
UV curable coatings, 149
UV curable ink jet ink(s)
  Adhesion, 140, 145
  Benzophenone use in, 140
  Chemical stability, 132, 135
  Conductive salts, 132, 135, 146
  Cure, 132, 135, 145
  Cycloaliphatic epoxide - use in, 32, 136-139, 145
  Dye selection, 142
  Dyes, 132, 135
  Effect of conductive salts on stability, 141
  Effect of ingredients, 132
  Epoxy resins use in, 140
  Formulations, 136, 138

Free radical initiated monomers in, 132
Mechanism of alternative cationic polymerisation, 136
Mechanism of pure cationic polymerisation, 136
Mechanisms, 136
Performance and reliability, 134
Physical and chemical properties, 134, 135
Principle of operation, 133, 134
Solvent resistance, 132, 138-140, 145
Typical specification, 134, 135
Viscosity increase on ageing, 142
UV curable media for
Decal production in the ceramics industry, 73-83
UV curable pigmented coatings
Formulation and processing, 111-118
UV curable pigmented systems, 165-167
Blue offset printing ink, 165, 166
Photoinitiator substituent effects, 172-174
White lacquers, 165-167
UV curable prepolymers in
Ink jet inks, 132-146
UV curable resins
Scope for different applications by solubilising, 279
UV cured coatings
Application areas, 1, 19, 21
Dental filling materials from, 383
Market volume in Eurpoe, 20
UV cured coatings for
Board, 4
Metal, 4
Paper, 4
Plastics, 4
Wood, 4
UV cured litho inks
Black, 129
Formulation, 124
Future, 130
White, 129
UV curing, 37
Advantages, 103
Cationic mechanism, 104
Disadvantage, 103
Free radical mechanism, 104
Initiation, 36, 37
Initiation by UV light, 36
Major restraint, 103
New applications areas, 19
Potential application in maintenance products, 19
Propagation, 36, 37
Protection against exposure, 433
Radical mechanism, 36
Termination, 36, 37
Thick pigmented systems, 69
Threshold Limit Values for dose levels, 432
Water based alternatives, 68
Wavelengths used, 432

UV curing cationic technology, 284
UV curing methodology, 180
UV curing of
  Black pigmented paint systems, 115, 117
  Coloured pigmented paint systems, 118, 119
  Red pigmented paint systems, 118, 119
  White pigmented paint systems, 113-116, 120-122
  Yellow pigmented paint systems, 118, 119
UV curing of paints
  Influence of pigments on, 104
UV curing of pigmented coatings
  Formulation key points, 118
  Processing, 120
UV curing of systems thro. monomer selection
  Performance of, 247-268
UV curing technology
  Cationic, 132
UV curing with
  Electrodesless lamps, 17
  Medium pressure lamps, 17
UV equipment, 61
  Air knife cooling, 63, 64
  Air-cooled, 61
  Control improvements, 66
  Cooling, 62-65, 69
  Designs for special applications, 67
  Dicroic reflectors, 64, 65
  Exposed machine part cooling, 62
  Extruded aluminium fabrication, 61
  Flameproof systems, 67
  Lamp cooling, 62
  Lamps, 65, 66
  Nitrogen purged systems, 67
  Ozone removal, 67
  Reflector cooling, 62
  Sheet metal fabrication, 61
  Substrate cooling, 62
  The future, 68, 69
  Trends in, 61-69
  Water cooled cylinders, 63, 64
  Water cooled platens, 63, 64, 67
  Water cooled refector, 63, 64, 67
  Water filtration system, 64, 65, 69
  Water-cooled, 61, 63
UV ink and lacquer formulations, 4
UV light sources
  Additive lamps, 66
  Capillary pressure mecury lamp, 18
  Electrodesless lamp, 65
  Evaluation of, 18
  Gallium doped lamps, 109, 112, 115, 118-121
  High pressure mercury lamp, 184, 200
  Low pressure mercury lamp, 18, 122
  Medium pressure mercury lamp, 62, 65, 105, 107,
    111-115, 118-122, 127, 128, 168-170, 199, 200, 252
  Metal halide lamps, 18, 66, 69
  Tungsten-halogen lamp, 184, 191, 192
  Xenon-mercury lamp, 201

UV photoinitiators in
  Pigmented systems, 216-243, 229, 230
UV radiation
  Most important effects, 216
UV sensitivity, 62
UV silicone coating, 67
UV spectroscopy, 188, 190, 369
UV technology
  Curing of colourless lacquers, 103
  Curing of printing inks, 103
  Production of thick pigmented coatings by, 103-123
UV-aerobic cure adhesives
  Compositional requirements, 155
  Conformal coated PCB for, 155
UV-anaerobic cure adhesives
  Air inhibition, 156
  Simplified depiction, 156
  Use in electronic and optical equipment, 156
UV-curable adhesives for opaque substrates
  Compositions, 157
  Important characteristics, 157
  Lap shear strengths, 157
  Potential approaches, 157
UV-curable clear coatings
  Photoinitiator performance in, 217
UV-Curestester
  Real-time, 381, 395-397
UV-moisture cure adhesives
  Acrylate coupling agents, 153, 154
  Acrylated trimethoxysilane coupling agent, 153, 154
  Compositional requirements, 153
  Isocyanates use in, 153
  Methacrylated silane coupling agents, 153, 154
  Potential advantage, 153
  Silanes use in, 153
  Trimethoxysilane coupling agents, 153, 154
  Trimethoxysilane groups, 153, 154
UV-pigmented paints systems
  Prerequisite for use of, 105
UV-thermal cure adhesives
  Composition, 150, 151
  Compositional requirements, 149
  Perester for, 152, 153
  Selected applications, 150, 151
  Surface mounted device (SMD), 151, 152
UV-Tronic computerised control system, 66
UV/visible spectrophotometry, 391
UVR 6110, 287
Vinyl ether oligomers, 200
Vinyl ethers, 211
  Heat of polymerisation of 201
  Hybrid systems containing, 285
Vinyl magnesium chloride, 361
Vinyl monomers, 48
5-Vinyl-2-norbornene
  Selective reaction with chlorodimethylsilane, 348

Vinylsilanes
  Incorporation into PSS fibres, 362
Viscometry
  Use in examining ink jet ink dyes and salts, 132,
  135
Viscosity
  On cured acrylates, 247, 250, 253, 255, 257, 261
Water thinnable oligomers, 269, 271, 283
White lacquers
  (2,4,6-Trimethyl-benzoyl)diphenylphosphine oxide
  (TMDPO) in, 170
  (4-Morpholino)acetophenone derivatives in, 169–171
  Alpha-amino ketones in, 169–171, 173
  Curing by medium pressure mercury lamps, 169–171
  Curing of, 169–171
  Formulation, 165
White pigmented coatings
  Acrylic modified binder, 121, 122
White pigments
  Absorption spectra of, 108
WIP electron beam source, 28
Zinc sulphide
  Absorption spectrum of, 108
Zirconium carboxylates, 155